I0038203

Linux Device Driver Development

Second Edition

Everything you need to start with device driver
development for Linux kernel and embedded Linux

John Madieu

Packt>

BIRMINGHAM—MUMBAI

Linux Device Driver Development
Second Edition

Copyright © 2022 Packt Publishing

All rights reserved. No part of this book may be reproduced, stored in a retrieval system, or transmitted in any form or by any means, without the prior written permission of the publisher, except in the case of brief quotations embedded in critical articles or reviews.

Every effort has been made in the preparation of this book to ensure the accuracy of the information presented. However, the information contained in this book is sold without warranty, either express or implied. Neither the author, nor Packt Publishing or its dealers and distributors, will be held liable for any damages caused or alleged to have been caused directly or indirectly by this book.

Packt Publishing has endeavored to provide trademark information about all of the companies and products mentioned in this book by the appropriate use of capitals. However, Packt Publishing cannot guarantee the accuracy of this information.

Group Product Manager: Rahul Nair

Publishing Product Manager: Rahul Nair

Senior Editor: Shazeen Iqbal

Content Development Editor: Romy Dias

Technical Editor: Nithik Cheruvakodan

Copy Editor: Safis Editing

Project Coordinator: Shagun Saini

Proofreader: Safis Editing

Indexer: Tejal Daruwale Soni

Production Designer: Alishon Mendonca

Marketing Coordinator: Hemangi Lotlikar

First published: October 2017

Second edition: March 2022

Production reference: 1140322

Published by Packt Publishing Ltd.

Livery Place

35 Livery Street

Birmingham

B3 2PB, UK.

ISBN 978-1-80324-006-0

www.packt.com

I dedicate this book to Laëlle Stella, my beautiful darling daughter.

Contributors

About the author

John Madieu is an embedded Linux and kernel engineer living in Paris, France. His main activity consists of developing device drivers and Board Support Packages (BSPs) for companies in domains including IoT, automation, transport, healthcare, energy, and the military. John is the founder and chief consultant of LABCSMART, a company that provides training and services for embedded Linux and Linux kernel engineering. He is an open source and embedded systems enthusiast, convinced that it is only by sharing knowledge that we can learn more. He is passionate about boxing, which he practiced for 6 years professionally, and continues to channel this passion through training sessions that he provides voluntarily.

I want to thank Jerôme Grard for his rather critical eye and his suggestions for improvement. I would also like to personally thank Pacôme Cyprien Nguefack, Claudia ATK, Elsie Zeufack, Loïca, and all those who participated directly or indirectly in the writing of this book, for accompanying me on this journey.

About the reviewer

Robertino Beniš has been involved in the embedded world for 15+ years, working on projects from smart homes and mobile devices (shipped 10+ million devices worldwide) to vehicle infotainment systems. He has worked with embedded Linux as well as with a number of real-time operating systems, including bare metal (does anyone still remember Qualcomm Brew?).

Currently, he is starting up NFTee Inc., California, in the blockchain engineering space, and studying Russian at Minsk State Linguistic University in Belarus.

Thanks to my (start-up) CEO, Kyle, for his understanding while I dedicated some time to reviewing this book, and to the city of Minsk as a whole, for enabling a clean, peaceful, and comfortable environment in which to study, work, and live.

Table of Contents

3
Dealing with Kernel Core Helpers

4
Writing Character Device Drivers

Section 2 - Linux Kernel Platform Abstraction and Device Drivers

5
Understanding and Leveraging the Device Tree

6
Introduction to Devices, Drivers, and Platform Abstraction

7
Understanding the Concept of Platform Devices and Drivers

Section 3 - Making the Most out of Your Hardware

10

Understanding the Linux Kernel Memory Allocation

11

Implementing Direct Memory Access (DMA) Support

12

Abstracting Memory Access – Introduction to the Regmap API: a Register Map Abstraction

13

Demystifying the Kernel IRQ Framework

14

Introduction to the Linux Device Model

Section 4 - Misc Kernel Subsystems for the Embedded World

15
Digging into the IIO Framework

16

Getting the Most Out of the Pin Controller and GPIO Subsystems

17

Leveraging the Linux Kernel Input Subsystem

Index

Other Books You May Enjoy

Preface

The Linux kernel is a complex, portable, modular, and widely used piece of software, running on around 80% of servers and embedded systems in more than half of the devices throughout the world. Device drivers play a critical role in the context of how well a Linux system operates. As Linux has turned out to be one of the most popular operating systems, interest in developing personal device drivers is also increasing steadily.

A device driver is the link between the user space and hardware devices, through the kernel.

This book will begin with two chapters that will help you understand the basics of drivers and prepare you for the long journey through the Linux kernel. This book will then cover driver development based on Linux subsystems, such as memory management, **industrial input/output (IIO)**, **general-purpose input/output (GPIO)**, **interrupt request (IRQ)** management, and **Inter-Integrated Circuit (I2C)** and **Serial Peripheral Interface (SPI)**. The book will also cover a practical approach to direct memory access and register map abstraction.

The source code in this book has been tested on both an x86 PC and UDOO QUAD from SECO, which is based on an ARM i.MX6 from NXP, with enough features and connections to allow us to cover all of the tests discussed in the book. Some drivers are also provided for testing purposes for inexpensive components, such as MCP23016 and 24LC512, which are an I2C GPIO controller and EEPROM memory, respectively.

By the end of this book, you will be comfortable with the concept of device driver development and will be able to write any device driver from scratch using the last stable kernel branch (v5.10.y at the time of writing).

Who this book is for

To make use of the content of this book, prior knowledge of basic C programming and Linux commands is expected. This book covers Linux driver development for widely used embedded devices, using kernel version v5.10. This book is essentially intended for embedded engineers, Linux system administrators, developers, and kernel hackers. Whether you are a software developer, a system architect, or a creator willing to dive into Linux driver development, this book is for you.

What this book covers

Chapter 1, Introduction to Kernel Development, introduces the Linux kernel development process. The chapter will discuss the downloading, configuring, and compiling steps of a kernel, for x86 as well as for ARM-based systems.

Chapter 2, Understanding Linux Kernel Module Basic Concepts, deals with Linux modularity by means of kernel modules and describes their loading/unloading. It also describes module architecture with some basic concepts.

Chapter 3, Dealing with Kernel Core Helpers, walks through frequently used kernel functions and mechanisms, such as the work queue, wait queue, mutexes, spinlock, and any other facilities that are useful for improved driver reliability.

Chapter 4, Writing Character Device Drivers, focuses on exporting device functionalities to the user space by means of character devices, as well as supporting custom commands using the ioctl interface.

Chapter 5, Understanding and Leveraging the Device Tree, discusses the mechanism to declare and describe devices to the kernel. This chapter explains device addressing, resource handling, and every data type supported in **device tree** (**DT**) and their kernel APIs.

Chapter 6, Introduction to Devices, Drivers, and Platform Abstraction, explains the general concept of platform devices, the concept of a pseudo-platform bus, as well as the device- and driver-matching mechanisms.

Chapter 7, Understanding the Concept of Platform Devices and Drivers, describes platform driver architecture in a general manner, and how to handle platform data.

Chapter 8, Writing I2C Device Drivers, dives into I2C device driver architecture, the data structures, and device addressing and accessing methods on the bus.

Chapter 9, Writing SPI Device Drivers, describes SPI-based device driver architecture, as well as the data structures involved. The chapter discusses each device's access method and specificities, as well as traps you should avoid. SPI DT binding is discussed too.

Chapter 10, *Understanding the Linux Kernel Memory Allocation*, first introduces the concept of virtual memory, to describe the whole kernel memory layout. This chapter then walks through the kernel memory management subsystem, discussing memory allocation and mapping, their APIs, and all devices involved in such mechanisms, as well as the kernel caching mechanism.

Chapter 11, *Implementing Direct Memory Access (DMA) Support*, introduces DMA and its new kernel API: the DMA Engine API. This chapter will talk about different DMA mappings and describe how to address cache coherency issues. In addition, the chapter summarizes all the concepts with a generic use case.

Chapter 12, *Abstracting Memory Access – Introduction to the Regmap API: a Register Map Abstraction*, provides an overview of the register map APIs and how they abstract the underlying SPI and I2C transactions. This chapter describes the generic API, as well as the dedicated API.

Chapter 13, *Demystifying the Kernel IRQ Framework*, demystifies the Linux IRQ core. This chapter walks through Linux IRQ management, starting from interrupt propagation over the system and moving to interrupt controller drivers, thus explaining the concept of IRQ multiplexing, using the Linux IRQ domain API.

Chapter 14, *Introduction to the Linux Device Model*, provides an overview of the heart of Linux, describing how objects are represented in the kernel, and how Linux is designed under the hood in a general manner, starting from `kobject` to devices, through to buses, classes, and device drivers.

Chapter 15, *Delving into the IIO Framework*, introduces the kernel data acquisition and measurement framework, to handle **Digital-to-Analog Converters** (**DACs**) and **Analog-to-Digital Converters** (**ADCs**). This chapter walks through the IIO APIs, both from kernel space and user space (thanks to `libiio`), dealing with triggered buffers and continuous data capture.

Chapter 16, *Getting the Most Out of the Pin Controller and GPIO Subsystems*, describes the kernel pin control infrastructure and APIs, as well as GPIO chip drivers and `gpiolib`, which is the kernel API to handle GPIO. This chapter also discusses the old and deprecated integer-based GPIO interface, as well as the descriptor-based interface, which is the new one, and finally, the way they can be configured from within the device tree. It also covers `libgpiod`, which is the official library for dealing with GPIO in user space.

Chapter 17, *Leveraging the Linux Kernel Input Subsystem*, provides a global view of input subsystems, dealing with both IRQ-based and polled input devices, and introducing both APIs. This chapter explains and shows how user space code deals with such devices.

To get the most out of this book

This book assumes a medium level of understanding of the Linux operating system and basic knowledge of C programming (at least data structures, pointer handling, and memory allocation). All code examples have been tested with Linux kernel v5.10. If additional skill is required for a given chapter, links to document references will be provided for you to quickly learn these skills.

Software/hardware covered in the book	Operating system requirements
A computer with good network bandwidth and enough disk space and RAM to download and build the Linux kernel	Preferably any Debian based distribution
Any cortex-A embedded board available on the market (for example, UDOO quad, Jetson nano, Raspberry Pi, Beagle bone)	Either a Yocto/Buildroot distribution or any embedded or vendor-specific OS (for example, Raspbian for raspberry Pi)

Other necessary packages are described in the dedicated chapter in the book. Internet connectivity is required for kernel source downloading.

If you are using the digital version of this book, we advise you to type the code yourself or access the code from the book's GitHub repository (a link is available in the next section). Doing so will help you avoid any potential errors related to the copying and pasting of code.

Download the example code files

You can download the example code files for this book from GitHub at `https://github.com/PacktPublishing/Linux-Device-Driver-Development-Second-Edition`. If there's an update to the code, it will be updated in the GitHub repository.

We also have other code bundles from our rich catalog of books and videos available at `https://github.com/PacktPublishing/`. Check them out!

Download the color images

We also provide a PDF file that has color images of the screenshots and diagrams used in this book. You can download it here: `https://static.packt-cdn.com/downloads/9781803240060_ColorImages.pdf`.

Conventions used

There are a number of text conventions used throughout this book.

`Code in text`: Indicates code words in text, database table names, folder names, filenames, file extensions, pathnames, dummy URLs, user input, and Twitter handles. Here is an example: "We can lock/unlock the spinlock using the `spin_lock()` and `spin_unlock()` inline functions, both defined in `include/linux/spinlock.h`."

A block of code is set as follows:

```
struct mutex {
    atomic_long_t owner;
    spinlock_t wait_lock;
#ifdef CONFIG_MUTEX_SPIN_ON_OWNER
    struct optimistic_spin_queue osq; /* Spinner MCS lock */
```

When we wish to draw your attention to a particular part of a code block, the relevant lines or items are set in bold:

```
struct fake_data {
    struct i2c_client *client;
    u16 reg_conf;
    struct mutex mutex;
};
```

Any command-line input or output is written as follows:

```
[342081.385491] Wait queue example
[342081.385505] Going to sleep my_init
[342081.385515] Waitqueue module handler work_handler
[342086.387017] Wake up the sleeping module
```

> **Tips or Important Notes**
> Appear like this.

Get in touch

Feedback from our readers is always welcome.

General feedback: If you have questions about any aspect of this book, email us at `customercare@packtpub.com` and mention the book title in the subject of your message.

Errata: Although we have taken every care to ensure the accuracy of our content, mistakes do happen. If you have found a mistake in this book, we would be grateful if you would report this to us. Please visit `www.packtpub.com/support/errata` and fill in the form.

Piracy: If you come across any illegal copies of our works in any form on the internet, we would be grateful if you would provide us with the location address or website name. Please contact us at `copyright@packt.com` with a link to the material.

If you are interested in becoming an author: If there is a topic that you have expertise in and you are interested in either writing or contributing to a book, please visit `authors.packtpub.com`.

Share Your Thoughts

Once you've read *Linux Device Driver Development - Second Edition*, we'd love to hear your thoughts! Scan the QR code below to go straight to the Amazon review page for this book and share your feedback.

https://packt.link/r/1803240067

Your review is important to us and the tech community and will help us make sure we're delivering excellent quality content.

Section 1 - Linux Kernel Development Basics

This section helps you make your first step into Linux kernel development. Here, we introduce the Linux kernel infrastructure (its structure and its build system), its compilation, and device driver development. As a mandatory step, we introduce the most used concepts kernel developers must know, such as sleeping, locking, basic work scheduling, and interrupt handling mechanisms. Last, we introduce the indispensable character device drivers, allowing interactions between kernel space and user space, via either standard system calls or an extended set of commands.

The following chapters will be covered in this section:

- *Chapter 1, Introduction to Kernel Development*
- *Chapter 2, Understanding Linux Kernel Module Basic Concepts*
- *Chapter 3, Dealing with Kernel Core Helpers*
- *Chapter 4, Writing Character Device Drivers*

1
Introduction to Kernel Development

Linux started as a hobby project in 1991 by a Finnish student, Linus Torvalds. The project has gradually grown and continues to do so, with roughly a thousand contributors around the world. Nowadays, Linux is a must, in embedded systems as well as on servers. A **kernel** is a central part of an operating system, and its development is not straightforward. Linux offers many advantages over other operating systems; it is free of charge, well documented with a large community, is portable across different platforms, provides access to the source code, and has a lot of free open source software.

This book will try to be as generic as possible. There is a special topic, known as the device tree, that is not a full **x86** feature yet. This topic will be dedicated to ARM processors, especially those that fully support the device tree. Why those architectures? Because they are mostly used on desktops and servers (for x86), as well as embedded systems (ARM).

In this chapter, we will cover the following topics:

- Setting up the development environment
- Understanding the kernel configuration process
- Building your kernel

Setting up the development environment

When you're working in embedded system fields, there are terms you must be familiar with, before even setting up your environment. They are as follows:

- **Target**: This is the machine that the binaries resulting from the build process are produced for. This is the machine that is going to run the binary.

- **Host**: This is the machine where the build process takes place.

- **Compilation**: This is also called native compilation or a **native build**. This happens when the target and the host are the same; that is, when you're building on machine A (the host) a binary that is going to be executed on the same machine (A, the target) or a machine of the same kind. Native compilation requires a native compiler. Therefore, a native compiler is one where the target and the host are the same.

- **Cross-compilation**: Here, the target and the host are different. It is where you build a binary from machine A (the host) that is going to be executed on machine B (the target). In this case, the host (machine A) must have installed the cross-compiler that supports the target architecture. Thus, a cross-compiler is a compiler where the target is different from the host.

Because embedded computers have limited or reduced resources (CPU, RAM, disk, and so on), it is common for the hosts to be x86 machines, which are much more powerful and have far more resources to speed up the development process. However, over the past few years, embedded computers have become more powerful, and they tend to be used for native compilation (thus used as the host). A typical example is the Raspberry Pi 4, which has a powerful quad-core CPU and up to 8 GB of RAM.

In this chapter, we will be using an x86 machine as the host, either to create a native build or for cross-compilation. So, any "native build" term will refer to an "x86 native build." Due to this, I'm running **Ubuntu 18.04**.

To quickly check this information, you can use the following command:

```
lsb_release -a
Distributor ID: Ubuntu
Description:    Ubuntu 18.04.5 LTS
Release:    18.04
Codename:    bionic
```

My computer is an **ASUS RoG**, with a 16 core AMD Ryzen CPU (you can use the `lscpu` command to pull this information out), 16 GB of RAM, 256 GB of SSD, and a 1 TB magnetic hard drive (information that you can obtain using the `df -h` command). That said, a quad-core CPU and 4 or 8 GB of RAM could be enough, but at the cost of an increased build duration. My favorite editor is **Vim**, but you are free to use the one you are most comfortable with. If you are using a desktop machine, you could use **Visual Studio Code** (**VS Code**), which is becoming widely used.

Now that we are familiar with the compilation-related keywords we will be using, we can start preparing the host machine.

Setting up the host machine

Before you can start the development process, you need to set up an **environment**. The environment that's dedicated to Linux development is quite simple – on **Debian**-based systems, at least (which is our case).

On the host machine, you need to install a few packages, as follows:

```
$ sudo apt update
$ sudo apt install gawk wget git diffstat unzip \
        texinfo gcc-multilib build-essential chrpath socat \
        libsdl1.2-dev xterm ncurses-dev lzop libelf-dev make
```

In the preceding code, we installed a few development tools and some mandatory libraries so that we have a nice user interface when we're configuring the Linux kernel.

Now, we need to install the compiler and the tools (linker, assembler, and so on) for the build process to work properly and produce the executable for the target. This set of tools is called **Binutils**, and the compiler + Binutils (+ other build-time dependency libraries if any) combo is called **toolchain**. So, you need to understand what is meant by *"I need a toolchain for <this> architecture"* or similar sentences.

Understanding and installing toolchains

Before we can start compiling, we need to install the necessary packages and tools for native or ARM cross-compiling; that is, the toolchains. GCC is the compiler that's supported by the Linux kernel. A lot of macros that are defined in the kernel are GCC-related. Due to this, we will use GCC as our (cross-)compiler.

For a native compilation, you can use the following toolchain installation command:

```
sudo apt install gcc binutils
```

When you need to cross-compile, you must identify and install the right toolchain. Compared to a native compiler, cross-compiler executables are prefixed by the name of the target operating system, architecture, and (sometimes) library. Thus, to identify architecture-specific toolchains, a naming convention has been defined: arch[-vendor][-os]-abi. Let's look at what the fields in the pattern mean:

- arch identifies the architecture; that is, arm, mips, x86, i686, and so on.
- vendor is the toolchain supplier (company); that is, Bootlin, Linaro, none (if there is no provider) or simply omitting the field, and so on.
- os is for the target operating system; that is, linux or none (bare metal). If omitted, bare metal is assumed.
- abi stands for application binary interface. It refers to what the underlying binary is going to look like, the function call convention, how parameters are passed, and more. Possible conventions include eabi, gnueabi, and gnueabihf. Let's look at these in more detail:

  - eabi means that the code that will be compiled will run on a bare metal ARM core.
  - gnueabi means that the code for Linux will be compiled.
  - gnueabihf is the same as gnueabi, but hf at the end means hard float, which indicates that the compiler and its underlying libraries are using hardware floating-point instructions rather than a software implementation of floating-point instructions, such as fixed-point software implementations. If no floating-point hardware is available, the instructions will be trapped and performed by a floating-point emulation module instead. When you're using software emulation, the only actual difference in functionality is slower execution.

The following are some toolchain names to illustrate the use of the pattern:

- arm-none-eabi: This is a toolchain that targets the ARM architecture. It has no vendor, targets a bare-metal system (does not target an operating system), and complies with the ARM EABI.

- `arm-none-linux-gnueabi` or `arm-linux-gnueabi`: This is a toolchain that produces objects for the ARM architecture to be run on Linux with the default configuration (ABI) provided by the toolchain. Note that `arm-none-linux-gnueabi` is the same as `arm-linux-gnueabi` because, as we have seen, when no vendor is specified, we assume there isn't one. The variant of this toolchain supporting hardware floating point would be `arm-linux-gnueabihf` or `arm-none-linux-gnueabihf`.

Now that we are familiar with toolchain naming conventions, we can determine which toolchain can be used to cross-compile for our target architecture.

To cross-compile for a 32-bit ARM machine, we would install the toolchain using the following command:

```
$ sudo apt install gcc-arm-linux-gnueabihf binutils-arm-linux-gnueabihf
```

Note that the 64-bit ARM backend/support in the Linux tree and GCC is called **aarch64**. So, the cross-compiler must be called something like `gcc-aarch64-linux-gnu*`, while Binutils must be called something like `binutils-aarch64-linux-gnu*`. Thus, for a 64-bit ARM toolchain, we would use the following command:

```
$ sudo apt install make gcc-aarch64-linux-gnu binutils-aarch64-linux-gnu
```

> **Note**
> Note that aarch64 only supports/provides hardware float aarch64 toolchains. Thus, there is no need to specify `hf` at the end.

Note that not all versions of the compiler can compile a given Linux kernel version. Thus, it is important to take care of both the Linux kernel version and the compiler (GCC) version. While the previous commands installed the latest version that's supported by your distribution, it is possible to target a particular version. To achieve this, you can use `gcc-<version>-<arch>-linux-gnu*`.

For example, to install version 8 of GCC for aarch64, you can use the following command:

```
sudo apt install gcc-8-aarch64-linux-gnu
```

Now that our toolchain has been installed, we can look at the version that was picked by our distribution package manager. For example, to check which version of the aarch64 cross-compiler was installed, we can use the following command:

```
$ aarch64-linux-gnu-gcc --version
aarch64-linux-gnu-gcc (Ubuntu/Linaro 7.5.0-3ubuntu1~18.04)
7.5.0
Copyright (C) 2017 Free Software Foundation, Inc.
[...]
```

For the 32-bit ARM variant, we can use the following command:

```
$ arm-linux-gnueabihf-gcc --version
arm-linux-gnueabihf-gcc (Ubuntu/Linaro 7.5.0-3ubuntu1~18.04)
7.5.0
Copyright (C) 2017 Free Software Foundation, Inc.
[...]
```

Finally, for the native version, we can use the following command:

```
$ gcc --version
gcc (Ubuntu 7.5.0-3ubuntu1~18.04) 7.5.0
Copyright (C) 2017 Free Software Foundation, Inc.
```

Now that we have set up our environment and made sure we are using the right tool versions, we can start downloading the Linux kernel sources and dig into them.

Getting the sources

In the early kernel days (until 2003), odd-even versioning styles were used, where odd numbers were stable and even numbers were unstable. When the 2.6 version was released, the versioning scheme switched to **X.Y.Z**. Let's look at this in more detail:

- **X**: This was the actual kernel's version, also called major. It was incremented when there were backward-incompatible API changes.

- **Y**: This was the minor revision. It was incremented after functionality was added in a backward-compatible manner.

- **Z**: This is also called PATCH and represented versions related to bug fixes.

This is called *semantic versioning* and was used until version *2.6.39*, when Linus Torvalds decided to bump the version to 3.0, which also meant the end of semantic versioning in 2011. At that point, an X.Y scheme was adopted.

When it came to version 3.20, Linus argued that he could no longer increase Y. Therefore, he decided to switch to an arbitrary versioning scheme, incrementing X whenever Y got so big that he ran out of fingers and toes to count it. This is the reason why the version has moved from 3.20 to 4.0 directly.

Now, the kernel uses an arbitrary **X.Y** versioning scheme, which has nothing to do with semantic versioning.

According to the Linux kernel release model, there are always two latest releases of the kernel out there: the stable release and the **long-term support** (**LTS**) release. All bug fixes and new features are collected and prepared by subsystem maintainers and then submitted to Linus Torvalds for inclusion into his Linux tree, which is called the mainline Linux tree, also known as the *master* Git repository. This is where every stable release originates from.

Before each new kernel version is released, it is submitted to the community through *release candidate* tags so that developers can test and polish all the new features. Based on the feedback he receives during this cycle, Linus decides whether the final version is ready to go. When Linus is convinced that the new kernel is ready to go, he makes the final release. We call this release "stable" to indicate that it's not a "release candidate:" those releases are *vX.Y* versions.

There is no strict timeline for making releases, but new mainline kernels are generally released every 2-3 months. Stable kernel releases are based on Linus releases; that is, the mainline tree releases.

Once a stable kernel is released by Linus, it also appears in the *linux-stable* tree (available at `https://git.kernel.org/pub/scm/linux/kernel/git/stable/linux.git/`), where it becomes a branch. Here, it can receive bug fixes. This tree is called a stable tree because it is used to track previously released stable kernels. It is maintained and curated by *Greg Kroah-Hartman*. However, all fixes must go into Linus's tree first, which is the mainline repository. Once the bug has been fixed in the mainline repository, it can be applied to previously released kernels that are still maintained by the kernel development community. All the fixes that have been backported to stable releases must meet a set of important criteria before they are considered – one of them is that they "must already exist in Linus's tree."

Note

Bugfix kernel releases are considered stable.

For example, when the 4.9 kernel is released by Linus, the stable kernel is released based on the kernel's numbering scheme; that is, 4.9.1, 4.9.2, 4.9.3, and so on. Such releases are called **bugfix kernel releases**, and the sequence is usually shortened with the number "4.9.y" when referring to their branch in the stable kernel release tree. Each stable kernel release tree is maintained by a single kernel developer, who is responsible for picking the necessary patches for the release and going through the review/release process. Usually, there are only a few bugfix kernel releases until the next mainline kernel becomes available – unless it is designated as a *long-term maintenance kernel*.

Every subsystem and kernel maintainer repository is hosted here: `https://git.kernel.org/pub/scm/linux/kernel/git/`. Here, we can also find either a Linus or a stable tree. In the Linus tree (`https://git.kernel.org/pub/scm/linux/kernel/git/torvalds/linux.git/`), there is only one branch; that is, the **master branch**. Its tags are either stable releases or release candidates. In the stable tree (`https://git.kernel.org/pub/scm/linux/kernel/git/stable/linux.git/`), there is one branch per stable kernel release (named *<A.B>.y*, where *<A.B>* is the release version in the Linus tree) and each branch contains its bugfix kernel releases.

Downloading the source and organizing it

In this book, we will be using Linus's tree, which can be downloaded using the following commands:

```
git clone https://git.kernel.org/pub/scm/linux/kernel/git/
torvalds/linux.git --depth 1
git checkout v5.10
ls
```

In the preceding commands we used `--depth 1` to avoid downloading the history (or rather, picking only the last commit history), which may considerably reduce the download size and save time. Since Git supports branching and tagging, the `checkout` command
allows you to switch to a specific tag or branch. In this example, we are switching to the `v5.10` tag.

> **Note**
> In this book, we will be dealing with Linux kernel v5.10.

Let's look at the content of the main source directory:

- `arch/`: To be as generic as possible, architecture-specific code is separated from the rest. This directory contains processor-specific code that's organized in a subdirectory per architecture, such as `alpha/`, `arm/`, `mips/`, `arm64/`, and so on.

- `block/`: This directory contains codes for block storage devices.

- `crypto/`: This directory contains the cryptographic API and the encryption algorithm's code.

- `certs/`: This directory contains certificates and sign files to enable a module signature to make the kernel load signed modules.

- `documentation/`: This directory contains the descriptions of the APIs that are used for different kernel frameworks and subsystems. You should look here before asking any questions on the public forums.

- `drivers/`: This is the heaviest directory since it is continuously growing as device drivers get merged. It contains every device driver, organized into various subdirectories.

- `fs/`: This directory contains the implementations of different filesystems that the kernel supports, such as NTFS, FAT, ETX{2,3,4}, sysfs, procfs, NFS, and so on.

- `include/`: This directory contains kernel header files.

- `init/`: This directory contains the initialization and startup code.

- `ipc/`: This directory contains the implementation of the **inter-process communication** (**IPC**) mechanisms, such as message queues, semaphores, and shared memory.

- `kernel/`: This directory contains architecture-independent portions of the base kernel.

- `lib/`: Library routines and some helper functions live here. This includes generic **kernel object** (**kobject**) handlers and **cyclic redundancy code** (**CRC**) computation functions.

- `mm/`: This directory contains memory management code.

- `net/`: This directory contains networking (whatever network type it is) protocol code.

- `samples/`: This directory contains device driver samples for various subsystems.

- `scripts/`: This directory contains scripts and tools that are used alongside the kernel. There are other useful tools here.

- `security/`: This directory contains the security framework code.

- `sound/`: Guess what falls here: audio subsystem code.

- `tools/`: This directory contains Linux kernel development and testing tools for various subsystems, such as USB, vhost test modules, GPIO, IIO, and SPI, among others.

- `usr/`: This directory currently contains the initramfs implementation.

- `virt/`: This is the virtualization directory, which contains the **kernel virtual machine** (**KVM**) module for a hypervisor.

To enforce portability, any architecture-specific code should be in the `arch` directory. Moreover, the kernel code that's related to the user space API does not change (system calls, `/proc`, `/sys`, and so on) as it would break the existing programs.

In this section, we have familiarized ourselves with the Linux kernel's source content. After going through all the sources, it seems quite natural to configure them to be able to compile a kernel. In the next section, we will learn how kernel configuration works.

Configuring and building the Linux kernel

There are numerous drivers/features and build options available in the Linux kernel sources. The configuration process consists of choosing what features/drivers are going to be part of the compilation process. Depending on whether we are going to perform native compilation or cross-compilation, there are environment variables that must be defined, even before the configuration process takes place.

Specifying compilation options

The compiler that's invoked by the kernel's `Makefile` is `$(CROSS_COMPILE)gcc`. That said, `CROSS_COMPILE` is the prefix of the cross-compiling tools (`gcc`, `as`, `ld`, `objcopy`, and so on) and must be specified when you're invoking `make` or must have been exported before any `make` command is executed. Only `gcc` and its related Binutils executables will be prefixed with `$(CROSS_COMPILE)`.

Note that various assumptions are made and options/features/flags are enabled by the Linux kernel build infrastructure based on the target architecture. To achieve that, in addition to the cross-compiler prefix, the architecture of the target must be specified as well. This can be done through the ARCH environment variable.

Thus, a typical Linux configuration or build command would look as follows:

```
ARCH=<XXXX> CROSS_COMPILE=<YYYY> make menuconfig
```

It can also look as follows:

```
ARCH=<XXXX> CROSS_COMPILE=<YYYY> make <make-target>
```

If you don't wish to specify these environment variables when you launch a command, you can export them into your current shell. The following is an example:

```
export CROSS_COMPILE=aarch64-linux-gnu-
export ARCH=aarch64
```

Remember that if these variables are not specified, the native host machine is going to be targeted; that is, if CROSS_COMPILE is omitted or not set, $(CROSS_COMPILE)gcc will result in gcc, and it will be the same for other tools that will be invoked (for example, $(CROSS_COMPILE)ld will result in ld).

In the same manner, if ARCH (the target architecture) is omitted or not set, it will default to the host where make is executed. It will default to $(uname -m).

As a result, you should leave CROSS_COMPILE and ARCH undefined to have the kernel natively compiled for the host architecture using gcc.

Understanding the kernel configuration process

The Linux kernel is a *Makefile-based* project that contains thousands of options and drivers. Each option that's enabled can make another one available or can pull specific code into the build. To configure the kernel, you can use `make menuconfig` for a ncurses-based interface or `make xconfig` for an X-based interface. The ncurses-based interface looks as follows:

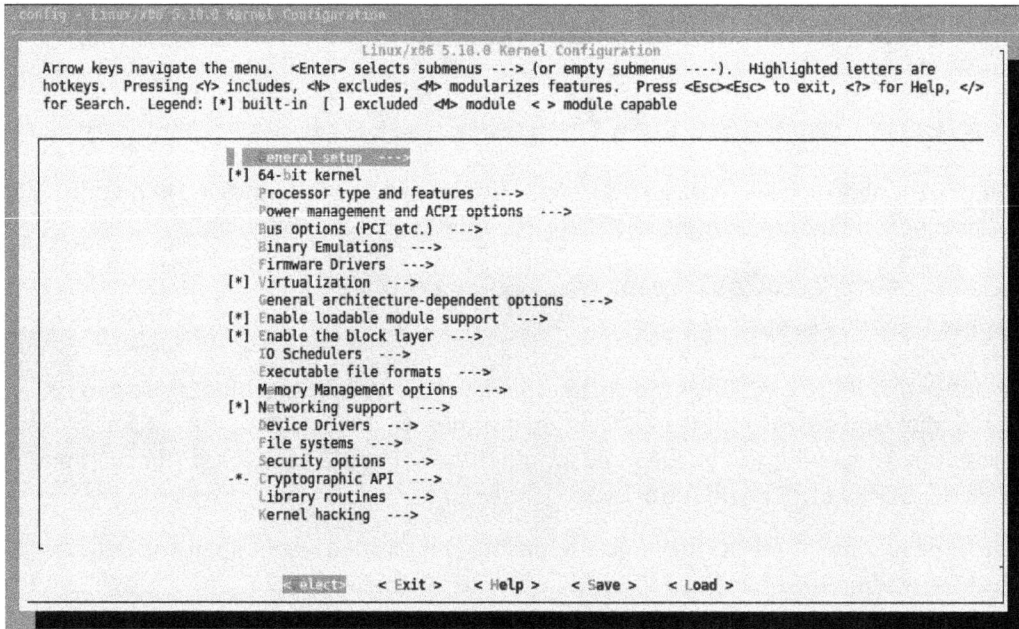

Figure 1.1 – Kernel configuration screen

For most options, you have three choices. However, we can enumerate five types of options while configuring the Linux kernel:

- Boolean options, for which you have two choices:

  - (`blank`), which leaves this feature out. Once this option is highlighted in the configuration menu, you can press the `<n>` key to leave the feature out. It is equivalent to false. When it's disabled, the resulting configuration option is commented out in the configuration file.

  - (`*`), which compiles it statically in the kernel. This means it will always be there when the kernel first loads. It is equivalent to true. You can enable a feature in the configuration menu by selecting it and pressing the `<y>` key. The resulting option will appear as `CONFIG_<OPTION>=y` in the configuration file; for example, `CONFIG_INPUT_EVDEV=y`.

- Tristate options, which, in addition to being able to take Boolean states, can take a third state, marked as `(M)` in the configuration windows. This results in `CONFIG_<OPTION>=m` in the configuration file; for example, `CONFIG_INPUT_EVDEV=m`. To produce a loadable module (provided that this option allows it), you can select the feature and press the `M` key.

- String options, which expect string values; for example, `CONFIG_CMDLINE="noinitrd console=ttymxc0,115200"`.

- Hex options, which expect hexadecimal values; for example, `CONFIG_PAGE_OFFSET=0x80000000`.

- Int options, which expect integer values; for example, `CONFIG_CONSOLE_LOGLEVEL_DEFAULT=7`.

The selected options will be stored in a `.config` file, at the root of the source tree.

It is very difficult to know which configuration is going to work on your platform. In most cases, there will be no need to start a configuration from scratch. There are default and functional configuration files available in each arch directory that you can use as a starting point (it is important to start with a configuration that already works):

```
ls arch/<your_arch>/configs/
```

For 32-bit ARM-based CPUs, these config files can be found in `arch/arm/configs/`. In this architecture, there is usually one default configuration per CPU family. For instance, for i.MX6-7 processors, the default config file is `arch/arm/configs/imx_v6_v7_defconfig`. However, on ARM 64-bit CPUs, there is only one big default configuration to customize; it is located in `arch/arm64/configs/` and is called `defconfig`. Similarly, for x86 processors, we can find the files in `arch/x86/configs/`. There will be two default configuration files here – `i386_defconfig` and `x86_64_defconfig`, for 32- and 64-bit x86 architectures, respectively.

The kernel configuration command, given a default configuration file, is as follows:

```
make <foo_defconfig>
```

This will generate a new `.config` file in the main (root) directory, while the old `.config` will be renamed `.config.old`. This can be useful to revert the previous configuration changes. Then, to customize the configuration, you can use the following command:

```
make menuconfig
```

Saving your changes will update your `.config` file. While you could share this config with your teammates, you are better off creating a default configuration file in the same minimal format as those shipped with the Linux kernel sources. To do that, you can use the following command:

```
make savedefconfig
```

This command will create a minimal (since it won't store non-default settings) configuration file. The generated default configuration file will be called `defconfig` and stored at the root of the source tree. You can store it in another location using the following command:

```
mv defconfig arch/<arch>/configs/myown_defconfig
```

This way, you can share a reference configuration inside the kernel sources and other developers can now get the same `.config` file as you by running the following command:

```
make myown_defconfig
```

> **Note**
>
> Note that, for cross-compilation, `ARCH` and `CROSS_COMPILE` must be set before you execute any `make` command, even for kernel configuration. Otherwise, you'll have unexpected changes in your configuration.

The followings are the various configuration commands you can use, depending on the target system:

- For a 64-bit x86 native compilation, it is quite straightforward (the compilation options can be omitted):

  ```
  make x86_64_defconfig
  make menuconfig
  ```

- Given a 32-bit ARM i.MX6-based board, you can execute the following command:

  ```
  ARCH=arm CROSS_COMPILE=arm-linux-gnueabihf- make imx_v6_
  v7_defconfig
  ARCH=arm CROSS_COMPILE=arm-linux-gnueabihf- make
  menuconfig
  ```

With the first command, you store the default options in the `.config` file, while with the latter, you can update (add/remove) various options, depending on your needs.

- For 64-bit ARM boards, you can execute the following commands:

```
ARCH=aarch64 CROSS_COMPILE=aarch64-linux-gnu- make
defconfig
ARCH=aarch64 CROSS_COMPILE=aarch64-linux-gnu- make
menuconfig
```

You may run into a Qt4 error with `xconfig`. In such a case, you should just use the following command to install the missing packages:

```
sudo apt install qt4-dev-tools qt4-qmake
```

> **Note**
>
> You may be switching from an old kernel to a new one. Given the old `.config` file, you can copy it into the new kernel source tree and run `make oldconfig`. If there are new options in the new kernel, you'll be prompted to include them or not. However, you may want to use the default values for those options. In this case, you should run `make olddefconfig`. Finally, to say no to every new option, you should run `make oldnoconfig`.

There may be a better option to find an initial configuration file, especially if your machine is already running. Debian and Ubuntu Linux distributions save the `.config` file in the `/boot` directory, so you can use the following command to copy this configuration file:

```
cp /boot/config-`uname -r` .config
```

The other distributions may not do this. So, I can recommend that you always enable the `IKCONFIG` and `IKCONFIG_PROC` kernel configuration options, which will enable access to `.config` through `/proc/configs.gz`. This is a standard method that also works with embedded distributions.

Some useful kernel configuration features

Now that we can configure the kernel, let's enumerate some useful configuration features that may be worth enabling in your kernel:

- IKCONFIG and IKCONFIG_PROC: These are the most important to me. It makes your kernel configuration available at runtime, in /proc/config.gz. It can be useful either to reuse this config on another system or simply look for the enabled state of a particular feature; for example:

```
# zcat /proc/config.gz | grep CONFIG_SOUND
CONFIG_SOUND=y
CONFIG_SOUND_OSS_CORE=y
CONFIG_SOUND_OSS_CORE_PRECLAIM=y
# CONFIG_SOUNDWIRE is not set
#
```

- CMDLINE_EXTEND and CMDLINE: The first option is a Boolean that allows you to extend the kernel command line from within the configuration, while the second option is a string containing the actual command-line extension value; for example, CMDLINE="noinitrd usbcore.authorized_default=0".

- CONFIG_KALLSYMS: This is a Boolean option that makes the kernel symbol table (the mapping between symbols and their addresses) available in /proc/kallsyms. This is very useful for tracers and other tools that need to map kernel symbols to addresses. It is used while you're printing oops messages. Without this, oops listings would produce hexadecimal output, which is difficult to interpret.

- CONFIG_PRINTK_TIME: This option shows timing information while printing messages from the kernel. It may be helpful to timestamp events that occurred at runtime.

- CONFIG_INPUT_EVBUG: This allows you to debug input devices.

- CONFIG_MAGIC_SYSRQ: This allows you to have some control (such as rebooting, dumping some status information, and so on) over the system, even after a crash, by simply using some combination keys.

- DEBUG_FS: This enables support for debug filesystems, where GPIO, CLOCK, DMA, REGMAP, IRQs, and many other subsystems can be debugged from.

- FTRACE and DYNAMIC_FTRACE: These options enable the powerful ftrace tracer, which can trace the whole system. Once ftrace has been enabled, some of its enumeration options can be enabled as well:

  - FUNCTION_TRACER: This allows you to trace any non-inline function in the kernel.

  - FUNCTION_GRAPH_TRACER: This does the same thing as the previous command, but it shows a call graph (the caller and the callee functions).

  - IRQSOFF_TRACER: This allows you to track off periods of IRQs in the kernel.

  - PREEMPT_TRACER: This allows you to measure preemption off latency.

  - SCHED_TRACER: This allows you to schedule latency tracing.

Now that the kernel has been configured, it must be built to generate a runnable kernel. In the next section, we will describe the kernel building process, as well as the expected build artifacts.

Building the Linux kernel

This step requires you to be in the same shell where you were during the configuration step; otherwise, you'll have to redefine the ARCH and CROSS_COMPILE environment variables.

Linux is a Makefile-based project. Building such a project requires using the make tool and executing the make command. Regarding the Linux kernel, this command must be executed from the main kernel source directory, as a normal user.

By default, if not specified, the make target is all. In the Linux kernel sources, for x86 architectures, this target points to (or depends on) vmlinux bzImage modules targets; for ARM or aarch64 architectures, it corresponds to vmlinux zImage modules dtbs targets.

In these targets, bzImage is an x86-specific make target that produces a binary with the same name, bzImage. vmlinux is a make target that produces a Linux image called vmlinux. zImage and dtbs are both ARM- and aarch64-specific make targets. The first produces a Linux image with the same name, while the second builds the device tree sources for the target CPU variant. modules is a make target that will build all the selected modules (marked with m in the configuration).

While building, you can leverage the host's CPU performance by running multiple jobs in parallel thanks to the `-j` make options. The following is an example:

```
make -j16
```

Most people define their `-j` number as 1.5x the number of cores. In my case, I always use `ncpus * 2`.

You can build the Linux kernel like so:

- For a native compilation, use the following command:

```
make -j16
```

- For a 32-bit ARM cross-compilation, use the following command:

```
ARCH=arm CROSS_COMPILE=arm-linux-gnueabihf- make -j16
```

Each make target can be invoked separately, like so:

```
ARCH=arm CROSS_COMPILE=arm-linux-gnueabihf- make dtbs
```

You can also do the following:

```
ARCH=arm CROSS_COMPILE=arm-linux-gnueabihf- make zImage -j16
```

Finally, you can also do the following:

```
make bzImage -j16
```

> **Note**
>
> I have used `-j16` in my commands because my host has an 8-core CPU. This number of jobs must be adapted according to your host configuration.

At the end of your 32-bit ARM cross-compilation jobs, you will see something like the following:

```
[...]
  LZO      arch/arm/boot/compressed/piggy_data
  CC       arch/arm/boot/compressed/misc.o
  CC       arch/arm/boot/compressed/decompress.o
  CC       arch/arm/boot/compressed/string.o
  SHIPPED  arch/arm/boot/compressed/hyp-stub.S
  SHIPPED  arch/arm/boot/compressed/lib1funcs.S
  SHIPPED  arch/arm/boot/compressed/ashldi3.S
  SHIPPED  arch/arm/boot/compressed/bswapsdi2.S
  AS       arch/arm/boot/compressed/hyp-stub.o
  AS       arch/arm/boot/compressed/lib1funcs.o
  AS       arch/arm/boot/compressed/ashldi3.o
  AS       arch/arm/boot/compressed/bswapsdi2.o
  AS       arch/arm/boot/compressed/piggy.o
  LD       arch/arm/boot/compressed/vmlinux
  OBJCOPY  arch/arm/boot/zImage
  Kernel: arch/arm/boot/zImage is ready
```

By using the default targets, various binaries will result from the build process, depending on the architecture. These are as follows:

- `arch/<arch>/boot/Image`: An uncompressed kernel image that can be booted
- `arch/<arch>/boot/*Image*`: A compressed kernel image that can also be booted:

  This is `bzImage` (which means "big zImage") for x86, `zImage` for ARM or aarch64, and `vary` for other architectures.
- `arch/<arch>/boot/dts/*.dtb`: This provides compiled device tree blobs for the selected CPU variant.
- `vmlinux`: This is a raw, uncompressed, and unstripped kernel image in ELF format. It's useful for debugging purposes but generally not used for booting purposes.

Now that we know how to (cross-)compile the Linux kernel, let's learn how to install it.

Installing the Linux kernel

The Linux kernel installation process differs in terms of native compilation or cross-compilation:

- In native installation (that is, you're installing host), you can simply run `sudo make install`. You must use `sudo` because the installation will take place in the `/boot` directory. If you're doing an x86 native installation, the following files are shipped:

  - `/boot/vmlinuz-<version>`: This is the compressed and stripped variant of `vmlinux`. It is the same kernel image as the one in `arch/<arch>/boot`.

  - `/boot/System.map-<version>`: This stores the kernel symbol table (the mapping between the kernel symbols and their addresses), not just for debugging, but also to allow some kernel modules to resolve their symbols and load properly. This file only contains a static kernel symbol table, while `/proc/kallsyms` (provided that `CONFIG_KALLSYMS` is enabled in the config) on the running kernel contains `System.map` and the loaded kernel module symbols.

  - `/boot/config-<version>`: This corresponds to the kernel configuration for the version that's been built.

- An embedded installation usually uses a single file kernel. Moreover, the target is not accessible, which makes manual installation preferred. Thus, embedded Linux build systems (such as Yocto or Buildroot) use internal scripts to make the kernel image available in the target root filesystem. While embedded installation may be straightforward thanks to the use of build systems, native installation (especially x86 native installation) can require running additional bootloader-related commands (such as `update-grub2`) to make the new kernel visible to the system.

Now that we are familiar with the kernel configuration, including the build and installation processes, let's look at kernel modules, which allow you to extend the kernel at runtime.

Building and installing modules

Modules can be built separately using the `modules` target. You can install them using the `modules_install` target. Modules are built in the same directory as their corresponding source. Thus, the resulting kernel objects are spread over the kernel source tree:

- For a native build and installation, you can use the following commands:

```
make modules
sudo make modules_install
```

  The resulting modules will be installed in `/lib/modules/$(uname -r)/kernel/`, in the same directory structure as their corresponding source. A custom install path can be specified using the `INSTALL_MOD_PATH` environment variable.

- When you're cross-compiling for embedded systems, as with all `make` commands, `ARCH` and `CROSS_COMPILE` must be specified. As it is not possible to install a directory in the target device filesystem, embedded Linux build systems (such as Yocto or Buildroot) set `INSTALL_MOD_PATH` to a path that corresponds to the target root filesystem so that the final root filesystem image contains the modules that have been built; otherwise, the modules will be installed on the host. The following is an example of a 32-bit ARM architecture:

```
ARCH=arm CROSS_COMPILE=arm-linux-gnueabihf- make modules
ARCH=arm CROSS_COMPILE=arm-linux-gnueabihf- INSTALL_MOD_
PATH=<dir> make modules_install
```

In addition to the `kernel` directory that is shipped with modules, the following files are installed in `/lib/modules/<version>` as well:

- `modules.builtin`: This lists all the kernel objects (`.ko`) that are built into the kernel. It is used by the module loading utility (`modprobe`, for example) so that it does not fail when it's trying to load something that's already built in. `modules.builtin.bin` is its binary counterpart.

- `modules.alias`: This contains the aliases for module loading utilities, which are used to match drivers and devices. This concept of module aliases will be explained in *Chapter 6, Introduction to Devices, Drivers, and Platform Abstraction*. `modules.alias.bin` is its binary equivalent.

- `modules.dep`: This lists modules, along with their dependencies. `modules.dep.bin` is its binary counterpart.

- `modules.symbols`: This tells us which module a given symbol belongs to. They are in the form of `alias symbol:<symbol> <modulename>`. An example is `alias symbol:v4l2_async_notifier_register videodev`. `modules.symbols.bin` is the binary counterpart of this file.

With that, we have installed the necessary modules. We've finished learning how to build and install Linux kernels and modules. We've also finished learning how to configure the Linux kernel and add the features we need.

Summary

In this chapter, you learned how to download the Linux source and process your first build. We also covered some common operations, such as configuring or selecting the appropriate toolchain. That said, this chapter was quite brief; it was just an introduction. This is why, in the next chapter, we will cover the kernel building process, how to compile a driver (either externally or as part of the kernel), and some basics that you should learn before you start the long journey that kernel development represents. Let's take a look!

2
Understanding Linux Kernel Module Basic Concepts

A kernel module is a piece of software whose aim is to extend the Linux kernel with a new feature. A kernel module can be a device driver, in which case it would control and manage a particular hardware device, hence the name **device driver**. A module can also add a framework support (for example **IIO**, the **Industrial Input Output** framework), extend an existing framework, or even a new filesystem or an extension of it. The thing to keep in mind is that kernel modules are not always device drivers, whereas device drivers are always kernel modules.

In opposition to kernel modules, there might be simple modules or user space modules, running in user space, with low privileges. This book, however, exclusively deals with kernel space modules, particularly Linux kernel modules.

That being said, this chapter will discuss the following topics:

- An introduction to the concept of modules
- Building a Linux kernel module
- Dealing with symbol exports and module dependencies
- Learning some Linux kernel programming tips

An introduction to the concept of modules

When building the Linux kernel, the resulting image is a single file made by the linking of all object files that correspond to features enabled in the configuration. As a result, all included features are therefore available as soon as the kernel starts, even if the filesystem is not yet ready or does not exist. These features are built-in, and the corresponding modules are called static modules. Such a module is available at any time in the kernel image and thus can't be unloaded, at the cost of extra size to the final kernel image. A static module is also known as a built-in module, since it is part of the final kernel image output. Any change in its code will require the whole kernel to be rebuilt.

Some features (such as device drivers, filesystems, and frameworks) can, however, be compiled as loadable modules. Such modules are separated from the final kernel image and are loaded on demand. These can be considered as plugins that can be loaded/unloaded dynamically to add or remove features (at runtime) to the kernel. Because each module is stored as a separate file on the filesystem, using loadable modules requires access to a filesystem.

To summarize, a module is to the Linux kernel what a plugin (add-on) is to user software (for example, Firefox). When it is statically linked to the resulting kernel image, it is said to be built in. When it is built as a separate file (which can be loaded/unloaded), it is loadable. It dynamically extends the kernel features without even the need to restart the machine.

To support module loading, the kernel must have been built with the following option enabled:

```
CONFIG_MODULES=y
```

Unloading modules is a kernel feature that can be enabled or disabled according to the CONFIG_MODULE_UNLOAD kernel configuration option. Without this option, we won't be able to unload any module. Thus, to be able to unload modules, the following feature must be enabled:

```
CONFIG_MODULE_UNLOAD=y
```

That said, the kernel is smart enough to prevent unloading modules that may probably break things (for example, because these are in use), even if it is asked to do so. This is because the kernel keeps a reference count of module usage so that it knows whether a module is currently in use or not. If the kernel believes it is unsafe to remove a module, it will not. We can, however, change this behavior with the following configuration feature:

```
MODULE_FORCE_UNLOAD=y
```

The preceding option allows us to force module unload.

Now that we are done with the main concepts behind modules, let's start practicing, first by introducing a module skeleton that will serve as a basis for this chapter.

Case study – module skeleton

Let's consider the following `hello-world` module. It will be the basis for our work throughout this chapter. Let's call its compilation unit `helloworld.c`, with the following content:

```
#include <linux/module.h>
#include <linux/init.h>

static int __init helloworld_init(void) {
    pr_info("Hello world initialization!\n");
    return 0;
}

static void __exit helloworld_exit(void) {
    pr_info("Hello world exit\n");
}

module_init(helloworld_init);
module_exit(helloworld_exit);

MODULE_LICENSE("GPL");
MODULE_AUTHOR("John Madieu <john.madieu@gmail.com>");
MODULE_DESCRIPTION("Linux kernel module skeleton");
```

In the preceding skeleton, headers are specific to the Linux kernel, hence the use of `linux/xxx.h`. The `module.h` header file is mandatory for all kernel modules, and `init.h` is needed for the `__init` and `__exit` macros. Other elements are described in the next sections. To build this skeleton module, we need to write a *special* makefile, which will be covered a little bit later in this chapter.

Module entry and exit points

The minimal requirement of a kernel module is an initialization method. This is a must. If the module can be built as a loadable module, then the `exit` method must be provided as well. The first method is the entry point and corresponds to the function called when the module is loaded (`modprobe` or `insmod`), and the latter is the cleanup and exit point and corresponds to the function executed at module unloading (at `rmmod` or `modprobe -r`).

All you need to do is to inform the kernel about which functions should be executed as an entry or exit point. The `helloworld_init` and `helloworld_exit` functions can be given any name. The only thing that is actually mandatory is to identify them as the corresponding initialization and exit functions, passing them as parameters to the `module_init()` and `module_exit()` macros.

To sum up, `module_init()` is used to declare the function that should be called when the module is loaded (with `insmod` or `modprobe` when the module is built as a loadable kernel module) or when the kernel reaches the run level corresponding to this module (when built-in). What is done in the initialization function will define the behavior of the module. `module_exit()` is used only when the module can be built as a loadable kernel module. It declares the function that should be called when the module is unloaded (with `rmmod`).

`init` or `exit` methods are invoked only once, whatever the number of devices currently handled by the module, provided the module is a device driver. It is common for modules that are platform (or alike) device drivers to register a platform driver and the associated `probe/remove` callback in their `init` functions, which this time will be invoked each time a device handled by the module is added or removed on the system. In such a case, they just unregister the platform driver from within their `exit` method.

__init and __exit attributes

`__init` and `__exit` are kernel macros, defined in `include/linux/init.h`, as shown here:

```
#define __init       __section(.init.text)
#define __exit       __section(.exit.text)
```

The `__init` keyword tells the linker to place the symbols (variables or functions) they prefix in a dedicated section in the resulting kernel object file. This section is known in advance to the kernel and freed when the module is loaded and the initialization function has finished. This applies only to built-in modules, not to loadable ones. The kernel will run the initialization function of the driver for the first time during its boot sequence. Since the driver cannot be unloaded, its initialization function will never be called again until the next reboot. There is no need to keep references on this initialization function anymore. It is the same for the `__exit` keyword and the `exit` method, whose corresponding code is omitted when the module is compiled statically into the kernel or when module unloading support is not enabled because, in both cases, the exit function is never called. `__exit` has no effect on loadable modules.

In conclusion, `__init` and `__exit` are Linux directives (macros) that wrap GNU C compiler attributes used for symbol placement. They instruct the compiler to put the code they prefix in the `.init.text` and `.exit.text` sections, respectively, even though the kernel can access different object sections.

Module information and metadata

Without having to read its code, it should be possible to gather some information (such as the author(s), module parameter descriptions, and the license) about a given module. A kernel module uses its `.modinfo` section to store information about the module. Any `MODULE_*` macro will update the content of this section with the values passed as parameters. Some of these macros are `MODULE_DESCRIPTION()`, `MODULE_AUTHOR()`, and `MODULE_LICENSE()`. That said, the real underlying macro provided by the kernel to add an entry to the module information section is `MODULE_INFO(tag, info)`, which adds generic information of the `tag = "info"` form. This means a driver author can add any freeform information they want, such as the following:

```
MODULE_INFO(my_field_name, "What easy value");
```

As well as the custom information we define, there is standard information we should provide, which the kernel provides macros for:

- MODULE_LICENSE: The license will define how your source code should be shared (or not) with other developers. MODULE_LICENSE() tells the kernel what license our module is under. It has an effect on your module behavior, since a license that is not compatible with GPL (General Public License) will result in your module not being able to see/use symbols exported by the kernel through the EXPORT_SYMBOL_GPL() macro, which shows the symbols for GPL-compatible modules only. This is the opposite of EXPORT_SYMBOL(), which exports functions for modules with any license. Loading a non-GPL-compatible module will also result in a tainted kernel; that means non-open source or untrusted code has been loaded, and you will likely have no support from the community. Remember that the module without MODULE_LICENSE() is not considered open source and will taint the kernel too. Available licenses can be found in include/linux/module.h, describing the license supported by the kernel.

- MODULE_AUTHOR() declares the module's author(s): MODULE_AUTHOR("John Madieu <john.madieu@foobar.com>");. It is possible to have more than one author. In this case, each author must be declared with MODULE_AUTHOR():

```
MODULE_AUTHOR("John Madieu <john.madieu@foobar.com>");
MODULE_AUTHOR("Lorem Ipsum <l.ipsum@foobar.com>");
```

- MODULE_DESCRIPTION() briefly describes what the module does: MODULE_DESCRIPTION("Hello, world! Module").

You can dump the content of the .modeinfo section of a kernel module using the objdump -d -j .modinfo command on the given module. For a cross-compiled module, you should use $(CORSS_COMPILE)objdump instead.

Now that we are done with providing module information and metadata, which are the last requirements when writing Linux kernel modules, let's learn how to build these modules.

Building a Linux kernel module

Two solutions exist for compiling a kernel module:

- The first solution is when code is outside of the kernel source tree, which is also known as out-of-tree building. The module source code is in a different directory. Building a module this way does not allow integration into the kernel configuration/compilation process, and the module needs to be built separately. It must be noted that with this solution, the module cannot be statically linked in the final kernel image – that is, it cannot be built in. Out-of-tree compilation only allows loadable kernel modules to be produced.

- The second solution is inside the kernel tree, which allows you to upstream your code, since it is well integrated into the kernel configuration/compilation process. This solution allows you to produce either a statically linked module (also known as built-in) or a loadable kernel module.

Now that we have enumerated and given the characteristics of the two possible solutions for building kernel modules, before studying each of them, let's first dig into the Linux kernel build process. This will help us to understand compilation prerequisites for each solution.

Understanding the Linux kernel build system

The Linux kernel maintains its own build system. It is called **kbuild** (note that the k is lower case). It allows you to configure the Linux kernel and compile it based on the configuration that has been done. It relies essentially on three files to achieve that. These are Kconfig, for feature selections, mainly used with in-kernel tree building, and Kbuild (note that the K is uppercase this time) or Makefile, for compilation rules.

The Kbuild or Makefile files

From within this build system, the makefile can be called either Makefile or Kbuild. If both files exist, only Kbuild will be used. That said, a makefile is a special file used to execute a set of actions, among which the most common is the compilation of programs. There is a dedicated tool to parse makefiles, called make. Using this tool, a kernel module build command pattern resembles the following:

```
make -C $KERNEL_SRC M=$(shell pwd) [target]
```

In the preceding pattern, $KERNEL_SRC refers to the path of the prebuilt kernel directory, -C $KERNEL_SRC instructs make to change into the specified directory when executing and change back when finished, and M=$(shell pwd) instructs the kernel build system to move back into this directory to find the module that is being built. The value given to M is the absolute path of the directory where the module sources (or the associated Kbuild file) are located. [target] corresponds to the subset of the make targets available when building an external module. These are as follows:

- modules: This is the default target for external modules. It has the same functionality as if no target was specified.

- modules_install: This installs the external module(s). The default location is /lib/modules/<kernel_release>/extra/. This path can be overridden.

- clean: This removes all generated files (in the module directory only).

However, we have not told the build system what object files to build or to link together. We must specify the name of the module(s) to be built, along with the list of requisite source files. It can be as simple as the following single line:

```
obj-<X> := <module_name>.o
```

In the preceding, the kernel build system will build <module_name>.o from <module_name>.c or <module_name>.S, and after linking, it will result in the <module_name>.ko kernel loadable module or will be part of the single-file kernel image. <X> can be either y, m, or left blank.

How and if mymodule.o will be built or linked depends on the value of <X>:

- If <X> is set to m, the obj-m variable is used, and mymodule.o will be built as a loadable kernel module.

- If <X> is set to y, the obj-y variable is used, and mymodule.o will be built as part of the kernel. You then say "foo is a built-in kernel module".

- If <X> is not set, the obj- variable is used, and mymodule.o will not be built at all.

However, the obj-$(CONFIG_XXX) pattern is often used, where CONFIG_XXX is a kernel configuration option, set or not, during the kernel configuration process. An example is the following:

```
obj-$(CONFIG_MYMODULE) += mymodule.o
```

$(CONFIG_MYMODULE) evaluates to either y, m, or nothing (blank), according to its value during the kernel configuration (displayed with menuconfig). If CONFIG_MYMODULE is neither y nor m, then the file will not be compiled nor linked. y means built-in (it stands for yes in the kernel configuration process), and m stands for a loadable module. $(CONFIG_MYMODULE) pulls the right answer from the normal config process.

So far, we have assumed the module is built from a single .c source file. When the module is built from multiple source files, an additional line is needed for listing these source files, as shown here:

```
<module_name>-y := <file1>.o <file2>.o
```

The preceding says that <module_name>.ko will be built from two files, file1.c and file2.c. However, if you wanted to build two modules, let's say foo.ko and bar.ko, the Makefile line would be as follows:

```
obj-m := foo.o bar.o
```

If foo.o and bar.o are made of source files other than foo.c and bar.c, you can specify the appropriate source files of each object file, as shown here:

```
obj-m := foo.o bar.o
foo-y := foo1.o foo2.o . . .
bar-y := bar1.o bar2.o bar3.o . . .
```

The following is another example of listing the requisite source files to build a given module:

```
obj-m := 8123.o
8123-y := 8123_if.o 8123_pci.o 8123_bin.o
```

The preceding example says that 8123 should be built as a loadable kernel module by building and linking 8123_if.c, 8123_pci.c, and 8123_bin.c together.

Apart from the files being part of the resulting build artifact, the Makefile file can also contain compiler and linker flags, such as the following:

```
ccflags-y := -I$(src)/include
ccflags-y += -I$(src)/src/hal/include
ldflags-y := -T$(src)foo_sections.lds
```

What is important here is the fact that such flags can be specified as well, not the values we have set in the example.

There is another use case of `obj-<X>`, described in the following:

```
obj-<X> += somedir/
```

This means that the kernel build system should go into the directory named `somedir` and look for any `Makefile` or `Kbuild` files inside, processing it in order to decide what objects should be built.

We can summarize what we just said with the following Makefile:

```
# kbuild part of makefile
obj-m := helloworld.o
#the following is just an example
#ldflags-y := -T foo_sections.lds

# normal makefile
KERNEL_SRC ?= /lib/modules/$(shell uname -r)/build

all default: modules
install: modules_install

modules modules_install help clean:
    $(MAKE) -C $(KERNEL_SRC) M=$(shell pwd) $@
```

The following describes this minimalist `Makefile` skeleton:

- `obj-m := helloworld.o`: `obj-m` lists modules we want to build. For each `<filename>.o`, the build system will look for `<filename>.c` or `<filename>.S` to build. `obj-m` is used to build a loadable kernel module, whereas `obj-y` will result in a built-in kernel module.

- `KERNEL_SRC= /lib/modules/$(shell uname -r)/build`: `KERNEL_SRC` is the location of the prebuilt kernel source. As we said earlier, we need a prebuilt kernel in order to build any module. If you have built your kernel from the source, you should set this variable with the absolute path of the built source directory. `-C` instructs the `make` utility to change into the specified directory reading the makefiles.

- `M=$(shell pwd)`: This is relevant to the kernel build system. The `Makefile` kernel uses this variable to locate the directory of an external module to build. Your `.c` files should be placed in that directory.

- `all default: modules`: This line instructs the `make` utility to execute the `modules` target as a dependency of `all` or `default` targets. In other words, `make default`, `make all`, or simple `make` commands will result in `make modules` being executed prior to execute any subsequent command if any.

- `modules modules_install help clean`: This line represents the list target that is valid in this makefile.

- `$(MAKE) -C $(KERNELDIR) M=$(shell pwd) $@`: This is the rule to be executed for each of the targets enumerated previously. `$@` will be replaced with the parameters given to `make`, which includes the target. Using this kind of magic word prevents us from writing as many (identical) lines as there are targets. In other words, if you run `make modules`, `$@` will be replaced with `modules`, and the rule will become `$(MAKE) -C $(KERNELDIR) M=$(shell pwd) modules`.

Now that we are familiar with the kernel build system requirements, let's see how modules are actually built.

Out-of-tree building

Before you can build an external module, you need to have a complete and precompiled kernel source tree. The prebuilt kernel version must be the same as the kernel you'll load and use your module with. There are two ways to obtain a prebuilt kernel version:

- Building it by yourself (which we discussed earlier): This can be used for both native and cross-compilation. Using a build system such as Yocto or Buildroot may help.

- Installing the `linux-headers-*` package from the distribution package feed: This applies only for x86 native compilations unless your embedded target runs a Linux distribution that maintains a package feed (such as Raspbian).

It must be noted that is it not possible to build a built-in kernel module with out-of-tree building. The reason is that building a Linux kernel module out of tree requires a prebuilt or prepared kernel.

Native and out-of-tree module compiling

With a native kernel module build, the easiest way is to install the prebuilt kernel headers and to use their directory path as the kernel directory in the makefile. Before we start doing so, headers can be installed with the following command:

```
sudo apt update
sudo apt install linux-headers-$(uname -r)
```

This will install preconfigured and prebuilt kernel headers (not the whole source tree) in /usr/src/linux-headers-$(uname -r). There will be a symbolic link, /lib/modules/$(uname -r)/build, pointing to the previously installed headers. It is the path you should specify as the kernel directory in Makefile. You should remember that $(uname -r) corresponds to the kernel version in use.

Now, when you are done with the makefile, still in your module source directory, run the make command or make modules:

```
$ make
make -C /lib/modules/ 5.11.0-37-generic/build \
    M=/home/john/driver/helloworld modules
make[1]: Entering directory '/usr/src/linux-headers- 5.11.0-37-
generic'
  CC [M]  /media/jma/DATA/work/tutos/sources/helloworld/
helloworld.o
  Building modules, stage 2.
  MODPOST 1 modules
  CC       /media/jma/DATA/work/tutos/sources/helloworld/
helloworld.mod.o
  LD [M]  /media/jma/DATA/work/tutos/sources/helloworld/
helloworld.ko
make[1]: Leaving directory '/usr/src/linux-headers- 5.11.0-37-
generic'
```

At the end of the build, you'll have the following:

```
$ ls
helloworld.c  helloworld.ko  helloworld.mod.c  helloworld.mod.o
helloworld.o  Makefile  modules.order  Module.symvers
```

To test, you can do the following:

```
$ sudo insmod  helloworld.ko
$ sudo rmmod helloworld
$ dmesg
[...]
[308342.285157] Hello world initialization!
[308372.084288] Hello world exit
```

The preceding example only deals with a native build, compiling on a machine running a standard distribution, allowing us to leverage its package repository to install prebuilt kernel headers. In the next section, we will discuss out-of-tree module cross-compilation.

Out-of-tree module cross-compiling

When it comes to cross-compiling an out-of-tree kernel module, there are essentially two variables that the kernel make command needs to be aware of. These are ARCH and CROSS_COMPILE, which respectively represent the target architecture and the cross-compiler prefix. Moreover, the location of a prebuilt kernel for the target architecture must be specified in the makefile. In our skeleton, we called it KERNEL_SRC.

When using a build system such as Yocto, the Linux kernel is first cross-compiled as a dependency before it starts cross-compiling the module. That said, I voluntarily used the KERNEL_SRC variable name for the prebuilt kernel directory, since this variable is automatically exported by Yocto for kernel module recipes. It is set to the value of STAGING_KERNEL_DIR within the module.bbclass class, inherited by all kernel module recipes.

That said, what changes between native compilation and cross-compilation of an out-of-tree kernel module is the final make command, which looks like the following for a 32-bit Arm architecture:

```
make ARCH=arm CROSS_COMPILE=arm-linux-gnueabihf-
```

For the 64-bit variant, it would look like the following:

```
make ARCH=aarch64 CROSS_COMPILE=aarch64-linux-gnu-
```

The previous commands assume the cross-compiled kernel source path has been specified in the makefile.

In-tree building

In-tree module building requires dealing with an additional file, Kconfig, which allows us to expose the module features in the configuration menu. That said, before you can build a module in the kernel tree, you should first identify which directory should host your source files. Given your filename, mychardev.c, which contains the source code of your special character driver, it should be changed to the drivers/char directory in the kernel source. Every subdirectory in the drivers has both Makefile and Kconfig files.

Add the following content to the Kconfig file of that directory:

```
config PACKT_MYCDEV
    tristate "Our packtpub special Character driver"
    default m
    help
      Say Y here to support /dev/mycdev char device.
      The /dev/mycdev is used to access packtpub.
```

In Makefile in that same directory, add the following line:

```
obj-$(CONFIG_PACKT_MYCDEV)    += mychardev.o
```

Be careful when updating Makefile – the .o file name must match the exact name of the .c file. If the source file is foobar.c, you must use foobar.o in Makefile. In order to have your module built as a loadable kernel module, add the following line to your defconfig board in the arch/arm/configs directory:

```
CONFIG_PACKT_MYCDEV=m
```

You can also run menuconfig to select it from the UI, run make to build the kernel, and then make modules to build modules (including yours). To make the driver built-in, just replace m with y:

```
CONFIG_PACKT_MYCDEV=y
```

Everything described here is what embedded board manufacturers do in order to provide a **Board Support Package** (**BSP**) with their board, with a kernel that already contains their custom drivers:

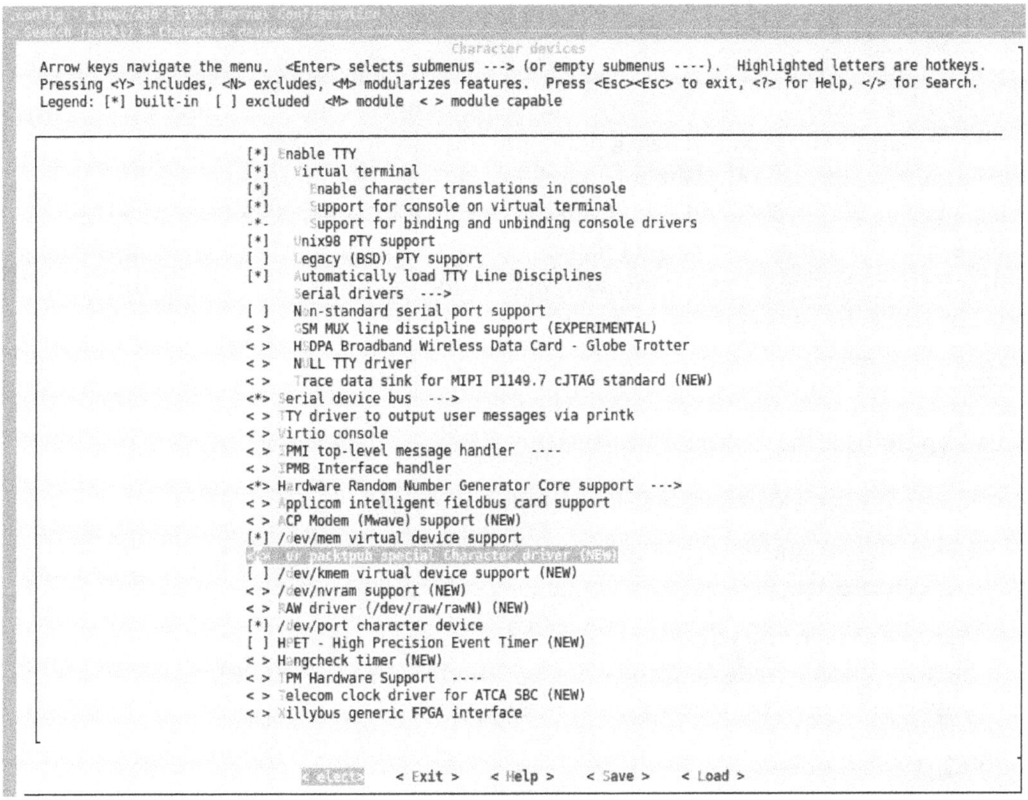

```
                                        Character devices
Arrow keys navigate the menu.  <Enter> selects submenus --->  (or empty submenus ----).  Highlighted letters are hotkeys.
Pressing <Y> includes, <N> excludes, <M> modularizes features.  Press <Esc><Esc> to exit, <?> for Help, </> for Search.
Legend: [*] built-in  [ ] excluded  <M> module  < > module capable

                        [*] Enable TTY
                        [*]   Virtual terminal
                        [*]     Enable character translations in console
                        [*]     Support for console on virtual terminal
                        -*-     Support for binding and unbinding console drivers
                        [*]   Unix98 PTY support
                        [ ]   Legacy (BSD) PTY support
                        [*]   Automatically load TTY Line Disciplines
                              Serial drivers  --->
                        [ ]   Non-standard serial port support
                        < >   GSM MUX line discipline support (EXPERIMENTAL)
                        < >   HSDPA Broadband Wireless Data Card - Globe Trotter
                        < >   NULL TTY driver
                        < >   Trace data sink for MIPI P1149.7 cJTAG standard (NEW)
                        <*> Serial device bus  --->
                        < > TTY driver to output user messages via printk
                        < > Virtio console
                        < > IPMI top-level message handler  ----
                        < > IPMB Interface handler
                        <*> Hardware Random Number Generator Core support  --->
                        < > Applicom intelligent fieldbus card support
                        < > ACP Modem (Mwave) support (NEW)
                        [*] /dev/mem virtual device support
                        <*> Our packtpub special character driver (NEW)
                        [ ] /dev/kmem virtual device support (NEW)
                        < > /dev/nvram support (NEW)
                        < > RAW driver (/dev/raw/rawN) (NEW)
                        [*] /dev/port character device
                        [ ] HPET - High Precision Event Timer (NEW)
                        < > Hangcheck timer (NEW)
                        < > TPM Hardware Support  ----
                        < > Telecom clock driver for ATCA SBC (NEW)
                        < > Xillybus generic FPGA interface

                   < Select >   < Exit >   < Help >   < Save >   < Load >
```

Figure 2.1 – The Packt_dev module in the kernel tree

Once configured, you can build the kernel with make, and build modules with make modules.

Modules included in the kernel source tree are installed in /lib/modules/$(unale -r)/kernel/. On your Linux system, it is /lib/modules/$(uname -r)/kernel/.

Now that we are familiar with out-of-tree or in-tree kernel module compilation, let's see how to handle module behavior adaptation by allowing parameters to be passed to this module.

Handling module parameters

Similar to a user program, a kernel module can accept arguments from the command line. This allows us to dynamically change the behavior of the module according to the given parameters, which can help a developer not have to indefinitely change/compile the module during a test/debug session. In order to set this up, we should first declare the variables that will hold the values of command-line arguments and use the `module_param()` macro on each of these. The macro is defined in `include/linux/moduleparam.h` (this should be included in the code too – `#include <linux/moduleparam.h>`) as follows:

```
module_param(name, type, perm);
```

This macro contains the following elements:

- `name`: The name of the variable used as the parameter.
- `type`: The parameter's type (`bool`, `charp`, `byte`, `short`, `ushort`, `int`, `uint`, `long`, and `ulong`), where `charp` stands for *character pointer*.
- `perm`: This represents the `/sys/module/<module>/parameters/<param>` file permissions. Some of them are `S_IWUSR`, `S_IRUSR`, `S_IXUSR`, `S_IRGRP`, `S_WGRP`, and `S_IRUGO`, where the following applies:
  - `S_I` is just a prefix.
  - `R` = read, `W` = write, and `X` = execute.
  - `USR` = user, `GRP` = group, and `UGO` = user, group, and others.

You can eventually use `|` (the `OR` operation) to set multiple permissions. If `perm` is `0`, the file parameter in Sysfs will not be created. You should use only `S_IRUGO` read-only parameters, which I highly recommend; by OR'ing with other properties, you can obtain fine-grained properties.

When using module parameters, `MODULE_PARM_DESC` can be used on a per-parameter basis to describe each of them. This macro will populate the module information section of each parameter's description. The following is a sample, from the `helloworld-params.c` source file provided with the code repository of this book:

```
#include <linux/moduleparam.h>
[...]

static char *mystr = "hello";
static int myint = 1;
```

```
static int myarr[3] = {0, 1, 2};

module_param(myint, int, S_IRUGO);
module_param(mystr, charp, S_IRUGO);
module_param_array(myarr, int,NULL, S_IWUSR|S_IRUSR);

MODULE_PARM_DESC(myint,"this is my int variable");
MODULE_PARM_DESC(mystr,"this is my char pointer variable");
MODULE_PARM_DESC(myarr,"this is my array of int");

static int foo()
{
    pr_info("mystring is a string: %s\n",
            mystr);
    pr_info("Array elements: %d\t%d\t%d",
            myarr[0], myarr[1], myarr[2]);
    return myint;
}
```

To load the module and feed our parameter, we do the following:

```
# insmod hellomodule-params.ko mystring="packtpub" myint=15
myArray=1,2,3
```

That said, we could have used modinfo prior to loading the module in order to display a description of parameters supported by the module:

```
$ modinfo ./helloworld-params.ko
filename:       /home/jma/work/tutos/sources/helloworld/./
helloworld-params.ko
license:        GPL
author:         John Madieu <john.madieu@gmail.com>
srcversion:     BBF43E098EAB5D2E2DD78C0
depends:
vermagic:       4.4.0-93-generic SMP mod_unload modversions
parm:           myint:this is my int variable (int)
parm:           mystr:this is my char pointer variable (charp)
parm:           myarr:this is my array of int (array of int)
```

It is also possible to find and edit the current values for the parameters of a loaded module from Sysfs in `/sys/module/<name>/parameters`. In that directory, there is one file per parameter, containing the parameter value. These parameter values can be changed if the associated files have write permissions (which depends on the module code).

The following is an example:

```
echo 0 > /sys/module/usbcore/parameters/authorized_default
```

Not just loadable kernel modules can accept parameters. Provided that a module is built in the kernel, you can specify parameters for this module from the Linux kernel command line (the one passed by the bootloader or the one that is provided by the CONFIG_ CMDLINE configuration option).

This has the following form:

```
[initial command line ...] my_module.param=value
```

In this example, `my_module` corresponds to the module name and `value` is the value assigned to this parameter.

Now that we are able to deal with module parameters, let's dive a bit deeper into a not-so-obvious scenario, where we will learn how the Linux kernel itself and its build system handles module dependencies.

Dealing with symbol exports and module dependencies

Only a limited number of kernel functions can be called from a kernel module. To be visible to a kernel module, functions and variables must be explicitly exported by the kernel. Thus, the Linux kernel exposes two macros that can be used to export functions and variables. These are the following:

- `EXPORT_SYMBOL(symbolname)`: This macro exports a function or variable to all modules.
- `EXPORT_SYMBOL_GPL(symbolname)`: This macro exports a function or variable only to GPL modules.

EXPORT_SYMBOL() or its GPL counterpart are Linux kernel macros that make a symbol available to loadable kernel modules or dynamically loaded modules (provided that said modules add an extern declaration – that is, include the headers corresponding to the compilation units that exported the symbols). EXPORT_SYMBOL() instructs the Kbuild mechanism to include the symbol passed as an argument in the global list of kernel symbols. As a result, kernel modules can access them. Code that is built into the kernel itself (as opposed to loadable kernel modules) can, of course, access any non-static symbol via an extern declaration, as with conventional C code.

These macros also allow us to export, from loadable kernel modules, symbols that can be accessed from other loadable kernel modules. An interesting thing is that a symbol thus exported by one module becomes accessible to another module that may depend on it! A normal driver should not need any non-exported function.

An introduction to the concept of module dependencies

A dependency of module B on module A is that module B is using one or more of the symbols exported by module A. Let's see in the next section how such dependencies are handled in the Linux kernel infrastructure.

The depmod utility

depmod is a tool that you run during the kernel build process to generate module dependency files. It does that by reading each module in /lib/modules/<kernel_release>/ to determine what symbols it should export and what symbols it needs. The result of that process is written to a modules.dep file, and its binary version, modules.dep.bin. It is a kind of module indexing.

Module loading and unloading

For a module to be operational, you should load it into the Linux kernel, either by using insmod and passing the module path as an argument, which is the preferred method during development, or by using modprobe, a clever command but which is preferable for use in production systems.

Manual loading

Manual loading needs the intervention of a user, which should have root access. The two classical methods to achieve this are modprobe and insmod, which are described as follows.

During development, you usually use insmod in order to load a module. insmod should be given the path of the module to load, as follows:

```
insmod /path/to/mydrv.ko
```

This is the low-level form of module loading, which forms the basis of the other module-loading method and the one we will use in this book. On the other hand, there is modprobe, mostly used by system admins or in a production system. modprobe is a clever command that parses the modules.dep file (discussed previously) in order to load dependencies first, prior to loading the given module. It automatically handles module dependencies, as a package manager does. It is invoked as follows:

```
modprobe mydrv
```

Whether we can use modprobe depends on depmod being aware of module installation.

Auto-loading

The depmod utility doesn't only build modules.dep and modules.dep.bin files; it does more than that. When kernel developers write drivers, they know exactly what hardware the drivers will support. They are then responsible for feeding the drivers with the product and vendor IDs of all devices supported by the driver. depmod also processes module files in order to extract and gather that information and generates a modules.alias file, located in /lib/modules/<kernel_release>/modules.alias, which maps devices to their drivers:

An excerpt of modules.alias is as follows:

```
alias usb:v0403pFF1Cd*dc*dsc*dp*ic*isc*ip*in* ftdi_sio
alias usb:v0403pFF18d*dc*dsc*dp*ic*isc*ip*in* ftdi_sio
alias usb:v0403pDAFFd*dc*dsc*dp*ic*isc*ip*in* ftdi_sio
alias usb:v0403pDAFEd*dc*dsc*dp*ic*isc*ip*in* ftdi_sio
alias usb:v0403pDAFDd*dc*dsc*dp*ic*isc*ip*in* ftdi_sio
alias usb:v0403pDAFCd*dc*dsc*dp*ic*isc*ip*in* ftdi_sio
[...]
```

At this step, you'll need a user space hotplug agent (or device manager), usually udev (or mdev), that will register with the kernel to get notified when a new device appears.

The notification is done by the kernel, sending the device's description (the product ID, the vendor ID, the class, the device class, the device subclass, the interface, and any other information that can identify a device) to the hotplug daemon, which in turn calls modprobe with this information. modprobe then parses the modules.alias file in order to match the driver associated with the device. Before loading the module, modprobe will look for its dependencies in module.dep. If it finds any, they will be loaded prior to the associated module loading; otherwise, the module is loaded directly.

There is another method for automatically loading a module, at boot time this time. This is achieved in /etc/modules-load.d/<filename>.conf. If you want some modules to be loaded at boot time, just create a /etc/modules-load.d/<filename>.conf file and add the module names that should be loaded, one per line. <filename> will be meaningful to you, and people usually use module: /etc/modules-load.d/modules.conf. You can create as many .conf files as you need.

An example of /etc/modules-load.d/mymodules.conf is as follows:

```
#This line is a comment
uio
iwlwifi
```

These configuration files are processed by systemd-modules-load.service, provided that systemd is the initialization manager on your machine. On SysVinit systems, these files are processed by the /etc/init.d/kmod script.

Module unloading

The usual command to unload a module is rmmod. This is preferable to unloading a module loaded with the insmod command. The command should be given the module name to unload as a parameter:

```
rmmod -f mymodule
```

On the other hand, a higher-level command to unload a module in a smart manner is modeprobe -r, which automatically unloads unused dependencies:

```
modeprobe -r mymodule
```

As you may have guessed, it is a helpful option for developers. Finally, we can check whether a module is loaded with the `lsmod` command, as follows:

```
$ lsmod
Module                Size   Used by
btrfs              1327104  0
blake2b_generic      20480  0
xor                  24576  1 btrfs
raid6_pq            114688  1 btrfs
ufs                  81920  0
[...]
```

The output includes the name of the module, the amount of memory it uses, the number of other modules that use it, and finally, the name of these. The output of `lsmod` is actually a nice formatting view of what you can see under `/proc /modules`, which is the file listing loaded modules:

```
$ cat /proc/modules
btrfs 1327104 0 - Live 0x0000000000000000
blake2b_generic 20480 0 - Live 0x0000000000000000
xor 24576 1 btrfs, Live 0x0000000000000000
raid6_pq 114688 1 btrfs, Live 0x0000000000000000
ufs 81920 0 - Live 0x0000000000000000
qnx4 16384 0 - Live 0x0000000000000000
```

The preceding output is raw and poorly formatted. Therefore, it is preferable to use `lsmod`.

Now that we are familiar with kernel module management, let's extend our kernel development skills by learning some tips that kernel developers have adopted.

Learning some Linux kernel programming tips

Linux kernel development is about learning from others and not reinventing the wheel. There is a set of rules to follow when doing kernel development. A whole chapter won't be enough to cover these rules. Thus, I picked two of the most relevant to me, those that are likely to change when programming for user space: error handling and message printing.

In user space, exiting from the main() method is enough to recover from all the errors that may have occurred. In the kernel, this is not the case, especially since it directly deals with the hardware. Things are different for message printing as well, and we will see that in this section.

Error handling

Returning the wrong error code for a given error can result in either the kernel or user space application misinterpreting and taking the wrong decision, producing unneeded behavior. To keep things clear, there are predefined errors in the kernel tree that cover almost every case you may face. Some of the errors (with their meaning) are defined in include/uapi/asm-generic/errno-base.h, and the rest of the list can be found in include/uapi/asm-generic/errno.h. The following is an excerpt of this list of errors, from include/uapi/asm-generic/errno-base.h:

```
#define EPERM     1    /* Operation not permitted */
#define ENOENT    2    /* No such file or directory */
#define ESRCH     3    /* No such process */
#define EINTR     4    /* Interrupted system call */
#define EIO       5    /* I/O error */
#define ENXIO     6    /* No such device or address */
#define E2BIG     7    /* Argument list too long */
#define ENOEXEC   8    /* Exec format error */
#define EBADF     9    /* Bad file number */
#define ECHILD    10   /* No child processes */
#define EAGAIN    11   /* Try again */
#define ENOMEM    12   /* Out of memory */
#define EACCES    13   /* Permission denied */
#define EFAULT    14   /* Bad address */
#define ENOTBLK   15   /* Block device required */
#define EBUSY     16   /* Device or resource busy */
#define EEXIST    17   /* File exists */
#define EXDEV     18   /* Cross-device link */
#define ENODEV    19   /* No such device */
[...]
```

Most of the time, the standard way to return an error is to do so in the form of return -ERROR, especially when it comes to answering system calls. For example, for an I/O error, the error code is EIO, and you should return -EIO, as follows:

```
dev = init(&ptr);
if(!dev)
    return -EIO
```

Errors sometimes cross the kernel space and propagate themselves to the user space. If the returned error is an answer to a system call (open, read, ioctl, or mmap), the value will be automatically assigned to the user space errno global variable, on which you can use strerror(errno) to translate the error into a readable string:

```
#include <errno.h>   /* to access errno global variable */
#include <string.h>
[...]
if(wite(fd, buf, 1) < 0) {
    printf("something gone wrong! %s\n", strerror(errno));
}
[...]
```

When you face an error, you must undo everything that has been set until the error occurred. The usual way to do that is to use the goto statement:

```
ret = 0;
ptr = kmalloc(sizeof (device_t));
if(!ptr) {
        ret = -ENOMEM
        goto err_alloc;
}
dev = init(&ptr);

if(!dev) {
        ret = -EIO
        goto err_init;
}
return 0;


err_init:
```

```
        free(ptr);
err_alloc:
        return ret;
```

The reason to use the `goto` statement is simple. When it comes to handling errors, let's say that at *step 5*, you have to clean up the previous operations (*steps 4, 3, 2*, and *1*), instead of doing lots of nested checking operations, as follows:

```
if (ops1() != ERR) {
    if (ops2() != ERR) {
        if (ops3() != ERR) {
            if (ops4() != ERR) {
```

This is less readable, error-prone, and confusing (readability also depends on indentation). By using the `goto` statement, we have straight control flow, as follows:

```
if (ops1() == ERR) // ||
    goto error1;   // ||
if (ops2() == ERR) // ||
    goto error2;   // ||
if (ops3() == ERR) // ||
    goto error3;   // ||
if (ops4() == ERR) // VV
    goto error4;
error5:
[...]
error4:
[...]
error3:
[...]
error2:
[...]
error1:
[...]
```

That said, you should only use `goto` to move forward in a function, not backward, nor to implement loops (as is the case in an assembler).

Handling null pointer errors

When it comes to returning an error from functions that are supposed to return a pointer, functions often return the NULL pointer. It is functional but it is a quite meaningless approach, since we do not exactly know why this NULL pointer is returned. For that purpose, the kernel provides three functions, ERR_PTR, IS_ERR, and PTR_ERR, defined as follows:

```
void *ERR_PTR(long error);
long IS_ERR(const void *ptr);
long PTR_ERR(const void *ptr);
```

The first macro returns the error value as a pointer. It can be seen as an *error value to pointer* macro. Given a function that is likely to return -ENOMEM after a failed memory allocation, we have to do something such as return ERR_PTR(-ENOMEM);. The second macro is used to check whether the returned value is a pointer error using if(IS_ERR(foo)). The last one returns the actual error code, return PTR_ERR(foo). It can be seen as a *pointer to error value* macro.

The following is an example of how to use ERR_PTR, IS_ERR, and PTR_ERR:

```
static struct iio_dev *indiodev_setup(){
    [...]
    struct iio_dev *indio_dev;
    indio_dev = devm_iio_device_alloc(&data->client->dev,
                                      sizeof(data));
    if (!indio_dev)
        return ERR_PTR(-ENOMEM);
    [...]
    return indio_dev;
}

static int foo_probe([...]){
    [...]
    struct iio_dev *my_indio_dev = indiodev_setup();
    if (IS_ERR(my_indio_dev))
        return PTR_ERR(data->acc_indio_dev);
    [...]
}
```

This is a plus with error handling, which is also an excerpt of the kernel coding style that states that if a function's name is an action or an imperative command, the function should return an integer error code. If, however, the function's name is a predicate, this function should return a Boolean to indicate the succeeded status of the operation.

`Add work`, for example, is a command, thus the `add_work()` function returns 0 for success or `-EBUSY` for failure. `PCI device present` is a predicate, and in the same way, this is why the `pci_dev_present()` function returns 1 if it succeeds in finding a matching device or 0 if it doesn't.

Message printing – goodbye printk, long life dev_*, pr_*, and net_* APIs

Apart from informing users of what is going on, printing is the first debugging technique. `printk()` is to the kernel what `printf()` is to the user space. `printk()` has, for a long time, ruled kernel message printing in a leveled manner. Written messages can be displayed using the `dmesg` command. Depending on how important the message to print was, `printk()` allowed you to choose between eight log-level messages, defined in `include/linux/kern_levels.h`, along with their meaning.

Nowadays, while `printk()` remains the low-level message printing API, the printk/log-level pair has been encoded into clearly named helpers, which are recommended for use in new drivers. These are as follows:

- `pr_<level>(...)`: This is used in regular modules that are not device drivers.
- `dev_<level>(struct device *dev, ...)`: This is to be used in device drivers that are not network devices (also known as `netdev` drivers).
- `netdev_<level>(struct net_device *dev, ...)`: This is used in `netdev` drivers exclusively.

In all these helpers, `<level>` represents the log level encoded into a quite meaningful name, as described in the following table:

Module helpers	Driver helpers	Netdev helper	Description	Log level
pr_debug, pr_devel	dev_dbg	netdev_ dbg	Used for debug messages. pr_ devel() is dead code. This means it is not compiled at all, so it's not present in the final binary unless DEBUG is defined. The preferred way to go is pr_debug.	7
pr_info	dev_info	netdev_ info	You can use this for informational purposes, such as start up information at a driver initialization.	6
pr_notice	dev_notice	netdev_ notice	This is a notice – nothing serious but notable, nevertheless. It is often used to report security events.	5
pr_warning	dev_warn	netdev_ warn	A warning that means nothing serious by itself but might indicate problems.	4
pr_err	dev_err	netdev_ err	An error condition, often used by drivers to indicate difficulties with hardware.	3
pr_crit	dev_crit	netdev_ crit	A critical condition occurred, such as a serious hardware/software failure.	2
pr_alert	dev_alert	netdev_ alert	Something bad happened and action must be taken immediately.	1
pr_emerg	dev_emerg	netdev_ emerg	Emergency messages – the system is about to crash or is unstable.	0

Table 2.1 – The Linux kernel printing API

Log levels work in a way that, whenever a message is printed, the kernel compares the message log level with the current console log level; if the former is higher (lower value) than the last, the message will be immediately printed to the console. You can check your log-level parameters with the following:

```
cat /proc/sys/kernel/printk
4        4        1        7
```

In the preceding output, the first value is the current log level (4). According to that, any message printed with higher importance (a lower log level) will be displayed in the console as well. The second value is the default log level, according to the `CONFIG_DEFAULT_MESSAGE_LOGLEVEL` option. Other values are not relevant for the purpose of this chapter, so let's ignore them.

The current log level can be changed with the following:

```
echo <level> > /proc/sys/kernel/printk
```

In addition, you can prefix the module output messages with a custom string. To achieve this, you should define the `pr_fmt` macro. It is common to define this message prefix with the module name, as follows:

```
#define pr_fmt(fmt)  KBUILD_MODNAME ": " fmt
```

For more concise log output, some overrides use the current function name as a prefix, as follows:

```
#define pr_fmt(fmt)  "%s: " fmt, __func__
```

If we consider the `net/bluetooth/lib.c` file in the kernel source tree, we can see the following in the first line:

```
#define pr_fmt(fmt)  "Bluetooth: " fmt
```

With that line, any `pr_<level>` (we are in a regular module, not a device driver) logging call will produce a log prefixed with `Bluetooth:`, similar to the following:

```
$ dmesg | grep Bluetooth
[ 3.294445] Bluetooth: Core ver 2.22
[ 3.294458] Bluetooth: HCI device and connection manager
initialized
[ 3.294460] Bluetooth: HCI socket layer initialized
[ 3.294462] Bluetooth: L2CAP socket layer initialized
[ 3.294465] Bluetooth: SCO socket layer initialized
[...]
```

This is all about message printing. We have learned how to choose and use the appropriate printing APIs according to the situation.

We are now done with our kernel module introduction series. At this stage, you should be able to download, configure, and (cross-)compile the Linux kernel, as well as write and build kernel modules against this kernel.

> **Note**
>
> `printk()` (or its encoded helpers) never blocks and is safe enough to be called even from atomic contexts. It tries to lock the console and print the message. If locking fails, the output will be written into a buffer and the function will return, never blocking. The current console holder will then be notified about new messages and will print them before releasing the console.

Summary

This chapter showed the basics of driver development and explained the concept of built-in and loadable kernel modules, as well as their loading and unloading. Even if you are not able to interact with the user space, you are ready to write a working module, print formatted messages, and understand the concept of init/exit.

The next chapter will deal with Linux kernel core functions, which, along with this chapter, form the Swiss army knife of Linux kernel development. In the next chapter, you will be able to target enhanced features, perform fancy operations that can impact the system, and interact with the core of the Linux kernel.

3
Dealing with Kernel Core Helpers

The Linux kernel is a standalone piece of software—as you'll see in this chapter—that does not depend on any external library as it implements any functionalities it needs to use (from list management to compression algorithms, everything is implemented from scratch). It implements any mechanism you may encounter in modern libraries and even more, such as compression, string functions, and so on. We will walk step by step through the most important aspects of such capabilities.

In this chapter, we will cover the following topics:

- Linux kernel locking mechanisms and shared resources
- Dealing with kernel waiting, sleeping, and delay mechanisms
- Understanding Linux kernel time management
- Implementing work-deferring mechanisms
- Kernel interrupt handling

Linux kernel locking mechanisms and shared resources

A resource is said to be shared when it is accessible by several contenders, whether exclusively or not. When it is exclusive, access must be synchronized so that only the allowed contender(s) may own the resource. Such resources might be memory locations or peripheral devices, and the contenders might be processors, processes, or threads. The operating system performs mutual exclusion by atomically modifying a variable that holds the current state of the resource, making this visible to all contenders that might access the variable at the same time. Atomicity guarantees the modification to be entirely successful, or not successful at all. Modern operating systems nowadays rely on hardware (which should allow atomic operations) to implement synchronization, though a simple system may ensure atomicity by disabling interrupts (and avoiding scheduling) around the critical code section.

We can enumerate two synchronization mechanisms, as follows:

- **Locks**: Used for mutual exclusion. When one contender holds the lock, no other can hold it (others are excluded). The most known locks in the kernel are spinlocks and mutexes.

- **Conditional variables**: For waiting for a change. These are implemented differently in the kernel, as we will see later.

When it comes to locking, it is up to the hardware to allow such synchronizations by means of atomic operations, which the kernel uses to implement locking facilities. Synchronization primitives are data structures used for coordinating access to shared resources. Because only one contender can hold the lock (and thus access the shared resource), it might perform an arbitrary operation on the resource associated with the lock, which would appear to be atomic to others.

Apart from dealing with the exclusive ownership of a given shared resource, there are situations where it is better to wait for the state of the resource to change—for example, waiting for a list to contain at least one object (its state then passes from empty to not empty) or for a task to complete (a **direct memory access** (**DMA**) transaction, for example). The Linux kernel does not implement conditional variables. From user space, we could think of conditional variables for both situations, but to achieve the same or even better, the kernel provides the following mechanisms:

- **Wait queue**: To wait for a change—designed to work in concert with locks

- **Completion queue**: To wait for the completion of a given computation, mostly used with DMAs

All the aforementioned mechanisms are supported by the Linux kernel and exposed to drivers by means of a reduced set of **application programming interfaces** (**APIs**) (which significantly ease their use for developers), which we will discuss in the coming sections.

Spinlocks

A spinlock is a hardware-based locking primitive that depends on hardware capabilities to provide atomic operations (such as `test_and_set`, which in a non-atomic implementation would result in read, modify, and write operations). It is the simplest and the base locking primitive, working as described in the following scenario.

When *CPUB* is running, and task *B* wants to acquire the spinlock (task *B* calls the spinlock's locking function), and this spinlock is already held by another **central processing unit** (**CPU**) (let's say *CPUA* running task *A*, which has already called this spinlock's locking function), then *CPUB* will simply spin around a `while` loop (thus blocking task *B*) until the other CPU releases the lock (task *A* calls the spinlock's release function). This spinning will only happen on multi-core machines (hence the use case described previously, involving more than one CPU) because, on a single-core machine, it cannot happen (the task either holds the spinlock and proceeds or never runs until the lock is released). A spinlock is said to be a lock held by a CPU, in contrast to a mutex (which we will discuss in the next section of this chapter), which is a lock held by a task.

A spinlock operates by disabling the scheduler on the local CPU (that is, the CPU running the task that called the spinlock's locking API). This also means that a task currently running on that CPU cannot be preempted except by **interrupt requests** (**IRQs**) if they are not disabled on the local CPU (more on this later). In other words, spinlocks protect resources that only one CPU can take/access at a time. This makes spinlocks suitable for **symmetrical multiprocessing** (**SMP**) safety and for executing atomic tasks.

Note

Not only do spinlocks take advantage of hardware atomic functions. In the Linux kernel, for example, preemption status depends on a per-CPU variable that, if it equals 0, means preemption is enabled; if it is greater than 0, this means preemption is disabled (`schedule()` becomes inoperative). Thus, disabling preemption (`preempt_disable()`) consists of adding 1 to the current per-CPU variable (`preempt_count`, actually), while `preempt_enable()` subtracts 1 from the variable, checks whether the new value is 0, and calls `schedule()`. Those addition/subtraction operations are atomic and thus rely on the CPU being able to provide atomic addition/subtraction functions.

A spinlock is created either statically using a `DEFINE_SPINLOCK` macro, as illustrated here, or dynamically by calling `spin_lock_init()` on an uninitialized spinlock:

```
static DEFINE_SPINLOCK(my_spinlock);
```

To understand how this works, we just must look at the definition of this macro in `include/linux/spinlock_types.h`, as follows:

```
#define DEFINE_SPINLOCK(x) spinlock_t x = \
                              __SPIN_LOCK_UNLOCKED(x)
```

This can be used as follows:

```
static DEFINE_SPINLOCK(foo_lock);
```

After this, the spinlock will be accessible through its name `foo_lock`, and its address would be `&foo_lock`.

However, for dynamic (runtime) allocation, it's better to embed the spinlock into a bigger structure, allocating memory for this structure and then calling `spin_lock_init()` on the spinlock element, as illustrated in the following code snippet:

```
struct bigger_struct {
    spinlock_t lock;
    unsigned int foo;
    [...]
};
static struct bigger_struct *fake_init_function()
{
    struct bigger_struct *bs;
    bs = kmalloc(sizeof(struct bigger_struct), GFP_KERNEL);
    if (!bs)
        return -ENOMEM;
    spin_lock_init(&bs->lock);
    return bs;
}
```

It's better to use DEFINE_SPINLOCK whenever possible. This offers compile-time initialization and requires fewer lines of code, with no real drawback. In this step, we can lock/unlock the spinlock using spin_lock() and spin_unlock() inline functions, both defined in include/linux/spinlock.h, as follows:

```
static __always_inline void spin_unlock(spinlock_t *lock)
static __always_inline void spin_lock(spinlock_t *lock)
```

That said, there are some known limitations in using spinlocks this way. Though a spinlock prevents preemption on the local CPU, it does not prevent this CPU from being hogged by an interrupt (thus executing this interrupt's handler). Imagine a situation where the CPU holds a "spinlock" on behalf of task A in order to protect a given resource, and an interrupt occurs. The CPU will stop its current task and branch to this interrupt handler. So far, so good. Now, imagine if this IRQ handler needs to acquire this same spinlock (you probably already guessed that the resource is shared with the interrupt handler). It will infinitely spin in place, trying to acquire a lock already locked by a task that it has preempted. This situation will result in a deadlock, for sure.

To address this issue, the Linux kernel provides _irq variant functions for spinlocks, which, in addition to disabling/enabling preemption, also disable/enable interrupts on the local CPU. These functions are spin_lock_irq() and spin_unlock_irq(), defined as follows:

```
static void spin_unlock_irq(spinlock_t *lock)
static void spin_lock_irq(spinlock_t *lock)
```

We might think that this solution is sufficient, but it isn't. The _irq variant partially solves the problem. Imagine interrupts are already disabled on the processor before your code starts locking; when you call spin_unlock_irq(), you will not just release the lock but enable interrupts also, but probably in an erroneous manner since there is no way for spin_unlock_irq() to know which interrupts were enabled before locking and which were not.

Let's consider the following example:

- Let's say interrupt x and y were disabled before a spinlock was acquired, and z was not.

- spin_lock_irq() will disable the interrupts (x, y, and z are now disabled) and take the lock.

- spin_unlock_irq() will enable the interrupts. x, y, and z find themselves all enabled, which was not the case before acquiring the lock. This is where the problem lies.

This makes `spin_lock_irq()` unsafe when called from IRQs off-context as its counterpart `spin_unlock_irq()` will dumbly enable IRQs, with the risk of enabling those that were not enabled while `spin_lock_irq()` was invoked. It makes sense to use `spin_lock_irq()` only when you know that interrupts are enabled—that is, you are sure nothing else might have disabled interrupts on the local CPU.

Now, imagine if you save the interrupts' status in a variable before acquiring the lock and restore them exactly as they were while releasing—there would be no further issues at all. To achieve this, the kernel provides `_irqsave` variant functions that behave exactly like the `_irq` ones, with saving and restoring interrupts status features in addition. These are `spin_lock_irqsave()` and `spin_lock_irqrestore()`, defined as follows:

```
spin_lock_irqsave(spinlock_t *lock, unsigned long flags)
spin_unlock_irqrestore(spinlock_t *lock, unsigned long flags)
```

> **Note**
>
> `spin_lock()` and all its variants automatically call `preempt_disable()`, which disables preemption on the local CPU, while `spin_unlock()` and its variants call `preempt_enable()`, which tries to enable preemption (Yes—tries!!! It depends on whether other spinlocks are locked, which would affect the value of the preemption counter), and which internally calls `schedule()` if enabled (depending on the current value of the counter, whose current value should be 0). `spin_unlock()` is then a preemption point and might re-enable preemption.

Disabling preemption versus disabling interrupts

Though disabling interrupts may prevent kernel preemption (scheduler tick disabled), nothing prevents the protected section from invoking the scheduler (`schedule()` function). A lot of kernel functions indirectly invoke the scheduler, such as those dealing with spinlocks. As a result, even a simple `printk()` function may invoke the scheduler since it deals with the spinlock that protects the kernel message buffer. The kernel disables or enables the scheduler (and, thus, preemption) by increasing or decreasing a kernel global and per-CPU variable (which defaults to 0, meaning *enabled*) called `preempt_count`. When this variable is greater than 0 (which is checked by the `schedule()` function), the scheduler simply returns and does nothing. This variable is incremented at each invocation of a `spin_lock*` family function. On the other side, releasing a spinlock (any `spin_unlock*` family function) decrements it from 1, and whenever it reaches 0, the scheduler is invoked, meaning that your critical section would not be that atomic.

Thus, only disabling interrupts protects you from kernel preemption only in cases where the protected code does not trigger preemption itself. That said, code that locked a spinlock may not sleep as there would be no way to wake it up (remember—timer interrupts and/or schedulers are disabled on the local CPU).

Mutexes

A mutex is the second and last locking primitive we will discuss in this chapter. It behaves exactly like a spinlock, with the only difference being that your code can sleep. If you try to lock a mutex that is already held by another task, your task will find itself suspended and woken up only when the mutex is released. No spinning this time, meaning that the CPU can do something else while your task waits in a sleeping state. As I mentioned previously, a spinlock is a lock held by a CPU. A mutex, on the other hand, is a lock held by a task.

A mutex is a simple data structure that embeds a wait queue (to put contenders to sleep) and a spinlock to protect access to this wait queue, as illustrated in the following code snippet:

```
struct mutex {
    atomic_long_t owner;
    spinlock_t wait_lock;
#ifdef CONFIG_MUTEX_SPIN_ON_OWNER
    struct optimistic_spin_queue osq; /* Spinner MCS lock */
#endif
    struct list_head wait_list;
[...]
};
```

In the preceding code snippet, elements used only in debugging mode have been removed for the sake of readability. However, as we can see, mutexes are built on top of spinlocks. owner represents the process that owns (holds) the lock. wait_list is the list in which the mutex's contenders are put to sleep. wait_lock is the spinlock that protects wait_list manipulation (removal or insertion of contenders to sleep in it). It helps to keep wait_list coherent on SMP systems.

The mutex APIs can be found in the include/linux/mutex.h header file. Prior to acquiring and releasing a mutex, it must be initialized. As for other kernel core data structures, there is a static initialization, as shown here:

```
static DEFINE_MUTEX(my_mutex);
```

Here is a definition of the DEFINE_MUTEX() macro:

```
#define DEFINE_MUTEX(mutexname) \
struct mutex mutexname = __MUTEX_INITIALIZER(mutexname)
```

A second approach the kernel offers is dynamic initialization, possible thanks to a call to a __mutex_init() low-level function, which is actually wrapped by a much more user-friendly macro, mutex_init(). You can see this in action in the following code snippet:

```
struct fake_data {
    struct i2c_client *client;
    u16 reg_conf;
    struct mutex mutex;
};
static int fake_probe(struct i2c_client *client)
{
[...]
    mutex_init(&data->mutex);
[...]
}
```

Acquiring (aka locking) a mutex is as simple as calling one of the following three functions:

```
void mutex_lock(struct mutex *lock);
int mutex_lock_interruptible(struct mutex *lock);
int mutex_lock_killable(struct mutex *lock);
```

If the mutex is free (unlocked), your task will immediately acquire it without going to sleep. Otherwise, your task will be put to sleep in a manner that depends on the locking function you use. With mutex_lock(), your task will be put in an uninterruptible sleep state (TASK_UNINTERRUPTIBLE) while waiting for the mutex to be released (if it is held by another task). mutex_lock_interruptible() will put your task in an interruptible sleep state, in which the sleep can be interrupted by any signal. mutex_lock_killable() will allow your sleeping task to be interrupted only by signals that actually kill the task. Each of these functions returns 0 if the lock has been acquired successfully. Moreover, interruptible variants return -EINTR when the locking attempt was interrupted by a signal.

Whichever locking function is used, the mutex owner (and only the owner) should release the mutex using `mutex_unlock()`, defined as follows:

```
void mutex_unlock(struct mutex *lock);
```

If it is worth checking the status of the mutex, you can use `mutex_is_locked()`, as follows:

```
static bool mutex_is_locked(struct mutex *lock)
```

This function simply checks if the mutex owner is NULL and returns `true` if so or `false` otherwise.

> **Note**
>
> It is recommended to use `mutex_lock()` only when you can guarantee the mutex will not be held for a long time. If not, you should use an interruptible variant instead.

There are specific rules while using mutexes. The most important ones are enumerated in the `include/linux/mutex.h` kernel mutex API header file, and some of these are outlined here:

- A mutex can be held by one and only one task at a time.
- Once held, the mutex can only be unlocked by the owner (that is, the task that locked it).
- Multiple, recursive, or nested locks/unlocks are not allowed.
- A mutex object must be initialized via the API. It must not be initialized by copying nor by using `memset`, just as held mutexes must not be reinitialized.
- A task that holds a mutex may not exit, just as memory areas where held locks reside must not be freed.
- Mutexes may not be used in hardware or software interrupt contexts such as tasklets and timers.

All this makes mutexes suitable for the following cases:

- Locking only in the user context
- If the protected resource is not accessed from an IRQ handler and the operations need not be atomic

However, it may be cheaper (in terms of CPU cycles) to use spinlocks for very small critical sections since the spinlock only suspends the scheduler and starts spinning, compared to the cost of using a mutex, which needs to suspend the current task and insert it into the mutex's wait queue, requiring the scheduler to switch to another task and rescheduling the sleeping task once the mutex is released.

Trylock methods

There are cases where we may need to acquire the lock only if it is not already held by another contender elsewhere. Such methods try to acquire the lock and immediately (without spinning if we are using a spinlock, nor sleeping if we are using a mutex) return a status value, showing whether the lock has been successfully locked or not.

Both spinlock and mutex APIs provide a trylock method. These are, respectively, spin_trylock() and mutex_trylock(), the latter of which you can see here. Both methods return 0 on failure (the lock is already locked) or 1 on success (lock acquired). Thus, it makes sense to use these functions along with an if statement:

```
int mutex_trylock(struct mutex *lock)
```

spin_trylock() actually targets spinlocks. It will lock the spinlock if it is not already locked, just as the spin_lock() method does. However, it immediately returns 0 without spinning in cases where the spinlock is already locked. You can see it in action here:

```
static DEFINE_SPINLOCK(foo_lock);
[...]
static void foo(void)
{
    [...]
    if (!spin_trylock(&foo_lock)) {
        /* Failure! the spinlock is already locked */
        [...]
        return;
    }
    /*
     * reaching this part of the code means that the
     * spinlock has been successfully locked
     */
    [...]
```

```
    spin_unlock(&foo_lock);
    [...]
}
```

On the other hand, `mutex_trylock()` targets mutexes. It will lock the mutex if it is not already locked, just as the `mutex_lock()` method does. However, it immediately returns 0 without sleeping in cases where the mutex is already locked. You can see an example of this in the following code snippet:

```
static DEFINE_MUTEX(bar_mutex);
[...]
static void bar (void)
{
    [...]
    if (!mutex_trylock(&bar_mutex)){
        /* Failure! the mutex is already locked */
        [...]
        return;
    }
    /*
     * reaching this part of the code means that the
     * mutex has been successfully acquired
     */
    [...]
    mutex_unlock(&bar_mutex);
    [...]
}
```

In the preceding excerpt, `mutex_trylock()` is used along with an `if` statement so that the driver can adapt its behavior.

Now that we are done with the trylock variant, let's switch to a totally different concept—learning how to explicitly delay execution.

Dealing with kernel waiting, sleeping, and delay mechanisms

The term *sleeping* in this section refers to a mechanism by which a task (on behalf of the running kernel code) voluntarily relaxes the processor, with the possibility of another task being scheduled. While simple sleeping would consist of a task sleeping and being awakened after a given duration (to passively delay an operation, for example), there are sleeping mechanisms based on external events (such as data availability). Simple sleeps are implemented in the kernel using dedicated APIs; waking up from such sleeps is implicit (handled by the kernel itself) after the duration expires. The other sleeping mechanism is conditioned on an event and the waking-up is explicit (another task must explicitly wake us up based on a condition, else we sleep forever) unless a sleeping timeout is specified. This mechanism is implemented in the kernel using the concept of wait queues. That said, both sleep APIs and wait queues implement what we can call passive waiting. The difference between the two is how the waking-up process occurs.

Wait queue

The kernel scheduler manages a list of tasks to run (tasks in a `TASK_RUNNING` state), known as a runqueue. On the other hand, sleeping tasks, whether interruptible or not (in a `TASK_INTERRUPTIBLE` or `TASK_UNINTERRUPTIBLE` state), have their own queues, known as wait queues.

Wait queues are a higher-level mechanism essentially used to process blocking **input/ output (I/O)**, to wait for a condition to be `true`, to wait for a given event to occur, or to sense data or resource availability. To understand how they work, let's have a look at the following structure in `include/linux/wait.h`:

```
struct wait_queue_head {
    spinlock_t lock;
    struct list_head head;
};
```

A wait queue is nothing but a list (with sleeping processes in it waiting to be awakened) and a spinlock to protect access to this list. We can use a wait queue when more than one process wants to sleep, waiting for one or more events to occur in order to be awakened. The head member is actually a list of processes waiting for the event(s). Each process that wants to sleep while waiting for the event to occur puts itself it this list before going to sleep. While a process is in the list, it is called wait queue entry. When an event occurs, one or more processes on the list are woken up and moved off the list.

We can declare and initialize a wait queue in two ways. The first method is to statically use `DECLARE_WAIT_QUEUE_HEAD`, as follows:

```
DECLARE_WAIT_QUEUE_HEAD(my_event);
```

Alternatively, we can dynamically use `init_waitqueue_head()`, as follows:

```
wait_queue_head_t my_event;
init_waitqueue_head(&my_event);
```

Any process that wants to sleep while waiting for my_event to occur can invoke either `wait_event_interruptible()` or `wait_event()`. Most of the time, the event is just the fact that a resource becomes available, thus it makes sense for a process to go to sleep only after a first check of the availability of that resource. To make things easy, these functions both take an expression in place of the second argument so that the process is put to sleep only if the expression evaluates `false`, as illustrated in the following code snippet:

```
wait_event(&my_event, (event_occured == 1));
/* or */
wait_event_interruptible(&my_event, (event_occured == 1));
```

`wait_event()` or `wait_event_interruptible()` simply evaluates the condition when called. If the condition is `false`, the process is put into either a TASK_UNINTERRUPTIBLE or a TASK_INTERRUPTIBLE (for the _interruptible variant) state and removed from the runqueue.

> **Note**
>
> `wait_event()` puts the process into an exclusive wait, aka uninterruptible sleep, and can't thus be interrupted by the signal. It should be used only for critical tasks. Interruptible functions are recommended in most situations.

There may be cases where you need not only the condition to be `true` but to time out after a certain waiting duration. You can address such cases using `wait_event_timeout()`, whose prototype is shown here:

```
wait_event_timeout(wq_head, condition, timeout)
```

This function has two behaviors, depending on the timeout having elapsed or not. These are outlined here:

- **Timeout elapsed**: The function returns 0 if the condition is evaluated to `false` or 1 if it is evaluated to `true`.

- **Timeout not elapsed yet**: The function returns the remaining time (in jiffies—at least 1) if the condition is evaluated to `true`.

The time unit for timeout is a jiffy. There are convenient APIs to convert convenient time units such as milliseconds and microseconds to jiffies, defined as follows:

```
unsigned long msecs_to_jiffies(const unsigned int m)
unsigned long usecs_to_jiffies(const unsigned int u)
```

After a change on any variable that could affect the result of the wait condition, you must call the appropriate `wake_up*` family function. That being said, in order to wake up a process sleeping on a wait queue, you should call either `wake_up()`, `wake_up_all()`, `wake_up_interruptible()`, or `wake_up_interruptible_all()`. Whenever you call any of these functions, the condition is re-evaluated again. If the condition is `true` at that time, then a process (or all processes for the `_all()` variant) in the wait queue will be awakened, and its (their) state set to `TASK_RUNNING`; otherwise, (the condition is `false`), nothing happens. The following code snippet illustrates this concept:

```
wake_up(&my_event);
wake_up_all(&my_event);
wake_up_interruptible(&my_event);
wake_up_interruptible_all(&my_event);
```

In the preceding code snippet, `wake_up()` will wake only one process from the wait queue, while `wake_up_all()` will wake all processes from the wait queue. On the other hand, `wake_up_interruptible()` will wake only one process from the wait queue that is in interruptible sleep, and `wake_up_interruptible_all()` will wake all processes from the wait queue that are in interruptible sleep.

Because they can be interrupted by signals, you should check the return value of the
_interruptible variants. A nonzero value means your sleep has been interrupted
by some sort of signal, and the driver should return ERESTARTSYS, as illustrated in the
following code snippet:

```c
#include <linux/module.h>
#include <linux/init.h>
#include <linux/sched.h>
#include <linux/time.h>
#include <linux/delay.h>
#include<linux/workqueue.h>

static DECLARE_WAIT_QUEUE_HEAD(my_wq);
static int condition = 0;
/* declare a work queue*/
static struct work_struct wrk;

static void work_handler(struct work_struct *work)
{
    pr_info("Waitqueue module handler %s\n", __FUNCTION__);
    msleep(5000);
    pr_info("Wake up the sleeping module\n");
    condition = 1;
    wake_up_interruptible(&my_wq);
}
static int __init my_init(void)
{
    pr_info("Wait queue example\n");
    INIT_WORK(&wrk, work_handler);
    schedule_work(&wrk);
    pr_info("Going to sleep %s\n", __FUNCTION__);
    if (wait_event_interruptible(my_wq, condition != 0)) {
        pr_info("Our sleep has been interrupted\n");
        return -ERESTARTSYS;
    }
    pr_info("woken up by the work job\n");
    return 0;
```

```
}
void my_exit(void)
{
    pr_info("waitqueue example cleanup\n");
}
module_init(my_init)
module_exit(my_exit);
MODULE_AUTHOR("John Madieu <john.madieu@gmail.com>");
MODULE_LICENSE("GPL");
```

In the preceding example, we have used the `msleep()` API, which will be explained shortly. Back to the behavior of the code—the current process (actually, `insmod`) will be put to sleep in the wait queue for 5 seconds and woken up by the work handler. The `dmesg` output is shown here:

```
[342081.385491] Wait queue example
[342081.385505] Going to sleep my_init
[342081.385515] Waitqueue module handler work_handler
[342086.387017] Wake up the sleeping module
[342086.387096] woken up by the work job
[342092.912033] waitqueue example cleanup
```

Now that we are comfortable with the concept of wait queue, which allows us to put processes to sleep and wait for these to be awakened, let's learn another simple sleeping mechanism that simply consists of delaying the execution flow unconditionally.

Simple sleeping in the kernel

This simple sleeping can also be referred to as **passive delay** because the task sleeps (allowing the CPU to schedule another task) while waiting, in contrast to active delay, which is a busy wait as the task waits by wasting the CPU clock and thus consuming resources. Before using sleeping APIs, the driver must include `#include <linux/delay>`, which would make the following function available:

```
usleep_range(unsigned long min, unsigned long max)
msleep(unsigned long msecs)
msleep(unsigned long msecs)
msleep_interruptible(unsigned long msecs)
```

In the preceding APIs, `msecs` is the number of milliseconds of sleep. `min` and `max` are the minimum and upper bounds of sleeping in microseconds.

The `usleep_range()` API relies on **high-resolution timers (hrtimers)**, and it is recommended to use this sleep for a few ~µsecs or small `msecs` (between 10 microseconds and 20 milliseconds), avoiding the busy-wait loop of `udelay()`.

`msleep*()` APIs are backed by `jiffies`/legacy timers. You should use this for larger milliseconds of sleep (10 milliseconds or more). This API sets the current task to `TASK_UNINTERRUPTIBLE`, whereas `msleep_interruptible()` sets the current task to `TASK_INTERRUPTIBLE` before scheduling the sleep. In short, the difference is whether the sleep can be ended early by a signal. It is recommended to use `msleep()` unless you have a need for the interruptible variant.

APIs in this section must be used for inserting delays in a non-atomic context exclusively.

Kernel delay or busy waiting

First, the term "delay" in this section can be considered as busy waiting as the task actively waits (corresponding to the `for()` or the `while()` loop), consuming CPU resources, in contrast to sleep, which is a passive delay as the task sleeps while waiting.

Even for busy loop waiting, the driver must include `#include <linux/delay>`, which would make the following APIs available as well:

```
ndelay(unsigned long nsecs)
udelay(unsigned long usecs)
mdelay(unsigned long msecs)
```

The advantage of such APIs is that they can be used in both atomic and non-atomic contexts.

The precision of `ndelay` depends on how accurate your timer is (not always the case on an embedded **system on a chip**, or **SoC**). `ndelay`-level precision may not actually exist on many non-PC devices. Instead, you are more likely to come across the following:

- `udelay`: This API is busy-wait loop-based. It will busy wait for enough loop cycles to achieve the desired delay. You should use this function if you need to sleep for a few µsecs (< ~10 microseconds). It is recommended to use this API even for sleeping less than 10 us because, on slower systems (some embedded SoCs), the overhead of setting up hrtimers for `usleep` may not be worth it. Such an evaluation will obviously depend on your specific situation, but it is something to be aware of.

- `mdelay`: This is a macro wrapper around `udelay` to account for possible overflow when passing large arguments to `udelay`. In general, the use of `mdelay` is discouraged, and code should be refactored to allow for the use of `msleep`.

At this step, we are done with Linux kernel sleeping or delay mechanisms. We should be able to design and implement a time-slice-managed execution flow. We can now get deeper into the way the Linux kernel manages time.

Understanding Linux kernel time management

Time is one of the most used resources in computer systems, right after memory. It is used to do almost everything: timer, sleep, scheduling, and many other tasks.

The Linux kernel includes software timer concepts to enable kernel functions to be invoked at a later time.

The concepts of clocksource, clockevent, and tick device

In the original Linux timer implementation, the main hardware timer was mainly used for timekeeping. It was also programmed to fire interrupts periodically at HZ frequency, whose corresponding period is called a **jiffy** (both are explained later in this chapter, in the *Jiffies and HZ* section). Each of these interrupts generated every 1/HZ second was (and still is) referred to as a tick. Throughout this section, the term *tick* will refer to the interrupt generated at a 1/HZ period.

The whole system time management (either from the kernel or user space) was bound to jiffies, which is also a global variable in the kernel, incremented at each tick. In addition to incrementing the value of `jiffies` (on top of which a timer wheel was implemented), the tick handler was also responsible for processes scheduling, statistic updating, and profiling.

Starting from kernel version 2.6.21, as a first improvement, hrtimers implementation was merged (and is now available through `CONFIG_HIGH_RES_TIMERS`). This feature was (and still is) transparent and has come with `hrtimer` timers as a functionality of its own, with a new data type, `ktime_t`, which is used to keep time value on a nanosecond basis. The former legacy (tick-based and low-res) timer implementation remained, however. The improvement converted the `nanosleep()`, **interval timers** (**itimers**), and **Portable Operating System Interface** (**POSIX**) timers to rely on high-resolution timers, resulting in better granularity and accuracy than with the current jiffy resolution. The unification of `nanosleep()` and `clock_nanosleep()` was made possible by the conversion of `nanosleep()` and POSIX timers. Without this improvement, the best accuracy that could be obtained for timer events would be 1 jiffy, whose duration depends on the value of `HZ` in the kernel.

> **Note**
> That said, high-resolution timer implementation is an independent feature. hrtimer timers can be enabled whatever the platform, but whether they work in high-resolution mode or not depends on the underlying hardware timer. Otherwise, the system is said to be in **low-resolution** (**low-res**) mode.

Improvements have been done all along the kernel evolutions until the generic clockevent interface came in, with the concepts of clock source, clock event, and tick devices. This completely changed time management in the kernel and influenced CPU power management.

The clocksource framework and clock source devices

A clock source is a monotonic, atomic, and free-running counter. This can be considered as a timer that acts as a free-running counter that provides time stamping and read access to the monotonically increasing time value. The common operation performed on clock source devices is reading the counter's value.

In the kernel, there is a `clocksource_list` global list that tracks the clock source devices registered with the system, enqueued ordered by rating. This allows the Linux kernel to know about all registered clock source devices and switch to a clock source with a better rating and features. For instance, the `__clocksource_select()` function is invoked after each registration of a new clock source, which ensures that the best clock source is always selected. Clock sources are registered with either `clocksource_mmio_init()` or `clocksource_register_hz()` (you can `grep` these words). However, clock source device drivers are in `drivers/clocksource/` in kernel sources.

On a running Linux system, the most intuitive way to list clock source devices that are registered with the framework is by looking for the word `clocksource` in the kernel log message buffer, as shown here:

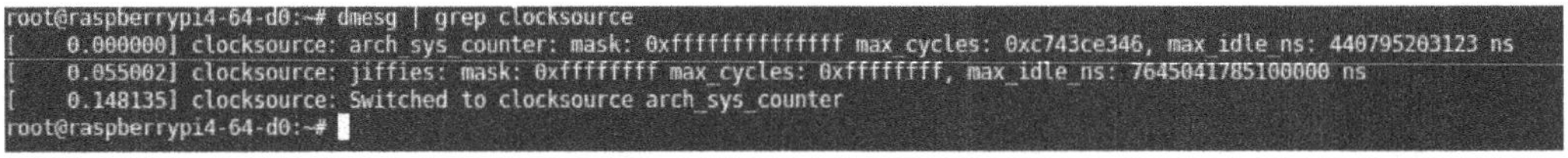

```
root@raspberrypi4-64-d0:~# dmesg | grep clocksource
[    0.000000] clocksource: arch_sys_counter: mask: 0xffffffffffffff max_cycles: 0xc743ce346, max_idle_ns: 440795203123 ns
[    0.055002] clocksource: jiffies: mask: 0xffffffff max_cycles: 0xffffffff, max_idle_ns: 7645041785100000 ns
[    0.148135] clocksource: Switched to clocksource arch_sys_counter
root@raspberrypi4-64-d0:~#
```

Figure 3.1 – System clocksource list

> **Note**
>
> In the preceding output logs (from a Pi 4), the `jiffies` clock source is a jiffy granularity-based and always provided clock source registered by the kernel as `clocksource_jiffies` in `kernel/time/jiffies.c`, with the lowest valid rating value (that is, used as the last resort). On the x86 platform, this clock source is refined and renamed into `refined-jiffies`—see the `register_refined_jiffies()` function call in `arch/x86/kernel/setup.c`.

However, the preferred way (especially where the `dmesg` buffer has rotated or has been cleared) to enumerate the available clock source on a running Linux system is by reading the content of the `available_clocksource` file in `/sys/devices/system/clocksource/clocksource0/`, as shown in the following code snippet (on a Pi 4):

```
root@raspberrypi4-64-d0:~# cat  /sys/devices/system/
clocksource/clocksource0/available_clocksource
arch_sys_counter
root@raspberrypi4-64-d0:~#
```

On an i.MX6 board, we have the following:

```
root@udoo-labcsmart:~# cat   /sys/devices/system/clocksource/
clocksource0/available_clocksource
mxc_timer1
root@udoo-labcsmart:~#
```

To check the currently used clock source, you can use the following code:

```
root@raspberrypi4-64-d0:~# cat  /sys/devices/system/
clocksource/clocksource0/current_clocksource
arch_sys_counter
root@raspberrypi4-64-d0:~#
```

On my x86 machine, we have the following for both available clock sources and the currently used one:

```
jma@labcsmart:~$ cat /sys/devices/system/clocksource/
clocksource0/available_clocksource
tsc hpet acpi_pm
jma@labcsmart:~$ cat /sys/devices/system/clocksource/
clocksource0/current_clocksource
tsc
jma@labcsmart:~$
```

To change the current clock source, you can echo the name of one of the available clock sources into the current_clocksource file, like this:

```
jma@labcsmart:~$ echo acpi_pm >  /sys/devices/system/
clocksource/clocksource0/current_clocksource
jma@labcsmart:~$
```

Changing the current clock source must be done with caution since the current clock source selected by the kernel during the boot is always the best one.

Linux kernel timekeeping

One of the main goals of the clock source device is feeding the timekeeper. There can be multiple clock sources in a system, but the timekeeper will choose the one with the highest precision to use. The timekeeper needs to obtain the value of the clock source periodically to update the system time, which is usually updated during the tick processing, as illustrated in the following diagram:

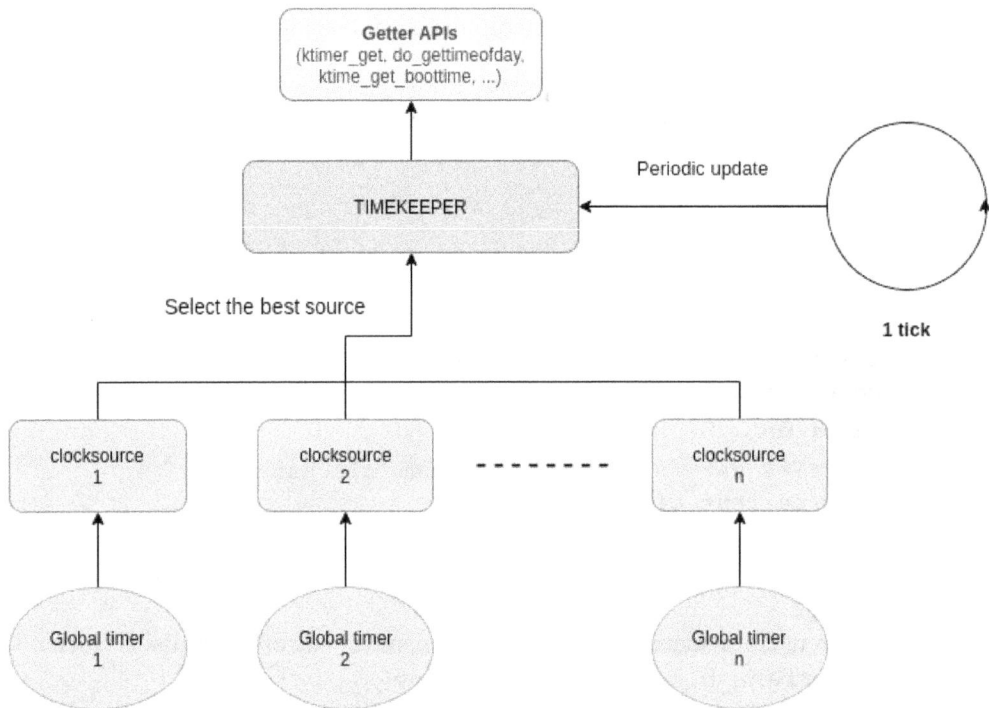

Figure 3.2 – Linux kernel timekeeper implementation

The timekeeper provides several types of time: **xtime, monotonic time, raw monotonic time**, and **boot time**, which are outlined in more detail here:

- **xtime**: This is wall (real) time, which represents the current time as given by the **Real-Time Clock (RTC)** chip.

- **Monotonic time**: The cumulative time since the system is turned on, but does not count the time the system sleeps.

- **Raw monotonic time**: This has the same meaning as monotonic time, but it is purer and will not be affected by **Network Time Protocol (NTP)** time adjustment.

- **Boot time**: This adds the time the system spent sleeping to the monotonic time, which gives the total time after the system is powered on.

The following table shows the different types of time and their kernel getter functions:

Time type	Precision	Access speed	Cumulative sleep time adjustments	Affected by NTP	Getter APIs
RTC	Low	Slow (**Inter-Integrated Circuit (I2C)** most of the time)	Yes	Yes	/ (NA or does not exist)
xtime	High	Quick	Yes	Yes	`do_gettimeofday()`, `ktime_get_real_ts()`, and `ktime_get_real()`
Monotonic	High	Quick	No	Yes	`ktime_get()` and `ktime_get_ts64()`
Raw monotonic	High	Quick	No	No	`ktime_get_raw()` and `getrawmonotonic64()`
Boot time	High	Quick	Yes	Yes	`ktime_get_boottime()`

Table 3.1 – Linux kernel timekeeping functions

Now that we are familiar with the Linux kernel timekeeping mechanisms and APIs, we are free to learn another concept involved in this time management—the clockevent framework.

The clockevent framework and clock event devices

Before the concept of clockevent was introduced, the locality of hardware timers was not considered. The clock source/event hardware was programmed to periodically generate HZ ticks (interrupts) per second, the interval between each tick being a jiffy. With the introduction of clockevent/source in the kernel, the interruption of the clock became abstracted as an event. The main function of the clockevent framework is to distribute the clock interrupts (events) and set the next trigger condition. It is a generic framework for next-event interrupt programming.

A clock event device is a device that can fire interrupts and allow us to program when the next interrupt (an event) will poke in the future. Each clock event device driver must provide a `set_next_event` function (or `set_next_ktime` in the case of an hrtimer-backed clock event device), which is used by the framework when it comes to using the underlying clock event device to program the next interrupt.

Clock event devices are orthogonal to clock source devices. This is probably why their drivers are in the same place (and, sometimes, in the same compilation unit) as clock source device drivers—that is, in `drivers/clocksource`. On most platforms, the same hardware and register range may be used for the clock event and for the clock source, but they are essentially different things. This is the case, for example, with the BCM2835 System Timer, which is a memory-mapped peripheral found on the BCM2835 used in the Raspberry Pi. It has a 64-bit free-running counter that runs at 1 **megahertz** (**MHz**), as well as four distinct "output compare registers" that can be used to schedule interrupts. In such cases, the driver usually registers the clock source and the clock event device in the same compilation unit.

On a running Linux system, the available clock event devices can by listed from the `/sys/devices/system/clockevents/` directory. Here is an example on a Pi 4:

```
root@raspberrypi4-64-d0:~# ls /sys/devices/system/clockevents/
broadcast     clockevent1   clockevent3   uevent
clockevent0   clockevent2   power
root@raspberrypi4-64-d0:~#
```

On a dual-core i.MX6 running system, we have the following:

```
root@udoo-labcsmart:~# ls /sys/devices/system/clockevents/
broadcast     clockevent0   clockevent1   consumers   power
suppliers     uevent
root@empair-labcsmart:~#
```

And finally, on my height core machine, we have the following:

```
jma@labcsmart:~$ ls /sys/devices/system/clockevents/
broadcast   clockevent0   clockevent1   clockevent2   clockevent3
clockevent4   clockevent5   clockevent6   clockevent7   power
uevent
jma@labcsmart:~$
```

From the preceding listings of available clock event devices on the system, we can say the following:

- There are as many clock event devices as CPUs on the system (allowing per-CPU clock devices, thus involving timer locality).

- There is always a strange directory, `broadcast`. We will discuss this particular timer in the next sections.

To know the underlying timer of a given clock event device, you can read the content of `current_device` in the clock event directory. We'll now look at some examples on three different machines.

On the i.MX 6 platform, we have the following:

```
root@udoo-labcsmart:~# cat /sys/devices/system/clockevents/
clockevent0/current_device
local_timer
root@udoo-labcsmart:~# cat /sys/devices/system/clockevents/
clockevent1/current_device
local_timer
```

On the Pi 4, we have the following:

```
root@raspberrypi4-64-d0:~# cat /sys/devices/system/clockevents/
clockevent2/current_device
arch_sys_timer
root@raspberrypi4-64-d0:~# cat /sys/devices/system/clockevents/
clockevent3/current_device
arch_sys_timer
```

On my x86 running machine, we have the following:

```
jma@labcsmart:~$ cat /sys/devices/system/clockevents/
clockevent0/current_device
lapic-deadline
jma@labcsmart:~$ cat /sys/devices/system/clockevents/
clockevent1/current_device
lapic-deadline
```

For the sake of readability, the choice has been made to read two entries only, and from what we have read, we can conclude the following:

- Clock event devices are backed by the same hardware timer, which is different from the hardware timer backing the clock source device.

- At least two hardware timers are needed to support the high-resolution timer interface, one playing the clock source role and one (ideally per-CPU) baking clock event devices.

A clock event device can be configured to work in either one-shot mode or in periodic mode, as outlined here:

- In periodic mode, it is configured to generate a tick every 1/HZ second and does all the things the legacy (low-resolution) timer-based tick did, such as updating jiffies, accounting CPU time, and so on. In other words, in periodic mode, it is used the same way as the legacy low-resolution timer was, but it is run out of the new infrastructure.

- One-shot mode makes the hardware generate a tick after a specific number of cycles from the current time. It is mostly used to program the next interrupt that will wake the CPU before it goes idle.

To track the operating mode of a clock event device, the concept of a tick device was introduced. This is further explained in the next section.

Tick devices

Tick devices are software extensions of clock event devices to provide a continuous stream of tick events that happen at regular time intervals. A tick device is automatically created by the kernel when a new clock event device is registered and always selects the best clock event device. It then goes without saying that a tick device is bound to a clock event device and that a tick device is backed by a clock event device.

The definition of a tick-device data structure is shown in the following code snippet:

```
struct tick_device {
    struct clock_event_device *evtdev;
    enum tick_device_mode mode;
};
```

In this data structure, evtdev is the clock event device that is abstracted by the tick device. mode is used to track the working mode of the underlying clock event. Therefore, when a tick device is said to be in periodic mode, it also means that the underlying clock event device is configured to work in this mode. The following diagram illustrates this:

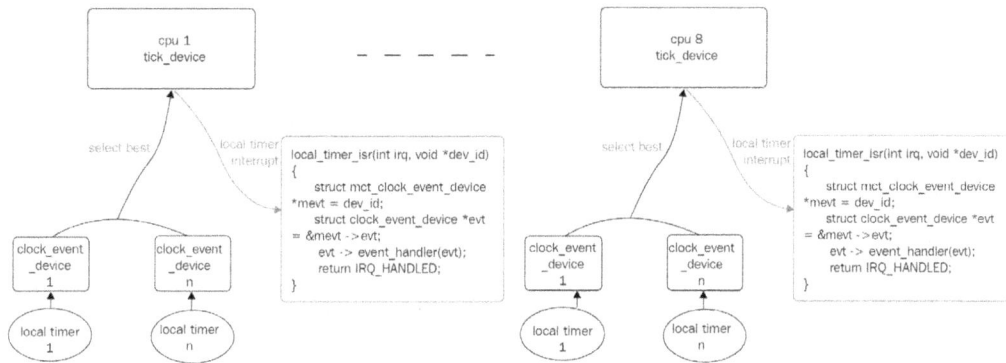

Figure 3.3 – Clockevent and tick-device correlation

A tick device can either be global to the system or local (per-CPU tick device). Whether a tick device must be global or not is decided by the framework, by selecting one local tick device based on the features of the underlying clock event device of this tick device. The descriptions of each type of tick device is as follows:

- A per-CPU tick device is used to provide local CPU functionality such as process accounting, profiling, and—obviously—CPU local periodic tick (in periodic mode) and CPU local next event interrupt (non-periodic mode), for CPU local hrtimers management (see the update_process_times() function to learn how all that is handled). In the timer core code, there is a tick_cpu_device per-CPU variable that represents the instance of the tick device for each CPU in the system.

- A global tick device is responsible for providing the period ticks that mainly run the do_timer() and update_wall_time() functions. Thus, the first one updates the global jiffies value and updates the system load average, while the latter updates the wall time (which is stored in xtime, which records the time difference from January 1, 1970, to now), and runs any dynamic timers that have expired (for instance, running local process timers). In the timer core code, there is the tick_do_timer_cpu global variable, which holds the CPU number whose tick device has the role of the global tick device—the one that executes do_timer(). There is another global variable, tick_next_period, which keeps track of the next time the global tick device will fire.

> **Note**
> This also means that the jiffies variable is always managed from one core at a time, but its function management affinity can jump from one core to the another as and when tick_do_timer_cpu changes.

From its interrupt routine, the driver of the underlying clock event device must invoke `evtdev->event_handler()`, which is the default handler of the clock device installed by the framework. While it is transparent for the device driver, this handler is set by the framework depending on other parameters, as follows: two kernel configuration options (`CONFIG_HIGH_RES_TIMERS` and `CONFIG_NO_HZ`), the underlying hardware timer resolution, and whether the tick device is operating in dynamic mode or one-shot mode.

`NO_HZ` is the kernel option enabling dynamic tick support, and `HIGH_RES_TIMERS` allows the use of hrtimer APIs. With hrtimers enabled, the base code is still tick-driven, but the periodic tick interrupt is replaced by timers under hrtimers (`softirq` is called in the timer softirq context). However, whether the hrtimers will work in high-resolution mode depends on the underlying hardware timer being high-resolution or not. If not, the hrtimers will be fed by the old low-resolution tick-based timer.

A tick device can operate in either one-shot mode or periodic mode. In periodic mode, the framework uses a per-CPU hrtimer via a control structure to emulate the ticks so that the base code is still tick-driven, but the periodic tick interrupt is replaced by timers under hrtimers embedded in the control structure. This control structure is a `tick_sched` struct, defined as follows:

```
struct tick_sched {
    struct hrtimer              sched_timer;
    enum tick_nohz_mode         nohz_mode;
[...]
};
```

Then, a per-CPU instance of this control structure is declared, as follows:

```
static DEFINE_PER_CPU(struct tick_sched, tick_cpu_sched);
```

This per-CPU instance allows a per-CPU tick emulation, which will drive the low-res timer processing via the `sched_timer` element, periodically reprogrammed to the next-low-res-timer-expires interval. This seems, however, obvious since each CPU has its own runqueue and ready processes list to manage.

A `tick_sched` element can be configured by the framework to work in three different modes, described as follows:

- `NOHZ_MODE_INACTIVE`: This mode means no dynamic tick and no hrtimers support. It is the state the system is in during initialization. In this mode, the local per-CPU tick-device event handler is `tick_handle_periodic()`, and entering `idle` will be interrupted by the tick timer interrupt.

- NOHZ_MODE_LOWRES: This is also the lowres mode, which means a dynamic tick enabled in low-resolution mode. It means no high-resolution hardware timer has been found on the system to allow hrtimers to work in high-resolution mode and that they work in low-precision mode, which has the same precision as the low-precision timer (**software (SW)** local timer), which is based on the tick. In this mode, the local per-CPU tick-device event handler is tick_nohz_handler(), and entering idle will not be interrupted by the tick timer interrupt.

- NOHZ_MODE_HIGHRES: This is also the highres mode. In this mode, both dynamic tick and hrtimer "high-resolution" modes are enabled. The local per-CPU tick-device event handler is hrtimer_interrupt(). Here, hrtimers work in high-precision mode, which has the same accuracy as the hardware timer (**hardware (HW)** local timer), which is much greater than the tick accuracy of the low-precision timer. In order to support the high-precision mode of hrtimers, hrtimers directly uses the one-shot mode of tick_device, and the conventional tick timer is converted into a sub-timer of hrtimer.

Tick core-related source code in the kernel is in the kernel/time/ directory, implemented in tick-*.c files. These are tick-broadcast.c, tick-common.c, tick-broadcast-hrtimer.c, tick-legacy.c, tick-oneshot.c, and tick-sched.c.

Broadcast tick device

On platforms implementing CPU power management, most (if not all) of the hardware timers backing clock event devices will be turned off in some CPUidle states. To keep the software timers functional, the kernel relies on an always-on clock event device (that is, backed by an always-on timer) to be present in the platform to relay the interrupt signaling when the timer expires. This always-on timer is called a **broadcast clock device**. In a nutshell, a tick broadcast is used to wake idle CPUs. It is represented by the broadcast directory in /sys/devices/system/clockevents/.

> **Note**
>
> A broadcast tick device is able to wake up any CPU by issuing an **inter-processor interrupt (IPI)** (discussed in *Chapter 13, Demystifying the Kernel IRQ Framework*) to that CPU. See wake_up_nohz_cpu(), which is used for this purpose.

To see the underlying timer backing the broadcast device, you can read the current_device variable in its directory.

On the x86 platform, we have the following output:

```
jma@labcsmart:~$ cat /sys/devices/system/clockevents/broadcast/
current_device
hpet
```

The Pi 4 output is shown here:

```
root@raspberrypi4-64-d0:~# cat /sys/devices/system/clockevents/
broadcast/current_device
bc_hrtimer
```

Finally, the i.MX 6 broadcast device is backed by the following timer:

```
root@udoo-labcsmart:~# cat /sys/devices/system/clockevents/
broadcast/current_device
mxc_timer1
```

From the preceding output showing the timer backing the broadcast device, we can conclude that clock source, clock event, and broadcast device timers are all different.

Whether a tick device can be used as a broadcast device or not is decided by the tick_install_broadcast_device() core function, invoked at each tick-device registration. This function will exclude tick devices with CLOCK_EVT_FEAT_C3STOP flags set (which means the underlying clock event device's timer stops in the C3 idle state) and rely on other criteria (such as supporting one-shot mode—that is, having the CLOCK_EVT_FEAT_ONESHOT flag set). Finally, the tick_broadcast_device global variable defined in kernel/time/tick-broadcast.c contains the tick device that has the role of a broadcast device. When tick device is selected as a broadcast device, its next event handler is set to tick_handle_periodic_broadcast(), instead of to tick_handle_periodic().

There are, however, platforms implementing CPU core gating that do not have an always-on hardware timer. For such platforms, the kernel provides a kernel hrtimer-based clock event device that is unconditionally registered upon boot (and can be chosen as a tick broadcast device) with the lowest possible rating value so that any broadcast-capable hardware clock event device present on the system will be chosen in preference to the tick broadcast device.

This hrtimer-backed clock event device relies on a dynamically chosen CPU (such that if there is a CPU about to enter into deep sleep with its wake-up time earlier than the hrtimer expiration time, this CPU becomes the new broadcast CPU) to be always powered up. This CPU will then relay the timer interrupt to CPUs in deep-idle states through its hardware local timer device. It is implemented in `kernel/time/tick-broadcast-hrtimer.c`, registered as `ce_broadcast_hrtimer` with the name field set to `bc_hrtimer`. It is, for instance, as you can see, the broadcast tick device used by the Pi 4 platform.

> **Note**
>
> It goes without saying that having an always-on CPU has implications for power management platform capabilities and makes `CPUidle` suboptimal since at least a CPU is kept always in a shallow idle state by the kernel to relay timer interrupts. It is a trade-off between CPU power management and a high-resolution timer interface, which at least leaves the kernel with a functional system with some working power management capabilities.

Understanding the sched_clock function

`sched_clock()` is a kernel timekeeping and timestamping function that returns the number of nanoseconds since the system started. It is weakly defined (to allow its overriding by architecture or platform code) in `kernel/sched/clock.c`, as follows:

```
unsigned long long __weak sched_clock(void)
```

It is, for instance, the function that provides a timestamp to `printk()` or is invoked when using `ktime_get_boottime()` or related kernel APIs. It defaults to a jiffy-backed implementation (which could affect scheduling accuracy). If overridden, the new implementation must return a 64-bit monotonic timestamp in nanoseconds that represents the number of nanoseconds since the last reboot. Most platforms achieve this by directly reading the timer registers. On platforms that lack timers, this feature is implemented with the same timer as the one used to back the main clock source device. This is the case on Raspberry Pi, for example. When this is the case, the registers to read the main clock source device value and the registers from where the `sched_clock()` value comes are the same: see `drivers/clocksource/bcm2835_timer.c`.

The timer driving `sched_clock()` has to be very fast compared with the clock source timers. As its name states, it is mainly used by the scheduler, which means it is called much more often. If you have to do trade-offs between accuracy compared to the clock source, you may sacrifice accuracy for speed in `sched_clock()`.

When `sched_clock()` is not overridden directly, the kernel time core provides a `sched_clock_register()` helper to supply a platform-dependent timer reading function as well as a rating value. Anyway, this timer reading function will end up in the cd kernel time framework variable, which is of type `struct clock_data` (assuming that the rate of the new underlying timer is greater than the rate of the timer driving the current function).

Dynamic tick/tickless kernel

Dynamic ticks are the logical consequence of migration to a high-resolution timer interface. Before they were introduced, periodic ticks periodically issued interrupts (`HZ` times per second) to drive the operating system. This kept the system awake even when there were no tasks or timer handlers to run. With this approach, long rests were impossible for the CPUs, which had to wake for no real purpose.

The dynamic tick mechanism came with a solution, allowing periodic ticks to be stopped during certain time intervals to save power. With this new approach, periodic ticks are enabled back only when some tasks need to be performed; otherwise, they are disabled.

How does it work? When a CPU has no more tasks to run (that is, the idle task is scheduled on this CPU), only two events can create new work to do after the CPU goes idle: the expiration of one of the internal kernel timers, which is predictable, or the completion of an I/O operation. When a CPU enters an idle state, the timer framework examines the next scheduled timer event and, if it is later than the next periodic tick, it reprograms the per-CPU clock event device to this later event. This will allow the idle CPU to enter longer idle sleeps without being interrupted unnecessarily by a periodic tick.

There are, however, systems with low power states where even the per-CPU clock event device would stop. On such platforms, it is the broadcast tick device that is programmed with the next future event.

Just before a CPU enters such an idle state (the `do_idle()` function), it calls into the tick broadcast framework (`tick_nohz_idle_enter()`), and the periodic tick of its `tick_device` variable is disabled (see `tick_nohz_idle_stop_tick()`). This CPU is then added to a list of CPUs to be woken up by setting the bit corresponding to this CPU in the `tick_broadcast_mask` "broadcast map" variable, which is a bitmap that represents a list of processors that are in a sleeping mode. Then, the framework calculates the time at which this CPU must be woken up (its next event time); if this time is earlier than the time at which `tick_broadcast_device` is currently programmed, the time at which `tick_broadcast_device` should interrupt is updated to reflect the new value, and this new value is programmed into the clock event device backing the broadcast tick device. The `tick_cpu_device` variable of the CPU that is about to enter a deep idle state is now put in shutdown mode, which means that it is no longer functional.

The foregoing procedures are repeated each time a CPU enters a deep idle state, and the `tick_broadcast_device` variable is programmed to fire at the earliest of the wake-up times of the CPUs in deep idle states.

When the tick broadcast device next event pokes, it will look into the bitmask of sleeping CPUs, looking for the CPU(s) owning the timer(s) that might have expired and will send an IPI to any remote CPU in this bitmask that might host an expired timer.

If, however, a CPU leaves the idle state upon an interrupt (the architecture code calls `handle_IRQ()`, which indirectly calls `tick_irq_enter()`), this CPU tick device is enabled (first in one-shot mode), and before it performs any task, the `tick_nohz_irq_enter()` function is called to ensure that `jiffies` are up to date so that the interrupt handler does not have to deal with a stale jiffy value, and then it resumes the periodic tick, which is kept active until the next call to `tick_nohz_idle_stop_tick()` (which is essentially called from `do_idle()`).

Using standard kernel low-precision (low-res) timers

Standard (now legacy and also referred to as low-resolution) timers are kernel timers operating on the granularity of `jiffies`. The resolution of these timers is bound to the resolution of the regular system tick, which depends on the architecture and configuration (that is, `CONFIG_HZ` or, simply, `HZ`) used by the kernel to control the scheduling and execution of a function at a certain point in the future (based on jiffies).

Jiffies and HZ

A jiffy is a kernel unit of time whose duration depends on the value of `HZ`, which represents the incrementation frequency of the `jiffies` variable in the kernel. Each increment event is called a tick. The clock source based on `jiffies` is the lowest common denominator clock source that should function on any system.

Since the `jiffies` variable is incremented `HZ` times every second, if `HZ = 1,000`, then it is incremented 1,000 times per second (that is, one tick every 1/1,000 seconds, or 1 millisecond). On most **Advanced RISC Machines (ARM)** kernel configurations, `HZ` defaults to `100`, while it defaults to `250` on x86, which would result in a resolution of 10 milliseconds or 4 milliseconds.

Here are different `HZ` values on two running systems:

```
jma@labcsmart:~$ grep 'CONFIG_HZ=' /boot/config-$(uname -r)
CONFIG_HZ=250
jma@labcsmart:~$
```

The preceding code has been executed on a running x86 machine. On an ARM running machine, we have the following:

```
root@udoo-labcsmart:~# zcat /proc/config.gz |grep CONFIG_HZ
CONFIG_HZ_100=y
root@udoo-labcsmart:~#
```

The preceding code says the current HZ value is 100.

Kernel timer APIs

A timer is represented in the kernel as an instance of struct timer_list, defined as follows:

```
struct timer_list {
    struct hlist_node entry;
    unsigned long expires;
    void (*function)(struct timer_list *);
    u32 flags;
);
```

In the preceding data structure, expires is an absolute value in jiffies that defines when this timer will expire in the future. entry is internally used by the kernel to track this timer in a per-CPU global list of timers. flags are OR'ed bitmasks that represent the timer flags, such as the way the timer is managed and the CPU on which the callback will be scheduled, and function is the callback to be executed when this timer expires.

You can dynamically define a timer using timer_setup() or statically create one with DEFINE_TIMER().

> **Note**
> In Linux kernel versions prior to 4.15, setup_timer() was used as the dynamic variant.

Here are the definitions of both macros:

```
void timer_setup( struct timer_list *timer,         \
            void (*function)( struct timer_list *), \
            unsigned int flags);


#define DEFINE_TIMER( _name, _function) [...]
```

After the timer has been initialized, you must set its expiration delay before starting it using one of the following APIs:

```
int mod_timer(struct timer_list *timer,
                unsigned long expires);
void add_timer(struct timer_list *timer)
```

The mod_timer() function is used to either set an initial expiration delay or to update its value on an active timer, which means that calling this function on an inactive timer will activate this timer.

> **Note**
> Activating a timer here means arming and queueing this timer. That said, when a timer is just armed, queued, and counting down, waiting for its expiration before running the callback function, it is said to be pending.

You should prefer this function over add_timer(), which is another function to start inactive timers exclusively. Before calling add_timer(), you must have set the timer expiration delay and the callback as follows:

```
my_timer.expires = jiffies + ((12 * HZ) / 10); /* 1.2s */
add_timer(&my_timer);
```

The value mod_timer() returns depends on the state of the timer prior to it being invoked. Calling mod_timer() on an inactive timer returns 0 on success, while it returns 1 when successfully invoked on a pending timer or a timer whose callback function is being executed. This means it is totally safe to execute mod_timer() from the timer callback. When invoked on an active timer, it is equivalent to del_timer(timer); timer->expires = expires; add_timer(timer);.

Once done with the timer, it can be released or cancelled using one of the following functions:

```
int del_timer(struct timer_list *timer);
int del_timer_sync(struct timer_list *timer);
```

`del_timer()` removes (dequeues) the `timer` object from the timer management queue. On success, it returns a different value depending on whether it is invoked on an inactive timer or on an active timer. In the first case, it returns 0, while it returns 1 in the latter case, even if the function callback of this timer is currently being executed.

Let's consider the following execution flow, where a timer is being deleted on a CPU while its callback is being executed on another CPU:

```
mainline (CPUx)                    handler(CPUy)

==============                     =============
                                   enter xxx_timer()
del_timer()
kfree(some_resource)
                                   access(some_resource)
```

In the preceding code snippet, using `del_timer()` does not guarantee that the callback is not running anymore. Here is another example:

```
mainline (CPUx)                    handler(CPUy)

==============                     =============
                                   enter xxx_timer()
 del_timer()
 kfree(timer)
                                   mod_timer(timer)
```

When `del_timer()` returns, it only guarantees that the timer is deactivated and unqueued, ensuring that it will not be executed in the future. However, on a multiprocessing machine, the timer function might already be executing on another processor. `del_timer_sync()` should be used in such cases, which will deactivate the timer and wait for any executing handler to exit before returning. This function will check each processor to make sure that the given timer is not currently running there. By using `del_timer_sync()` in the preceding race condition examples, `kfree()` could be invoked without worrying about the resource being used in the callback or not. You should almost always use `del_timer_sync()` instead of `del_timer()`. The driver must not hold a lock preventing the handler's completion; otherwise, it will result in a deadlock. This makes the `del_timer()` context agnostic as it is asynchronous, while `del_timer_sync()` is to be used in a non-atomic context exclusively.

Moreover, for sanity purposes, we can independently check whether the timer is pending or not using the following API:

```
int timer_pending(const struct timer_list *timer);
```

This function checks whether this timer is armed and pending.

The following code snippet shows a basic usage of standard kernel timers:

```
#include <linux/init.h>
#include <linux/kernel.h>
#include <linux/module.h>
#include <linux/timer.h>


static struct timer_list my_timer;

void my_timer_callback(struct timer_list *t)
{
    pr_info("Timer callback&; called\n");
}

static int __init my_init(void)
{
    int retval;
    pr_info("Timer module loaded\n");

    timer_setup(&my_timer, my_timer_callback, 0);
    pr_info("Setup timer to fire in 500ms (%ld)\n",
                jiffies);
    retval = mod_timer(&my_timer,
                        jiffies + msecs_to_jiffies(500));
    if (retval)
        pr_info("Timer firing failed\n");

    return 0;
}

static void my_exit(void)
```

```
{
    int retval;
    retval = del_timer(&my_timer);
    /* Is timer still active (1) or no (0) */
    if (retval)
        pr_info("The timer is still in use...\n");

    pr_info("Timer module unloaded\n");
}
```

```
module_init(my_init);
module_exit(my_exit);
MODULE_AUTHOR("John Madieu <john.madieu@gmail.com>");
MODULE_DESCRIPTION("Standard timer example");
MODULE_LICENSE("GPL");
```

In the preceding example, we demonstrated a basic usage of standard timers. We request 500 milliseconds of timeout. That said, the unit of time of this kind of timer is a jiffy. So, in order to pass a timeout value in a human format (seconds or milliseconds), you must use conversion helpers. You can see some here:

```
unsigned long msecs_to_jiffies(const unsigned int m)
unsigned long usecs_to_jiffies(const unsigned int u)
unsigned long timespec64_to_jiffies(
                const struct timespec64 *value);
```

With the preceding helper functions, you should not expect any accuracy better than a jiffy. For example, using usecs_to_jiffies(100) will return a jiffy. The returned value is rounded up to the closest jiffy value.

In order to pass additional arguments to the timer callback, the preferred way is to embed them as elements into a structure together with the timer and use the from_timer() macro on the element to retrieve the bigger structure, from which you can access each element. This macro is defined as follows:

```
#define from_timer(var, callback_timer, timer_fieldname) \
    container_of(callback_timer, typeof(*var), timer_fieldname)
```

As an example, let's consider we need to pass two elements to the timer callback: the first one is of type `struct sometype` and the second one is an integer. In order to pass arguments, we define an additional structure, as follows:

```
struct fake_data {
    struct timer_list timer;
    struct sometype foo;
    int bar;
};
```

After this, we pass the embedded timer to the setup function, as follows:

```
struct fake_data *fd = alloc_init_fake_data();
timer_setup(&fd->timer, timer_callback, 0);
```

Later in the callback, you must use the `from_timer` variable to retrieve the bigger structure from which you can access arguments. Here is an example of this in use:

```
void timer_callback(struct timer_list *t)
{
    struct fake_data *fd = from_timer(fd, t, timer);
    sometype data = fd->data;
    int var = fd->bar;
[...]
}
```

In the preceding code snippet, we described how to pass data to the timer callback and how to get this data, using a container structure. By default, a pointer to `timer_list` is passed to the callback function, instead of the `unsigned long` data type in versions prior to 4.15.

High-resolution timers (hrtimers)

While the legacy timer implementation is bound to ticks, high-precision timers provide us with nanosecond-level and tick-agnostic timing accuracy to meet the urgent need for precise time applications or kernel drivers. This has been introduced in kernel v2.6.16 and can be enabled in the build using the `CONFIG_HIGH_RES_TIMERS` option in the kernel configuration.

While the standard timer interface keeps/represents time values in jiffies, the high-resolution timer interface came with a new data type, allowing us to keep the time value: ktime_t, which is a simple 64-bit scalar.

> **Note**
>
> Prior to kernel 3.17, the ktime_t type was represented differently on 32- or 64-bit machines. On 64-bit CPUs, it was represented as a plain 64-bit nanosecond value as it is nowadays all over the kernel, while it was represented as a two-32-bit-fields data structure ([seconds − nanoseconds] pair) on 32-bit CPUs.

Hrtimer APIs require a #include <linux/hrtimer.h> header. That said, in the header file, the structure that characterizes a high-resolution timer is defined as follows:

```
struct hrtimer {
    ktime_t                 _softexpires;
    enum hrtimer_restart    (*function)(struct hrtimer *);
    struct hrtimer_clock_base    *base;
    u8                      state;
[...]
};
```

The elements in the data structure have been shortened to the strict minimum to cover the needs of the book. For the rest, we'll now look at their meaning.

Before using the hrtimer, it must be initialized with hrtimer_init(), defined as follows:

```
void hrtimer_init(struct hrtimer *timer,
                clockid_t which_clock,
                enum hrtimer_mode mode);
```

In the preceding function, timer is a pointer to the hrtimer to initialize. clock_id tells which type of clock must be used to feed this hrtimer. The following are common options:

- CLOCK_REALTIME: This selects the real-time time—that is, the wall time. If the system time changes, it can affect this timer.

- CLOCK_MONOTONIC: This is an incremental time, not affected by system changes. However, it stops incrementing when the system goes to sleep or suspends.

- CLOCK_BOOTTIME: The running time of the system. Similar to CLOCK_ MONOTONIC, the difference is that it includes sleep time. When suspended, it will still increase.

In the preceding code snippet, the mode parameter tells how the hrtimer should be working. Here are some possible options:

- HRTIMER_MODE_ABS: This means that this timer expires after an absolute specified time.

- HRTIMER_MODE_REL: This timer expires after a specified relative from now.

- HRTIMER_MODE_PINNED: This hrtimer is bound to a CPU. This is only considered when starting the hrtimer so that the hrtimer fires and executes the callback on the same CPU on which it is queued.

- HRTIMER_MODE_ABS_PINNED: This is a combination of the first and the third flags.

- HRTIMER_MODE_REL_PINNED: This is a combination of the second and third flags.

After the hrtimer has been initialized, it must be assigned a callback function that will be executed upon the timer expiration. The following code snippet shows the expected prototype:

```
enum hrtimer_restart callback(struct hrtimer *h);
```

hrtimer_restart is the type to be returned by the callback. It must be either HRTIMER_NORESTART to indicate that the timer must not be restarted (used to perform a one-shot operation) or HRTIMER_RESTART to indicate that the timer must be restarted (to simulate periodical mode). In the first case, when returning HRTIMER_NORESTART, the driver will have to explicitly restart the timer (using hrtimer_start(), for example) if need be. When returning HRTIMER_RESTART, the timer restart is implicit as it will be handled by the kernel. However, the driver needs to reset the timeout before returning from the callback. In order to do so, the driver can use hrtimer_forward(), defined as follows:

```
u64 hrtimer_forward(struct hrtimer *timer,
                    ktime_t now, ktime_t interval)
```

In the preceding code snippet, timer is the hrtimer to forward, now is the point from where the timer must be forwarded, and interval is how long in the future the timer must be forwarded. Do, however, note that this only updates the timer expiry value and does not requeue the timer.

The now parameter can be obtained in different ways, either by using `ktime_get()`, which would return the current monotonic clock time or with `hrtimer_get_expires()`, which would return the time when the timer is supposed to expire before forwarding. This is illustrated in the following code snippet:

```
hrtimer_forward(hrtimer, ktime_get(), ms_to_ktime(500));
/* or */
hrtimer_forward(handle, hrtimer_get_expires(handle),
                ns_to_ktime(450));
```

In the first line of the preceding example, the hrtimer is forwarded 500 milliseconds from the current time, while in the second line, it is forwarded 450 nanoseconds from the time when it was supposed to expire. The first line in the example is equivalent to `hrtimer_forward_now()`, which forwards the hrtimer to a specified time from the current time (from now). It is declared as follows:

```
u64 hrtimer_forward_now(struct hrtimer *timer,
                        ktime_t interval)
```

Now that the timer has been set up and its callback defined, it can be armed (started) using `hrtimer_start()`, which has the following prototype:

```
int hrtimer_start(struct hrtimer *timer, ktime_t time,
                  const enum hrtimer_mode mode);
```

In the preceding code snippet, mode represents the timer expiry mode, and it should be either HRTIMER_MODE_ABS for an absolute time value or HRTIMER_MODE_REL for a time value relative to now. This parameter must be consistent with the initialization mode parameter. The timer parameter is a pointer to the initialized hrtimer. Finally, time is the expiry time of the hrtimer. Since it is of type `ktime_t`, various helper functions allow us to generate a `ktime_t` element from various input time units. These are shown here:

```
ktime_t ktime_set(const s64 secs,
                  const unsigned long nsecs);
ktime_t ns_to_ktime(u64 ns);
ktime_t ms_to_ktime(u64 ms);
```

In the preceding list, `ktime_set()` generates a `ktime_t` element from a given number of seconds and nanoseconds. `ns_to_ktime()` or `ms_to_ktime()` generate a `ktime_t` element from a given number of nanoseconds or milliseconds, respectively.

You may also be interested in returning the number of nano-/microseconds, given a `ktime_t` input element using the following functions:

```
s64 ktime_to_ns(const ktime_t kt)
s64 ktime_to_us(const ktime_t kt)
```

Moreover, given one or two `ktime_t` elements, you can perform some arithmetical operations using the following helpers:

```
ktime_t ktime_sub(const ktime_t lhs, const ktime_t rhs);
ktime_t ktime_sub(const ktime_t lhs, const ktime_t rhs);
ktime_t ktime_add(const ktime_t add1, const ktime_t add2);
ktime_t ktime_add_ns(const ktime_t kt, u64 nsec);
```

To subtract or add `ktime` objects, you can use `ktime_sub()` and `ktime_add()`, respectively. `ktime_add_ns()` increments a `ktime_t` element by a specified number of nanoseconds. `ktime_add_us()` is another variant for microseconds. For subtraction, `ktime_sub_ns()` and `ktime_sub_us()` can be used.

After calling `hrtimer_start()`, the hrtimer will be armed (activated) and enqueued in a (time-ordered) per-CPU bucket, waiting for its expiration. This bucket will be local to the CPU on which `hrtimer_start()` has been invoked, but it is not guaranteed that the callback will run on this CPU (migration might happen). For a CPU-bound hrtimer, you should use a `*_PINNED` mode variant when you initialize the hrtimer.

An enqueued hrtimer is always started. Once the timer expires, its callback is invoked, and depending on the return value, the hrtimer can be requeued or not. In order to cancel a timer, drivers can use `hrtimer_cancel()` or `hrtimer_try_to_cancel()`, declared as follows:

```
int hrtimer_cancel(struct hrtimer *timer);
int hrtimer_try_to_cancel(struct hrtimer *timer);
```

Both functions return 0 when the timer is not active during the call. `hrtimer_try_to_cancel()` will return 1 if the timer is active (running but not executing the callback function) and has been successfully canceled or will fail, returning -1 if the callback function is being executed. On the other hand, `hrtimer_cancel()` will cancel the timer if the callback function is not running yet or will wait for it to finish if it is being executed. When `hrtimer_cancel()` returns, the caller can be guaranteed that the timer is no longer active and that its expiration function is not running.

Drivers can, however, independently check whether the hrtimer callback is still running with the following code:

```
int hrtimer_callback_running(struct hrtimer *timer);
```

For instance, hrtimer_try_to_cancel() internally calls hrtimer_callback_running() and returns -1 if the callback is running.

Let's write a module example to put our hrtimer knowledge into practice. We first start by writing the callback function, as follows:

```
#include <linux/module.h>
#include <linux/kernel.h>
#include <linux/hrtimer.h>
#include <linux/ktime.h>

static struct hrtimer hr_timer;

static enum hrtimer_restart timer_callback(struct hrtimer
*timer)
{
    pr_info("Hello from timer!\n");
#ifdef PERIODIC_MS_500
    hrtimer_forward_now(timer, ms_to_ktime(500));
    return HRTIMER_RESTART;
#else
    return HRTIMER_NORESTART;
#endif
}
```

In the preceding hrtimer callback function, we can decide to run in one-shot mode or periodic mode. For periodic mode, the user must define PERIODIC_MS_500, in which case the timer will be forwarded 500 milliseconds in the future from the current hrtimer clock base time before being requeued.

Then, the rest of the module implementation looks like this:

```
static int __init hrtimer_module_init(void)
{;
    ktime_t init_time;
```

```
    init_time = ktime_set(1, 1000);
    hrtimer_init(&hr_timer, CLOCK_MONOTONIC,
                    HRTIMER_MODE_REL);
    hr_timer.function = &timer_callback;
    hrtimer_start(&hr_timer, init_time, HRTIMER_MODE_REL);
    return 0;
}

static void __exit hrtimer_module_exit(void) {
    int ret;
    ret = hrtimer_cancel(&hr_timer);
    if (ret)
        pr_info("Our timer is still in use...\n");
    pr_info("Uninstalling hrtimer module\n");
}

module_init(hrtimer_module_init);
module_exit(hrtimer_module_exit);
```

In the preceding implementation, we generated an initial `ktime_t` element of 1 second and 1,000 nanoseconds—that is, 1 second and 1 millisecond, which is used as initial expiration duration. When the hrtimer expires for the first time, our callback is invoked. If `PERIODIC_MS_500` is defined, the hrtimer will be forwarded to 500 milliseconds later, and the callback will be periodically invoked (every 500 milliseconds) after the initial invocation; otherwise, it is a one-shot invocation.

Implementing work-deferring mechanisms

Deferring is a method by which you schedule a piece of work to be executed in the future. It's a way to report an action later. Obviously, the kernel provides facilities to implement such a mechanism; it allows you to defer functions, whatever their type, to be called and executed later. There are three of them in the kernel, as outlined here:

- **Softirqs**: Executed in an atomic context
- **Tasklets**: Executed in an atomic context
- **Workqueues**: Executed in a process context

In the next three sections, we will learn in detail the implementation of each of them.

Softirqs

As the name suggests, **softirq** stands for **software interrupt**. Such a handler can preempt
all other tasks on the system but the hardware IRQ handlers since it is executed with IRQs
enabled. Softirqs are intended to be used for high-frequency threaded job scheduling.
Network and block devices are the only two subsystems in the kernel that make direct use
of softirqs. Even though softirq handlers run with interrupts enabled, they cannot sleep,
and any shared data needs proper locking. The softirq APIs are implemented in `kernel/`
`softirq.c` in the kernel source tree, and drivers that wish to use this API need to
include `<linux/interrupt.h>`.

Softirqs are represented by `struct softirq_action` structures and are defined as
follows:

```
struct softirq_action {
    void (*action)(struct softirq_action *);
};
```

This structure embeds a pointer to the function to run when the softirq is raised. Thus, the
prototype of your softirq handler should look like this:

```
void softirq_handler(struct softirq_action *h)
```

Running a softirq handler results in executing this action function, which has only
one parameter: a pointer to the corresponding `softirq_action` structure. You can
register the softirq handler at runtime by means of the `open_softirq()` function, as
illustrated here:

```
void open_softirq(int nr,
                  void (*action)(struct softirq_action *))
```

`nr` represents the softirq index, which is also considered as the softirq priority (where 0 is
the highest). `action` is a pointer to the softirq handler. Possible indexes are enumerated
in the following code snippet:

```
enum
{
    HI_SOFTIRQ=0,    /* High-priority tasklets */
    TIMER_SOFTIRQ,   /* Timers */
    NET_TX_SOFTIRQ,  /* Send network packets */
    NET_RX_SOFTIRQ,  /* Receive network packets */
    BLOCK_SOFTIRQ,   /* Block devices */
```

```
    BLOCK_IOPOLL_SOFTIRQ, /* Block devices with I/O polling
                          * blocked on other CPUs */
    TASKLET_SOFTIRQ,/* Normal Priority tasklets */
    SCHED_SOFTIRQ,   /* Scheduler */
    HRTIMER_SOFTIRQ,/* High-resolution timers */
    RCU_SOFTIRQ,     /* RCU locking */
    NR_SOFTIRQS      /* This only represent the number
                     * of softirqs type, 10 actually */
};
```

Softirqs with lower indexes (highest priority) run before those with higher indexes (lowest priority). The name of all the available softirqs in the kernel are listed in the following array:

```
const char * const softirq_to_name[NR_SOFTIRQS] = {
    "HI", "TIMER", "NET_TX", "NET_RX", "BLOCK", "BLOCK_IOPOLL",
    "TASKLET", "SCHED", "HRTIMER", "RCU"
};
```

It's easy to check some in the output of the /proc/softirqs virtual file, as follows:

```
root@udoo-labcsmart:~# cat /proc/softirqs
                  CPU0        CPU1
        HI:       3535          1
     TIMER:    4211589    4748893
    NET_TX:    1277827         39
    NET_RX:    1665450          0
     BLOCK:       1978        201
  IRQ_POLL:          0          0
   TASKLET:     455761         33
     SCHED:    4212802    4750408
   HRTIMER:          3          0
       RCU:     438826     286874
root@udoo-labcsmart:~#
```

A NR_SOFTIRQS-entry array of struct softirq_action is declared in kernel/
softirq.c, as follows:

```
static struct softirq_action softirq_vec[NR_SOFTIRQS] ;
```

Each entry in this array may contain one—and only one—softirq. Consequently, there can
be a maximum of NR_SOFTIRQS (actually, 10 in v5.10, the last stable version at the time
of writing) registered softirqs. The following code snippet from kernel/softirq.c
illustrates this:

```
void open_softirq(int nr,
                      void (*action)(struct softirq_action *))
{
    softirq_vec[nr].action = action;
}
```

A concrete example is the network subsystem, which registers the softirqs it needs (in
net/core/dev.c), as follows:

```
open_softirq(NET_TX_SOFTIRQ, net_tx_action);
open_softirq(NET_RX_SOFTIRQ, net_rx_action);
```

Before a registered softirq can execute, it should be activated/scheduled. In order to do so,
you have to call raise_softirq() or raise_softirq_irqoff() (if interrupts are
already off), as illustrated in the following code snippet:

```
void __raise_softirq_irqoff(unsigned int nr)
void raise_softirq_irqoff(unsigned int nr)
void raise_softirq(unsigned int nr)
```

The first function simply sets the appropriate bit in the per-CPU softirq bitmap (the __
softirq_pending field in the struct irq_cpustat_t data structure allocated per
CPU in kernel/softirq.c), as follows:

```
irq_cpustat_t irq_stat[NR_CPUS] ____cacheline_aligned;
EXPORT_SYMBOL(irq_stat);
```

When the flag is checked, this allows it to run. This function is described here for study
purposes and should not be used directly.

`raise_softirq_irqoff` needs to be called with interrupts disabled. First, it internally calls `__raise_softirq_irqoff()`, described previously, to activate the softirq. Afterward, it checks whether it has been called from within an interrupt (either hardirq or softirq) context by mean of an `in_interrupt()` macro (which simply returns the value of `current_thread_info( )->preempt_count`, where 0 means preemption-enabled, stating that we are not in an interrupt context, and a > 0 value means we are in an interrupt context). If `in_interrupt() > 0`, this does nothing when we are in an interrupt context because softirq flags are checked on the exit path of any I/O IRQ handler (see `asm_do_IRQ()` for ARM or `do_IRQ()` for x86 platforms, which makes a call to `irq_exit()`). Here, softirqs run in an interrupt context. However, if `in_interrupt() == 0`, it invokes `wakeup_softirqd()`, responsible for waking the local CPU `ksoftirqd` thread up (it schedules it, actually) in order to ensure the softirq runs soon but in a process context this time.

`raise_softirq`, on the other hand, first calls `local_irq_save()` (which disables interrupts on the local processor after saving its current interrupt flags). It then calls `raise_softirq_irqoff()`, described previously, in order to schedule the softirq on the local CPU (remember—this function must be invoked with IRQs disabled on the local CPU). Finally, it calls `local_irq_restore()` in order to restore the previously saved interrupt flags.

Here are a few things to remember about softirqs:

- A softirq can never preempt another softirq. Only hardware interrupts can. Softirqs are executed at a high priority, with scheduler preemption disabled but IRQs enabled. This makes softirqs suitable for the most time-critical and important deferred processing on the system.

- While a handler runs on a CPU, other softirqs on this CPU are disabled. Softirqs run concurrently, however. While a softirq is running, another softirq (even the same one) can run on another processor. This is one of the main advantages of softirqs over hardirqs and is the reason why they are used in the networking subsystem, which may require heavy CPU power.

- Softirqs are mostly scheduled in the return path of hardware interrupt handlers. If any is scheduled out of the interrupt context, it will run in a process context if it is still pending when the local `ksoftirqd` thread is given to the CPU. Their execution may be triggered in the following cases:

  - By the local per-CPU timer interrupt (on SMP system only, with `CONFIG_SMP` enabled). See `timer_tick()`, `update_process_times()`, and `run_local_timers()`.

  - By a call to the `local_bh_enable()` function (mostly invoked by the network subsystem for handling packet-receiving/-transmitting softirqs).

  - On the exit path of any I/O IRQ handler (see `do_IRQ`, which makes a call to `irq_exit()`, in turn invoking `invoke_softirq()`.

  - When the local `ksoftirqd` thread is given to the CPU (aka awakened).

The actual kernel function responsible for walking through the softirqs' pending bitmap and running them is `__do_softirq()`, defined in `kernel/softirq.c`. This function is always invoked with interrupts disabled on the local CPU. It does the following tasks:

- Once invoked, the function first saves the current per-CPU pending softirqs' bitmap in a variable named `pending`, and locally disables softirqs by means of `__local_bh_disable_ip`.

- It then resets the current per-CPU pending bitmask (which has already been saved) and then re-enables interrupts (softirqs run with interrupts enabled).

- After this, it enters a `while` loop, checking for pending softirqs in the saved bitmap. If there is no softirq pending, it will execute the handler of each pending softirq, taking care to increment their execution statistics.

- After all pending IRQ handlers have been executed (we are out of the `while` loop), `__do_softirq()` again reads the per-CPU pending bitmask in order to check if any softirqs were scheduled again when it was in the `while` loop. If there are any pending softirqs, the whole process will restart (based on a `goto` loop), starting from *Step 2*. This helps in handling, for example, softirqs that rescheduled themselves.

However, `__do_softirq()` will not repeat if one of the following conditions occurs:

- It has already repeated up to `MAX_SOFTIRQ_RESTART` times, which is set to 10 in `kernel/softirq.c`. This is the limit of the softirqs' processing loop, not the upper bound of the previously described `while` loop.

- It has hogged the CPU more than `MAX_SOFTIRQ_TIME`, which is set to 2 milliseconds (`msecs_to_jiffies(2)`) in `kernel/softirq.c`, since this would prevent the scheduler from being enabled.

If one of the two aforementioned situations occurs, `__do_softirq()` will break its loop and call `wakeup_softirqd()` in order to wake the local `ksoftirqd` thread, which will later execute the pending softirqs in a process context. Since `do_softirq` is called at many points in the kernel, it is likely that another invocation of `__do_softirqs` handles pending softirqs before the `ksoftirqd` thread has a chance to run.

Some words about ksoftirqd

`ksoftirqd` is a per-CPU kernel thread raised to handle unserviced software interrupts. It is spawned early during the kernel boot process, as stated in `kernel/softirq.c` and shown here:

```
static __init int spawn_ksoftirqd(void)
{
    cpuhp_setup_state_nocalls(CPUHP_SOFTIRQ_DEAD,
"softirq:dead",
                        NULL, takeover_tasklets);
    BUG_ON(smpboot_register_percpu_thread(&softirq_threads));
    return 0;
}
early_initcall(spawn_ksoftirqd);
```

After running the top command in the preceding code snippet, we can see some `ksoftirqd/<n>` entries, where `<n>` is the logical CPU index of the CPU running the `ksoftirqd` thread. As `ksoftirqd` threads run in a process context, they are equal to classic processes/threads, so they compete for the CPU. `ksoftirqd` threads hogging CPUs for a long time may indicate a system under heavy load.

Tasklets

Before starting to discuss about **tasklets**, you must notice that these are scheduled for removal in the Linux kernel, thus the purpose of this section is purely for pedagogic reasons, to help you understanding their use in older kernel modules. Consequently, you must not use these in your developments.

Tasklets are bottom halves that are built on top of `HI_SOFTIRQ` and `TASKLET_SOFTIRQ` **softirqs,** with the exception that the `HI_SOFTIRQ`-based tasklets run before `TASKLET_SOFTIRQ`-based ones. Simply put, tasklets are softirqs and obey the same rules. Unlike softirqs, however, two of the same tasklets never run concurrently. The tasklet API is quite basic and intuitive.

Tasklets are represented by a `struct tasklet_struct` structure defined in `<linux/interrupt.h>`. Each instance of this structure represents a unique tasklet, as illustrated in the following code snippet:

```
struct tasklet_struct
{
    struct tasklet_struct *next;
    unsigned long state;
    atomic_t count;
    bool use_callback;
    union {
        void (*func)(unsigned long data);
        void (*callback)(struct tasklet_struct *t);
    };
    unsigned long data;
};
```

Though this API is scheduled for removal, it has been slightly modernized as compared to its legacy implementation. The callback function is stored in the `callback()` field rather than `func()`, which is kept for compatibility with the old implementation. This new callback simply takes a pointer to the `tasklet_struct` structure as its one argument. The handler will be executed by the underlying softirq. It is the equivalent of `action` to a softirq, with the same prototype and the same argument meaning. `data` will be passed as its sole argument.

Whether `callback()` handler or `func()` handler is executed depends on the way the tasklet is initialized. A tasklet can be statically initialized using either `DECLARE_TASKLET()` macro or `DECLARE_TASKLET_OLD()` macro. These macros are defined as follows:

```
#define DECLARE_TASKLET_OLD(name, _func)        \
    struct tasklet_struct name = {              \
    .count = ATOMIC_INIT(0),                    \
    .func = _func,                              \
```

```
}
#define DECLARE_TASKLET(name, _callback)         \
    struct tasklet_struct name = {               \
    .count = ATOMIC_INIT(0),                      \
    .callback = _callback,                        \
    .use_callback = true,                         \
}
```

From what we can see, by using `DECLARE_TASKLET_OLD()`, the legacy implementation is kept and `func()` is used as the callback. Therefore, the prototype of the provided handler must be as follows:

```
void foo(unsigned long data);
```

By using `DECLARE_TASKLET()`, the `callback` field is used as the handler and `use_callback` filed is set to `true` (this is because the tasklet core checks this value to determine the handler that must be invoked). In this case, the protype of the callback is as follows:

```
void foo(struct tasklet_struct *t)
```

In the previous snipped, `t` pointer is passed by the tasklet core while invoking the handler. It will point to your tasklet. Since a pointer to the tasklet is passed as argument to the callback, it is common to embed the tasklet object within a larger, user-specific structure, the pointer to which can be obtained with the `container_of()` macro. In order to do so, you should rather use the dynamic initialization, which can be achieved thanks to `tasklet_setup()` function, defined as follows:

```
void tasklet_setup(struct tasklet_struct *t,
        void (*callback)(struct tasklet_struct *));
```

According to the previous prototype, we can guess that by using the dynamic initialization, we have no choice but to use the new implementation where `callback` field is used as the tasklet handler.

Using static or dynamic method depends on what you need to achieve, for example, if you want the tasklet to be unique for the whole module or to be private per probed device, or even more, if you need to have a direct or indirect reference to the tasklet.

By default, an initialized tasklet is runnable when it is scheduled: *it is said to be enabled*. DECLARE_TASKLET_DISABLED is an alternative to statically initialize default-disabled tasklets. There is no such alternative for a dynamically initialized tasklet, unless you invoke tasklet_disable() on this tasklet after it has been dynamically initialized. A disabled tasklet will require the tasklet_enable() function to be invoked to make this tasklet runnable. Tasklets are scheduled (similar to raising the softirq) via the tasklet_schedule() and tasklet_hi_schedule() functions. You can use tasklet_disable() API to disable a scheduled or running tasklet. This function disables the tasklet and returns only when the tasklet has terminated its execution (assuming it was running). After this, the tasklet can still be scheduled, but it will not run on the CPU until it is enabled again. The asynchronous variant tasklet_disable_nosync() can be used too, which returns immediately, even if the termination has not occurred. Moreover, a tasklet that has been disabled several times should be enabled the same number of times (this is enforced and verified by the kernel thanks to the count field in the struct tasklet object). The definition of the previously mentioned tasklet APIs is illustrated in the following snippet:

```
DECLARE_TASKLET(name, _callback)
DECLARE_TASKLET_DISABLED(name, _callback);
DECLARE_TASKLET_OLD(name, func);
void tasklet_setup(struct tasklet_struct *t,
    void (*callback)(struct tasklet_struct *));
void tasklet_enable(struct tasklet_struct *t);
void tasklet_disable(struct tasklet_struct *t);
void tasklet_schedule(struct tasklet_struct *t);
void tasklet_hi_schedule(struct tasklet_struct *t);
```

The kernel maintains normal-priority and high-priority tasklets in two per-CPU queues (each CPU maintains its low- and high-priority queue pair). tasklet_schedule() adds the tasklet into the normal priority list of the CPU on which it is invoked, scheduling the associated softirq with a TASKLET_SOFTIRQ flag. With tasklet_hi_schedule(), the tasklet is added into the high-priority list (still of the list on which it is invoked), scheduling the associated softirq with a HI_SOFTIRQ flag. When the tasklet is scheduled, its TASKLET_STATE_SCHED flag is set, and the tasklet is queued for execution. At the time of execution, a TASKLET_STATE_RUN flag is set, and the TASKLET_STATE_SCHED state is removed, thus making the tasklet re-schedulable during its execution, either by the tasklet itself or from within an interrupt handler.

High priority tasklets are meant to be used for soft interrupt handlers with low latency requirements. Calling `tasklet_schedule()` on a tasklet already scheduled and whose execution has not started yet will do nothing, resulting in the tasklet being executed only once. A tasklet can reschedule itself, and you can safely call `tasklet_schedule()` in a tasklet. High priority tasklets are always executed before normal ones and should then be used carefully, else you may increase system latency. Stopping a tasklet is as simple as calling `tasklet_kill()`, as illustrated in the following code snippet, which will prevent the tasklet from running again, or waiting for its completion before killing it if the tasklet is currently scheduled to run. If a tasklet re-schedules itself, you should first prevent the tasklet from re-scheduling itself prior to calling this function:

```
void tasklet_kill(struct tasklet_struct *t);
```

Writing your tasklet handler

All that being said, let's see some use cases, as follows:

```
# #include <linux/init.h>
#include <linux/module.h>
#include <linux/kernel.h>
#include <linux/interrupt.h>     /* for tasklets api */


/* Tasklet handler, that just prints the handler name */
void tasklet_function(struct tasklet_struct *t)
{
    pr_info("running %s\n", __func__);
}


DECLARE_TASKLET(my_tasklet, tasklet_function);

static int __init my_init(void)
{
    /* Schedule the handler */
    tasklet_schedule(&my_tasklet);
    pr_info("tasklet example\n");
    return 0;
}

void my_exit( void )
```

```
{
    /* Stop the tasklet before we exit */
    tasklet_kill(&my_tasklet);
    pr_info("tasklet example cleanup\n");
    return;
}

module_init(my_init);
module_exit(my_exit);
MODULE_AUTHOR("John Madieu <john.madieu@gmail.com>");
MODULE_LICENSE("GPL");
```

In the preceding code snippet, we statically declare our `my_tasklet` tasklet and the function supposed to be invoked when this tasklet is scheduled. Since we did not used the `_OLD` variant, we defined the handler prototype the same as `callback` field in the tasklet object.

> **Note**
> Tasklet API is deprecated, and you should consider using threaded
> IRQs instead.

Workqueues

Added since Linux kernel 2.6, the most used and simple deferring mechanism is a workqueue. As a deferring mechanism, it takes an opposite approach to the others we've seen, running only in a preemptible context. It is the only choice when sleeping is required unless you implicitly create a kernel thread or unless you are using a threaded interrupt. That said, workqueues are built on top of kernel threads, and for this simple reason, we will not implicitly cover kernel threads in this book.

At the core of the workqueue subsystem, there are two data structures that fairly well explain the concept behind this, as follows:

- The work to be deferred (referred to as a work item), represented in the kernel by instances of struct work_struct, which indicates the handler function to run. If you need a delay before the work runs after it has been submitted to the workqueue, the kernel provides a struct delayed_work instance instead. A work item is a simple structure that only contains a pointer to the function that is to be scheduled for asynchronous execution. To summarize, we can enumerate two types of work item structures, as follows:

  - A work_struct structure, which schedules a task to run as soon as possible when the system allows it

  - A delayed_work structure, which schedules a task to run after at least a given time interval

- The workqueue itself, which is represented by a struct workqueue_struct instance and is the structure onto which work is placed. It is a queue of work items.

Apart from these data structures, there are two generic terms you should be familiar with, as follows:

- **Worker threads**, which are dedicated threads that execute and pull the functions off the queue, one by one, one after the other.

- **Worker pools**: This is a collection of worker threads (a thread pool) that are used to better manage the worker threads.

The first step in using workqueues consists of creating a work item, represented by struct work_struct or struct delayed_work for the delayed variant, and defined in linux/workqueue.h. The kernel provides either a DECLARE_WORK macro to statically declare and initialize a work structure or dynamically uses an INIT_WORK macro. If you need delayed work, you can use the INIT_DELAYED_WORK macro for dynamic allocation and initialization or DECLARE_DELAYED_WORK for a static one. You can see the macros in action in the following code snippet:

```
DECLARE_WORK(name, function)
DECLARE_DELAYED_WORK(name, function)
INIT_WORK(work, func );
INIT_DELAYED_WORK( work, func);
```

Here is what our work item structure looks like:

```
struct work_struct {
    atomic_long_t data;
    struct list_head entry;
    work_func_t func;
};

struct delayed_work {
    struct work_struct work;
    struct timer_list timer;
    struct workqueue_struct *wq;
    int cpu;
};
```

The func field, which is of type work_func_t, tells us a bit more about the header of a work function, as illustrated here:

```
typedef void (*work_func_t)(struct work_struct *work);
```

work is an input parameter that corresponds to the work structure to be scheduled. If you submitted delayed work, it would correspond to the delayed_work.work field. It would then be necessary to use the to_delayed_work() function in order to get the underlying delayed work structure, as illustrated in the following code snippet:

```
struct delayed_work *to_delayed_work(
                struct work_struct *work)
```

Workqueue infrastructure allows drivers to create a dedicated kernel thread (a workqueue) called a worker thread to run work functions. A new workqueue can be created with the following functions:

```
struct workqueue_struct *create_workqueue(const char *name)
struct workqueue_struct *create_singlethread_workqueue(
                                    const char *name)
```

`create_workqueue()` creates a dedicated thread (a worker thread) per CPU on the system. For example, on an 8-core system, it will result in 8 kernel threads created to run the works submitted to your workqueue. Unless you have strong reasons to create a thread per CPU, you should prefer the single thread variant. In most cases, a single system kernel thread should be enough. In this case, you should use `create_singlethread_workqueue()` instead, which creates—as its name states—a single-threaded workqueue. Either normal or delayed works can be enqueued onto the same queue. In order to schedule works on your created workqueue, you can use either `queue_work()` or `queue_delayed_work()`, depending on the nature of the work. These functions are defined as follows:

```
bool queue_work(struct workqueue_struct *wq,
                struct work_struct *work)
bool queue_delayed_work(struct workqueue_struct *wq,
                        struct delayed_work *dwork,
                        unsigned long delay)
```

These functions return `false` if the work was already on a queue and `true` otherwise. `queue_dalayed_work()` is to be used to schedule a work item (a delayed one) for later execution after a given delay. The time unit for the delay is a jiffy. There are, however, APIs to convert milliseconds and microseconds to jiffies, defined as follows:

```
unsigned long msecs_to_jiffies(const unsigned int m)
unsigned long usecs_to_jiffies(const unsigned int u)
```

The following example uses 200 milliseconds as a delay:

```
queue_delayed_work(my_wq, &drvdata->tx_work,
                   usecs_to_jiffies(200));
```

You should not expect this delay to be accurate as the delay will be rounded up to the closest jiffy value. Thus, even when requesting 200 us, you should expect a jiffy. Submitted work items can be canceled by calling either `cancel_delayed_work()`, `cancel_delayed_work_sync()`, or `cancel_work_sync()`. These cancelation functions are defined as follows:

```
bool cancel_work_sync(struct work_struct *work)
bool cancel_delayed_work(struct delayed_work *dwork)
bool cancel_delayed_work_sync(struct delayed_work *dwork)
```

`cancel_work_sync()` synchronously cancels the given work—in other words, it cancels work and waits for its execution to finish. The kernel guarantees `work` not to be pending or executing on any CPU on return from this function, even if the work migrates to another workqueue or requeues itself. It returns `true` if `work` was pending and `false` otherwise.

`cancel_delayed_work()` asynchronously cancels a delayed entry. It returns `true` (actually, a nonzero value) if `dwork` was pending and canceled and `false` if it wasn't pending, probably because it is actually running, and thus might still be running after `cancel_delayed_work()` has returned. To make sure the work really ran to its end, you may want to use `flush_workqueue()`, which flushes every work item in the given queue, or `cancel_delayed_work_sync()`, which is the synchronous version of `cancel_delayed_work()`.

When you are done with a workqueue, you should destroy it with `destroy_workqueue()`, as illustrated here:

```
void flush_workqueue(struct worksqueue_struct * queue);
void destroy_workqueue(structure workqueque_struct *queue);
```

While waiting for any pending work to execute, the `_sync` variant functions sleep and thus can be called only from a process context.

Kernel-global workqueue – the shared queue

In most situations, your code does not necessarily need the performance of its own dedicated set of threads, and because `create_workqueue()` creates one worker thread for each CPU, it may be a bad idea to use it on very large multi-CPU systems. You may then want to use the kernel shared queue, which already has its set of kernel threads pre-allocated (early during the boot, via the `workqueue_init_early()` function) for running works.

This kernel-global workqueue is the so-called `system_wq` workqueue, defined in `kernel/workqueue.c`. There is actually one instance per CPU, each backed by a dedicated thread named `events/n`, where n is the processor number (or index) to which the thread is bound.

You can queue a work item in the default system workqueue using one of the following functions:

```
int schedule_work(struct work_struct *work);
int schedule_delayed_work(struct delayed_work *dwork,
                          unsigned long delay);
```

```
int schedule_work_on(int cpu,
                struct work_struct *work);
int schedule_delayed_work_on(int cpu,
                struct delayed_work *dwork,
                unsigned long delay);
```

schedule_work() immediately schedules the work that will be executed as soon as possible after the worker thread on the current processor wakes up. With schedule_delayed_work(), the work will be put in the queue in the future, after the delay timer ticks. _on variants are used to schedule the work on a specific CPU (which does not absolutely need to be the current one). Each of these functions queue works on the system's shared workqueue, system_wq, defined in kernel/workqueue.c as follows:

```
struct workqueue_struct *system_wq __read_mostly;
EXPORT_SYMBOL(system_wq);
```

You should also note that since this system workqueue is shared, you should not queue works which can run for too long, otherwise it may slow down other contender works, which could then wait for longer than they should before being executed.

In order to flush the kernel-global workqueue—that is, ensure a given batch of work is completed—we can use flush_scheduled_work(), as follows:

```
void flush_scheduled_work(void);
```

flush_scheduled_work() is a wrapper that calls flush_workqueue() on system_wq. Note that there may be works in the system_wq workqueue that you have not submitted and have no control over. Flushing this workqueue entirely is thus overkill, and it is recommended to run cancel_delayed_work_sync() or cancel_work_sync() instead.

> **Note**
> Unless you have a strong reason for creating a dedicated thread, the default (kernel-global) thread is preferred.

Workqueues' new generation

The original (now legacy) workqueue implementation used two kinds of workqueues: those with a single thread system-wide, and those with a thread per CPU. However, for a growing number of CPUs, this led to some limitations, as outlined here:

- On very large systems, the kernel could run out of **process identifiers** (**PIDs**) (defaulting to 32,000) just at the boot, even before the init process started.

- Multithreaded workqueues provided poor concurrency management as their threads compete for the CPU with each other threads on the system. As there were more CPU contenders, this introduced some overhead—that is, more context switches than necessary.

- The consumption of far more resources than were really needed.

Moreover, subsystems that needed a dynamic or fine-grained level of concurrency had to implement their own thread pools. As a result of this, a new workqueue API has been designed, and the legacy workqueue API (create_workqueue(), create_singlethread_workqueue(), and create_freezable_workqueue()) is scheduled for removal, though they are actually wrappers around the new one, the so-called **Concurrency Managed Workqueue** (**cmwq**), using per-CPU worker pools shared by all workqueues in order to automatically provide a dynamic and flexible level of concurrency, abstracting such details for the API users.

cmwq

cmwq is an upgrade of workqueue APIs. Using this new API implies you are choosing between the alloc_workqueue() function and the alloc_ordered_workqueue() macro to create a workqueue. They both allocate a workqueue and return a pointer to it on success, and NULL on failure. The returned workqueue can be freed using the destroy_workqueue() function. You can see an illustration of the code in the following snippet:

```
struct workqueue_struct *alloc_workqueue(const char *fmt,
                            unsigned int flags,
                            int max_active, ...);
#define alloc_ordered_workqueue(fmt, flags, args...) [...]
void destroy_workqueue(struct workqueue_struct *wq)
```

fmt is the printf format for the name of the workqueue, and args... are arguments for fmt.

destroy_workqueue() is to be called on the workqueue once you are done with it. All works currently pending will be done first before the kernel truly destroys the workqueue. alloc_workqueue() creates a workqueue based on max_active, which defines the concurrency level by limiting the number of works (tasks, actually) that can be executing (workers in a runnable state) simultaneously from this workqueue on any given CPU. For example, a max_active value of 5 would mean at most 5 work items of this workqueue can be executing at the same time per CPU. On the other hand, alloc_ordered_ workqueue() creates a workqueue that processes each work item one by one in queued order (that is, **first-in, first-out (FIFO)** order).

flags controls how and when work items are queued, assigned execution resources, scheduled, and executed. There are various flags used in this new API on which we should spend some time, as follows:

- WQ_UNBOUND: Legacy workqueues had a worker thread per CPU and were designed to run tasks on the CPU where they were submitted. The kernel scheduler had no choice but to always schedule a worker on the CPU on which it was defined. With this approach, even a single workqueue was able to prevent a CPU from idling and being turned off, which leads to increased power consumption or poor scheduling policies. WQ_UNBOUND turns off the previously-described behavior. Work items are not bound to the CPU anymore, hence the name **unbound workqueues**. There is no more locality, and the scheduler can reschedule workers on any CPU as it sees fit. The scheduler has the last word now and can balance CPU load, especially for long and sometimes CPU-intensive works.

- WQ_MEM_RECLAIM: This flag is to be set for workqueues that need to guarantee forward progress during the memory reclaim path (when free memory is running dangerously low; the system is said to be under memory pressure). In this case, GFP_KERNEL allocations may block and deadlock the entire workqueue. The workqueue is then guaranteed to have at least a ready-to-use worker thread—a so-called rescuer thread—reserved for it, regardless of memory pressure, so that it can progress forward. There's one rescuer thread allocated for each workqueue that has this flag set.

Let's consider a situation where we have three work items (w1, w2, and w3) in our workqueue W. w1 does some work and then waits for w3 to complete (let's say it depends on the computation result of w3). Afterward, w2 (which is independent of the others) does some `kmalloc()` allocation (`GFP_KERNEL`) and, oops—there's not enough memory. While w2 is blocked, it still occupies W's workqueue. This results in w3 not being able to run, even though there is no dependency between w2 and w3. As there is not enough memory available, there would be no way of allocating a new thread to run w3. A pre-allocated thread would for sure solve this problem, not by magically allocating the memory for w2, but by running w3 so that w1 can continue its job, and so on. w2 will continue its progression as soon as possible when there is enough available memory to allocate. This pre-allocated thread is a so-called **rescuer thread**. You must set the `WQ_MEM_RECLAIM` flag if you think the workqueue is likely to be used in the memory reclaim path. This flag replaces the legacy `WQ_RESCUER` flag as of this commit: `https://git.kernel.org/pub/scm/linux/kernel/git/torvalds/linux.git/commit/?id=493008a8e475771a2126e0ce95a73e35b371d277`.

> **Note**
>
> Memory reclaim is a Linux mechanism on a memory allocation path that consists of allocating memory only after throwing the current content of that memory somewhere else.

- `WQ_FREEZABLE`: This flag is used for power-management purposes. A workqueue with this flag set will be frozen when the system is suspended or hibernates. On the freezing path, all current work items of the worker(s) will be processed. When the freeze is complete, no new work item will be executed until the system is unfrozen. Filesystem-related workqueue(s) may use this flag (that is, to ensure that modifications made on files are pushed to the disk or create a hibernation image on the freezing path and that no modification is made on-disk after the hibernation image has been created. In this situation, non-freezable items or doing things differently could lead to filesystem corruption. As an example, all **Extents File System (XFS)** internal workqueues have this flag set (see `fs/xfs/xfs_super.c`) to ensure no further changes are made on-disk once the freezer infrastructure freezes kernel threads and creates a hibernation image. You should definitely not have this flag set if your workqueue can run tasks as part of the hibernation/suspend/resume process of the system. More information on this topic can be found in `Documentation/power/freezing-of-tasks.txt` and by having a look at the `freeze_workqueues_begin()` and `thaw_workqueues()` internal kernel functions.

- `WQ_HIGHPRI`: Tasks with this flag set run immediately and do not wait for the CPU to become available. This flag is used for workqueues that queue work items requiring a high priority for execution. Such workqueues have worker threads with a high priority level (lower nice value). In the earlier days of the cmwq, high-priority work items were just queued at the head of a global normal priority worklist so that they could immediately run. Nowadays, there are no interactions between normal-priority and high-priority workqueues as each has its own worklist and its own worker pool. Work items of high-priority workqueues are queued to the high-priority worker pool of the target CPU. Tasks in this workqueue should not block much. Use this flag if you do not want your work item competing for CPU with normal- or lower-priority tasks. Crypto and block devices subsystems use this, for example.

- `WQ_CPU_INTENSIVE`: Work items of CPU-intensive workqueues may burn a lot of CPU cycles and do not participate in workqueue concurrency management. Instead, as with any other task, their execution is regulated by the system scheduler, which makes this flag handy for bound work items that may consume a lot of CPU time. Though the system scheduler controls their execution, concurrency management controls the start of their execution, and runnable non-CPU-intensive work items might cause CPU-intensive work items to be delayed. The `crypto` and `dm-crypt` subsystems use such workqueues. To prevent such tasks from delaying the execution of other non-CPU-intensive work items, they will not be considered when the workqueue code determines whether the CPU is available or not.

To be compliant and feature-compatible with the old workqueue API, the following mappings are done to keep this API compatible with the original one:

- `create_workqueue(name)` is mapped onto `alloc_workqueue(name,WQ_MEM_RECLAIM, 1)`

- `create_singlethread_workqueue(name)` is mapped onto `alloc_ordered_workqueue(name, WQ_MEM_RECLAIM)`

- `create_freezable_workqueue(name)` is mapped onto `alloc_workqueue(name,WQ_FREEZABLE | WQ_UNBOUND|WQ_MEM_RECLAIM, 1)`

To summarize, `alloc_ordered_workqueue()` actually replaces `create_freezable_workqueue()` and `create_singlethread_workqueue()` (as per this commit: `https://git.kernel.org/pub/scm/linux/kernel/git/next/linux-next.git/commit/?id=81dcaf6516d8`). Workqueues allocated with `alloc_ordered_workqueue()` are unbound and have `max_active` set to 1.

When it comes to scheduling items in a workqueue, the work items that have been queued to a specific CPU using `queue_work_on()` will execute on that CPU. Work items queued via `queue_work()` will prefer the queueing CPU, but locality is not guaranteed.

> **Note**
>
> Note that `schedule_work()` is a wrapper that calls `queue_work()` on the system workqueue (`system_wq`), while `schedule_work_on()` is a wrapper around `queue_work_on()`. Also, keep in mind the following: `system_wq = alloc_workqueue("events", 0, 0);`. You can have a look at the `workqueue_init_early()` function in `kernel/workqueue.c` in kernel sources to see how other system-wide workqueues are created.

We are done with the new Linux kernel workqueue management implementation—that is, cmwq. Because workqueues can be used to defer works from interrupt handlers, we can move on to the next section and learn how to handle interrupts from the Linux kernel.

Kernel interrupt handling

Apart from servicing processes and user requests, another job of the Linux kernel is managing and speaking with hardware. This is either from the CPU to the device or from the device to the CPU and is achieved by means of interrupts. An interrupt is a signal sent to the processor by an external hardware device requesting immediate attention. Prior to an interrupt being visible to the CPU, this interrupt should be enabled by the interrupt controller, which is a device on its own whose main job consists of routing interrupts to CPUs.

The Linux kernel allows the provision of handlers for interrupts we are interested in so that when those interrupts are triggered, our handlers are executed.

An interrupt is how a device halts the kernel, telling it that something interesting or important has happened. These are called IRQs on Linux systems. The main advantage interrupts offer is to avoid device polling. It is up to the device to tell if there is a change in its state; it is not up to us to poll it.

To be notified when an interrupt occurs, you need to register with that IRQ, providing a function called an interrupt handler that will be called every time that interrupt is raised.

Designing and registering an interrupt handler

When an interrupt handler is executed, it runs with interrupts disabled on the local CPU. This involves respecting certain constraints while designing an **interrupt service routine** (**ISR**), as outlined here:

- **Execution time**: As IRQ handlers run with interrupts disabled on the local CPU, the code must be as short and small as possible and fast enough to assure a fast re-enabling of the previously disabled CPU-local interrupts in order not to miss any further occurring IRQs. Time-consuming IRQ handlers may considerably alter the real-time properties of the system and slow it down.

- **Execution context**: Since interrupt handlers are executed in an atomic context, sleeping (or other mechanisms that may sleep—such as mutexes—copying data from kernel to user space or vice versa, and so on) is forbidden. Any part of the code requiring or involving sleeping must be deferred into another, safer context (that is, a process context).

An IRQ handler needs to be given two arguments: the interrupt line to install the handler for, and a **unique device ID** (**UDI**) of the peripheral (mostly used as a context data structure—that is, a pointer to the per-device or private structure of the associated hardware device)—as illustrated here:

```
typedef irqreturn_t (*irq_handler_t)(int, void *);
```

The device driver wishing to register an interrupt handler for a given IRQ should call devm_request_irq(), defined in <linux/interrupt.h> as follows:

```
devm_request_irq(struct device *dev, unsigned int irq,
                 irq_handler_t handler,
                 unsigned long irqflags,
                 onst char *devname, void *dev_id)
```

The preceding function argument list, `dev`, is the device responsible for the IRQ line, `irq` represents the interrupt line (that is, the interrupt number of the issuing device) to register `handler` for. Prior to validating the request, the kernel will make sure the requested interrupt is valid and that it is not already assigned to another device unless both devices request this `irq` line to be shared (with help of `flags`). `handler` is the function pointer to the interrupt handler, and `flags` represent the interrupt flags. `name` is an **American Standard Code for Information Interchange (ASCII)** string that should namely describe the interrupt, and `dev` should be unique to each registered handler and cannot be `NULL` for shared IRQs since it is used by the kernel IRQ core to identify the device. A common way of using it is to provide a pointer to the device structure or a pointer to any per-device (and potentially useful to the handler) data structure, since when an interrupt occurs, both the interrupt line (`irq`) and this parameter will be passed to the registered handler, which can use this data as context data for further processing.

`flags` mangles the state or the behavior of the IRQ line or its handler by means of the following masks, which can be OR'ed to form a final desired bitmask according to your needs:

```
#define IRQF_SHARED 0x00000080
#define IRQF_PROBE_SHARED 0x00000100
#define IRQF_NOBALANCING 0x00000800
#define IRQF_IRQPOLL 0x00001000
#define IRQF_ONESHOT 0x00002000
#define IRQF_NO_SUSPEND 0x00004000
#define IRQF_FORCE_RESUME 0x00008000
#define IRQF_NO_THREAD 0x00010000
#define IRQF_EARLY_RESUME 0x000200002
#define IRQF_COND_SUSPEND 0x00040000
```

Note that `flags` can be 0 as well. Let's now explain some important flags—we leave the rest for the user to explore in `include/linux/interrupt.h`, but these are the ones we'll look at in more detail:

- `IRQF_NOBALANCING` excludes the interrupt from IRQ balancing, which is a mechanism that consists of distributing/relocating interrupts across CPUs, with a goal of increasing performance. It is kind of preventing the CPU affinity of that IRQ from being changed. This flag may be useful to provide a flexible setup for clock source or clock event devices, to prevent misattribution of the event to the wrong core. This flag is meaningful on multi-core systems only.

- `IRQF_IRQPOLL`: This flag allows the implementation of an `irqpoll` mechanism, intended to fix interrupt problems, meaning this handler should be added to the list of known interrupt handlers that can be looked for when a given interrupt is not handled.

- `IRQF_ONESHOT`: Normally, the actual interrupt line being serviced is re-enabled after its hardirq handler completes, whether it awakes a threaded handler or not. This flag keeps the interrupt line disabled after the hardirq handler finishes. It must be set on threaded interrupts (we will discuss this later) for which the interrupt line must remain disabled until the threaded handler has completed, after which it will be re-enabled.

- `IRQF_NO_SUSPEND` does not disable the IRQ during system hibernation/suspension. It does mean the interrupt is able to wake up the system from a suspended state. Such IRQs may be timer interrupts that may trigger and need to be handled even during system suspension. The whole IRQ line is affected by this flag such that if the IRQ is shared, every registered handler for this shared line will be executed, not only the one that installed this flag. You should avoid as much as possible using `IRQF_NO_SUSPEND` and `IRQF_SHARED` at the same time.

- `IRQF_FORCE_RESUME` enables the IRQ in the system resume path even if `IRQF_NO_SUSPEND` is set.

- `IRQF_NO_THREAD` prevents the interrupt handler from being threaded. This flag overrides the kernel `threadirqs` command-line option that forces every interrupt to be threaded. This flag has been introduced to address the non-threadability of some interrupts (for example, timers, which cannot be threaded even when all interrupt handlers are forced to be threaded).

- `IRQF_TIMER` marks this handler as being specific to system timer interrupts. It helps not to disable the timer IRQ during system suspension to ensure normal resumption and not thread them when full preemption (that is, `PREEMPT_RT`) is enabled. It is just an alias for `IRQF_NO_SUSPEND | IRQF_NO_THREAD`.

- `IRQF_EARLY_RESUME` resumes IRQ early at resume time of **system core** (**syscore**) operations instead of at device resume time. The following link points to the message of the commit introducing its support: `https://lkml.org/lkml/2013/11/20/89`.

- `IRQF_SHARED` allows for the sharing of the interrupt line among several devices. However, each device driver that needs to share the given interrupt line must set with this flag; otherwise, the handler registration will fail.

We must also consider the `irqreturn_t` return type of interrupt handlers since it may involve further actions after the return of the handler. Possible return values are listed here:

- `IRQ_NONE`: On a shared interrupt line, once the interrupt occurs, the kernel IRQ core successively walks through handlers registered for this line and executes them in the order they have been registered. The driver then has the responsibility to check whether it is its device that issued the interrupt. If the interrupt does not come from its device, it must return `IRQ_NONE` to instruct the kernel to call the next registered interrupt handler. This return value is mostly used on shared interrupt lines since it informs the kernel that the interrupt does not come from our device. However, if 99,900 of the previous 100,000 interrupts of a given IRQ line have not been handled, the kernel then assumes that this IRQ is stuck in some manner, drops a diagnostic, and tries to turn the IRQ off. For more information on this, you can have a look at the `__report_bad_irq()` function in the kernel source tree.

- `IRQ_HANDLED`: This value should be returned if the interrupt has been handled successfully. On a threaded IRQ, this value does acknowledge the interrupt (at a controller level) without waking the thread handler up.

- `IRQ_WAKE_THREAD`: On a thread IRQ handler, this value must be returned by the hard-IRQ handler to wake the handler thread. In this case, `IRQ_HANDLED` must only be returned by that threaded handler, previously registered with `devm_request_threaded_irq()`. We will discuss this later in the chapter.

> **Note**
> You should never re-enable IRQs from within your IRQ handler as this would involve allowing "interrupt reentrancy".

`devm_request_irq()` is the managed version of `request_irq()`, defined as follows:

```
int request_irq(unsigned int irq, irq_handler_t handler,
                unsigned long flags, const char *name,
                void *dev)
```

They both have the same variable meanings. If the driver used the managed version, the IRQ core will take care of releasing the resources. In other cases, such as at the unloading path or when the device leaves, the driver will have to release the IRQ resources by unregistering the interrupt handler using `free_irq()`, declared as follows:

```
void free_irq(unsigned int irq, void *dev_id)
```

`free_irq()` removes the handler (identified by `dev_id` when it comes to shared interrupts) and disables the line. If the interrupt line is shared, the handler is just removed from the list of handlers for this `irq`, and the interrupt line is disabled in the future when the last handler is removed. Moreover, if possible, your code must make sure the interrupt is really disabled on the card it drives before calling this function, since omitting this may lead to spurious IRQ.

There are a few things worth mentioning here about interrupts that you should never forget, as follows:

- On Linux systems, when the handler of an IRQ is being executed by a CPU, all interrupts are disabled on that CPU and the interrupt being serviced is masked on all the other cores. This means interrupt handlers need not be reentrant because the same interrupt will never be received until the current handler has completed. However, all other interrupts but the serviced one remain enabled (or, should we say, unchanged) on other cores, so other interrupts keep being serviced, though the current line is always disabled, as well as further interrupts on the local CPU. As a result, the same interrupt handler is never invoked concurrently to service a nested interrupt. This makes writing your interrupt handler a lot easier.

- Critical regions that need to run with interrupts disabled should be limited as much as possible. To remember this, tell yourself that your interrupt handler has interrupted other code and needs to give the CPU back.

- The interrupt context has its own (fixed and quite low) stack size. It thus totally makes sense to disable IRQs while running an ISR as reentrancy could cause stack overflow if too many preemptions happen.

- Interrupt handlers cannot block; they do not run in a process context. Thus, you may not perform the following operations from within an interrupt handler:

  - You cannot transfer data to/from user space since this may block.

  - You cannot sleep or rely on code that may lead to sleep, such as invoking `wait_event()`, memory allocation with any flag other than `GFP_ATOMIC`, or using a mutex/semaphore. The threaded handler can handle this.

  - You cannot trigger or call `schedule()`.

> **Note**
>
> If a device issues an IRQ while this IRQ is disabled (or masked) at a controller level, it will not be processed at all (masked in the flow handler), but an interrupt will instantaneously occur if it is still pending (at a device level) when the IRQ is enabled (or unmasked).
>
> The concept of non-reentrancy for an interrupt means that, if an interrupt is already in an active state, it cannot enter it again until the active status is cleared.

Understanding the concept of top and bottom halves

External devices send interrupt requests to the CPU either to signal a particular event or request a service. As stated in the previous section, bad interrupt management may considerably increase a system's latency and decrease its real-time properties. We also stated that interrupt processing—that is, the hard-IRQ handler at least—must be very fast not only to keep the system responsive but also not to miss other interrupt events.

The idea here is to split the interrupt handler into two parts. The first part (a function, actually) will run in a so-called hard-IRQ context with interrupts disabled, and will perform the minimum required work (such as doing some quick sanity checks— essentially, time-sensitive tasks, read/write hardware registers, and fast processing of this data and acknowledging interrupts to the device that raised it). This first part is the so-called **top half** on Linux systems. The top-half would then schedule a thread handler, which would run a so-called **bottom-half** function, with interrupts re-enabled, and which is the second part of the interrupt. The bottom half could then perform time-consuming operations (such as buffer processing) and tasks that may sleep, as it runs in a thread.

This splitting would considerably increase system responsiveness as the time spent with IRQs disabled is reduced to its minimum, and since bottom halves are run in kernel threads, they compete for the CPU with other processes on the runqueue. Moreover, they may have their real-time properties set. The top half is actually the handler registered using devm_request_irq(). When using devm_request_threaded_irq(), as we will see in the next section, the top half is the first handler given to the function.

As described previously in the *Implementing work-deferring mechanisms* section, a bottom half represents nowadays any task (or work) scheduled from within an interrupt handler. Bottom halves are designed using work-deferring mechanisms, which we have seen previously.

Depending on which one you choose, it may run in a (software) interrupt context or in a process context. These are softirqs, tasklets, workqueues, and threaded IRQs.

> **Note**
> Tasklets and softirqs have nothing to do with the "thread interrupt" mechanism since they run in their own special (atomic) contexts.

Since softirq handlers run at a high priority with the scheduler preemption disabled, not relinquishing CPU to processes/threads until they complete, care must be taken while using them for bottom-half delegation. As nowadays the quantum allocated for a particular process may vary, there is no strict rule for how long the softirq handler should take to complete in order not to slow the system down as the kernel would not be able to give CPU time to other processes. I would say no longer than half a jiffy.

The hard IRQ (top half, actually) must be as fast as possible, and most of the time, just reading and writing in I/O memory. Any other computation should be deferred in the bottom half, whose main goal is to perform any time-consuming and not interrupt-related work not performed by the top half. There is no clear guideline on the repartition of work between the top and bottom halves. Here is some advice:

- Hardware-related or time-sensitive work can be performed in the top half.

- If the work really need not be interrupted, it can be performed in the top half.

- From my point of view, everything else can be deferred and thus performed in the bottom half, which will run with interrupts enabled and when the system is less busy.

- If the hard IRQ is fast enough to process and acknowledge interrupts within a few microseconds consistently, then there is absolutely no need to use bottom-half delegations.

Working with threaded IRQ handlers

Threaded interrupt handlers were introduced to reduce the time spent in interrupt handlers and defer the rest of the work (that is, processing) out into kernel threads. So, the top half (hard IRQ) would consist of quick sanity checks such as ensuring whether the interrupt comes from its device and waking the bottom half accordingly. A threaded interrupt handler runs in its own thread, either in the thread of their parent (if they have one) or in a separate kernel thread. Moreover, the dedicated kernel thread can have its real-time priority set, though it runs at normal real-time priority (that is, `MAX_USER_RT_PRIO/2`, as you can see in the `setup_irq_thread()` function in `kernel/irq/manage.c`).

The general rule behind threaded interrupts is simple: keep the hard-IRQ handler as minimal as possible and defer as much work to the kernel thread as possible (preferably, all work). You should use devm_request_threaded_irq() if you wish to request a threaded interrupt handling. Here is its prototype:

```
devm_request_threaded_irq(struct device *dev, unsigned int irq,
                irq_handler_t handler, irq_handler_t thread_
fn,
                unsigned long irqflags, const char *devname,
                void *dev_id);
```

This function accepts two special parameters on which we should spend some time, handler, and thread_fn. They are outlined in more detail here:

- handler immediately runs when the interrupt occurs, in an interrupt context, and acts as a hard-IRQ handler. Its job usually consists of reading the interrupt cause (in the device's status register) to determine whether or how to handle the interrupt (this is frequent on **memory-mapped I/O (MMIO)** devices). If the interrupt does not come from our device, this function should return IRQ_NONE. This return value usually only makes sense on shared interrupt lines.

  If this hard-IRQ handler can finish interrupt processing fast enough (this is not a universal rule, but let's say no longer than a half of jiffy—that is, not longer than 500 µs if CONFIG_HZ, which defines the value of a jiffy, is set to 1000) for some set of interrupt causes, it should return IRQ_HANDLED after processing in order to acknowledge the interrupts. Interrupt processing that does not fall in this time lapse should be deferred in the thread IRQ handlers. In this case, the hard-IRQ handler should return IRQ_WAKE_THREAD to awake the threaded handler. Returning IRQ_WAKE_THREAD makes sense only when the thread_fn handler is also provided.

- thread_fn is the threaded handler added to the scheduler runqueue when the hard-IRQ handler function returns IRQ_WAKE_THREAD. If thread_fn is NULL while the handler is set and returns IRQ_WAKE_THREAD, nothing happens at the return path of the hard-IRQ handler but a simple warning message (we can see that in the __irq_wake_thread() function in the kernel sources). As thread_fn competes for the CPU with other processes on the runqueue, it may be executed immediately or later in the future when the system has less load. This function should return IRQ_HANDLED when it has completed the interrupt handling. After that, the associated kernel thread will be taken off the runqueue and put in a blocked state until woken up again by the hard-IRQ function.

A default hard-IRQ handler will be installed by the kernel if `handler` is `NULL` and `thread_fn != NULL`. This is the default primary handler. It does nothing but return `IRQ_WAKE_THREAD` to wake up the associated kernel thread that will execute the `thread_fn` handler.

It is implemented as follows:

```
/* Default primary interrupt handler for threaded
 * interrupts. Assigned as primary handler when
 * request_threaded_irq is called with handler == NULL.
 * Useful for oneshot interrupts.
 */
static irqreturn_t irq_default_primary_handler(int irq,
                                               void *dev_id)
{
    return IRQ_WAKE_THREAD;
}
int request_threaded_irq(unsigned int irq,
        irq_handler_t handler, irq_handler_t thread_fn,
        unsigned long irqflags, const char *devname,
        void *dev_id)
{
[...]
    if (!handler) {
        if (!thread_fn)
            return -EINVAL;
        handler = irq_default_primary_handler;
    }
[...]
}
EXPORT_SYMBOL(request_threaded_irq);
```

This makes it possible to move the execution of interrupt handlers entirely to the process context, thus preventing buggy drivers (buggy IRQ handlers, actually) from breaking the whole system and reducing interrupt latency.

In new kernel releases, `request_irq()` simply wraps `request_threaded_irq()` with the `thread_fn` parameter set to `NULL` (the same goes for the `devm_` variant).

Note that the interrupt is acknowledged at an interrupt controller level when you return from the hard-IRQ handler (whatever the return value is), thus allowing you to take other interrupts into account. In such a situation, if the interrupt hasn't been acknowledged at the device level, the interrupt will fire again and again, resulting in stack overflows (or being stuck in the hard-IRQ handler forever) for level-triggered interrupts since the issuing device will still have the interrupt line asserted.

For threaded interrupt implementation, when drivers needed to run the bottom half in a thread, they had to mask the interrupt at device level from the hard-interrupt handler. This required accessing the issuing device, which is not, however, always possible for devices sitting on slow buses (such I2C or **Serial Peripheral Interface** (**SPI**) buses, for example) because such buses require a thread context. With the introduction of IRQF_ONESHOT, this operation is not mandatory anymore as it helps keep the IRQ disabled at a controller level even when the threaded handler runs. Drivers must, however, clear the device interrupt in the threaded handler before it completes.

Using devm_request_threaded() (or the non-managed variant), it is possible to request an exclusively threaded IRQ by omitting the hard-interrupt handler. In this case, it is mandatory to set the IRQF_ONESHOT flag, else the kernel will complain because the threaded handler would run with the interrupt unmasked at both device and controller levels.

Here is an example of this:

```
static irqreturn_t data_event_handler(int irq,
                                      void *dev_id)
{
    struct big_structure *bs = dev_id;
    clear_device_interupt(bs);
    process_data(bs->buffer);
    return IRQ_HANDLED;
}
static int my_probe(struct i2c_client *client)
{
[...]
    if (client->irq > 0) {
        ret = request_threaded_irq(client->irq, NULL,
                &data_event_handler,
                IRQF_TRIGGER_LOW | IRQF_ONESHOT,
                id->name, private);
```

```
        if (ret)
            goto error_irq;
    }
...
    return 0;
error_irq:
    do_cleanup();
    return ret;
}
```

In the preceding example, our device sits on an I2C bus, so accessing the device may put the underlying task to sleep. Such an operation must never be performed in the hard-interrupt handler.

Here is an excerpt from the message in the link that introduced the `IRQF_ONESHOT` flag and which explains what it does (the whole message can be found via this link: `http://lkml.iu.edu/hypermail/linux/kernel/0908.1/02114.html`):

"*It allows drivers to request that the interrupt is not unmasked (at controller level) after the hard interrupt context handler has been executed and the thread has been woken. The interrupt line is unmasked after the thread handler function has been executed.*"

If one driver has set either `IRQF_SHARED` or `IRQF_ONESHOT` flags on a given IRQ, then the other driver sharing the IRQ must set the same flags. The `/proc/interrupts` file lists IRQs with their number of processing per CPU, the IRQ name as given during the requesting step, and a comma-separated list of drivers that registered a handler for that interrupt.

Threading the IRQs is the best choice for interrupt processing that can hog too many CPU cycles (exceeding a jiffy, for example), such as bulk data processing. Threading IRQs allows the priority and CPU affinity of their associated threads to be managed individually. As this concept comes from the real-time kernel tree, it fulfills many requirements of a real-time system, such as allowing a fine-grained priority model and reducing interrupt latency in the kernel. You can have a look at `/proc/irq/<IRQ>/smp_affinity`, which can be used to get or set the corresponding `<IRQ>` affinity. This file returns and accepts a bitmask that represents which processors can handle ISRs registered for this IRQ. This way, you can—for example—decide to set the affinity of the hard-interrupt handler to one CPU, while setting the affinity of the threaded handler to another CPU.

Requesting a context-agnostic IRQ

A driver requesting an IRQ must know in advance the nature of the interrupt and decide whether its handler can run in a hard-IRQ context or not, which may influence the choice between `devm_request_irq()` and `devm_request_threaded_irq()`.

The problem with those approaches is that sometimes, a driver requesting an IRQ does not know about the nature of the interrupt controller that provides this IRQ line, especially when the interrupt controller is a discrete chip (typically, a **general-purpose I/O (GPIO)** expander connected over SPI or I2C buses). Now comes the `request_any_context_irq()`, function with which drivers requesting an IRQ will know whether the handler will run in a thread context or not, and call `request_threaded_irq()` or `request_irq()` accordingly. This means that whether the IRQ associated with our device comes from an interrupt controller that may not sleep (a memory-mapped one) or from one that can sleep (behind an I2C/SPI bus), there will be no need to change the code. Its prototype looks like this:

```
int request_any_context_irq(unsigned int irq,
                    irq_handler_t handler, unsigned long flags,
                    const char *name, void *dev_id)
```

`devm_request_any_context_irq()` and `devm_request_irq()` have the same interface but different semantics. Depending on the underlying context (the hardware platform), `devm_request_any_context_irq()` selects either a hard-interrupt handling using `request_irq()` or a threaded handling method using `request_threaded_irq()`. It returns a negative error value on failure, while on success, it returns either `IRQC_IS_HARDIRQ` (meaning a hard-interrupt handling method is used) or `IRQC_IS_NESTED` (meaning a threaded one is used). With this function, the behavior of the interrupt handler is decided at runtime. For more information, you can have a look at the commit introducing it in the kernel by following this link: `https://git.kernel.org/pub/scm/linux/kernel/git/next/linux-next.git/commit/?id=ae731f8d0785`.

The advantage of using `devm_request_any_context_irq()` is that the driver does not need to care about what can be done in the IRQ handler, as the context in which the handler will run depends on the interrupt controller that provides the IRQ line. For example, for a GPIO-IRQ based device driver, if the GPIO belongs to a controller that sits on an I2C or SPI bus (GPIO access may sleep), the handler will be threaded. Otherwise (that is, the GPIO access does not sleep and is memory-mapped as it is part of the SoC), the handler will run in a hard-IRQ handler.

In the following example, the device expects an IRQ line mapped to a GPIO. The driver cannot assume that the given GPIO line will be memory-mapped, coming from the SoC. It may come from a discrete I2C or SPI GPIO controller as well. A good practice would be to use `request_any_context_irq()` here:

```
static irqreturn_t packt_btn_interrupt(int irq,
                                       void *dev_id)
{
    struct btn_data *priv = dev_id;
    input_report_key(priv->i_dev, BTN_0,
        gpiod_get_value(priv->btn_gpiod) & 1);
    input_sync(priv->i_dev);
    return IRQ_HANDLED;
}
static int btn_probe(struct platform_device *pdev)
{
    struct gpio_desc *gpiod;
    int ret, irq;
    gpiod = gpiod_get(&pdev->dev, "button", GPIOD_IN);
    if (IS_ERR(gpiod))
        return -ENODEV;
    priv->irq = gpiod_to_irq(priv->btn_gpiod);
    priv->btn_gpiod = gpiod;
[...]
    ret = request_any_context_irq(priv->irq,
            packt_btn_interrupt,
            (IRQF_TRIGGER_FALLING | IRQF_TRIGGER_RISING),
            "packt-input-button", priv);
    if (ret < 0)
        goto err_btn;

    return 0;
err_btn:
    do_cleanup();
    return ret;
}
```

The preceding code is simple enough but quite safe since `devm_request_any_context_irq()` does the job, which prevents it from mistaking the type of the underlying GPIO. The advantage of this approach is that you do not need to care about the nature of the interrupt controller that provides the IRQ line. In our example, if the GPIO belongs to a controller sitting on an I2C or SPI bus, the handler will be threaded. Otherwise (memory-mapped), the handler will run in a hard-IRQ context.

Using a workqueue to defer the bottom half

As we have already discussed the workqueue API in a dedicated section, it is now preferable to give an example here. This example is not error-free and has not been tested. It is just a demonstration to highlight the concept of bottom-half deferring by means of a workqueue.

Let's start by defining a data structure that will hold the elements we need for further development, as follows:

```
struct private_struct {
    int counter;
    struct work_struct my_work;
    void __iomem *reg_base;
    spinlock_t lock;
    int irq;
    /* Other fields */
    [...]
};
```

In the preceding data structure, our work structure is represented by the `my_work` element. We do not use a pointer here because we will need to use a `container_of()` macro in order to grab back the pointer to the initial data structure. Next, we can define a method that will be invoked in the worker thread, as follows:

```
static void work_handler(struct work_struct *work)
{
    int i;
    unsigned long flags;
    struct private_data *my_data =
            container_of(work, struct private_data, my_work);
    /*
     * Processing at least half of MIN_REQUIRED_FIFO_SIZE
```

```
    * prior to re-enabling the irq at device level,
    * so that buffer can receive further data
    */
    for (i = 0, i < MIN_REQUIRED_FIFO_SIZE, i++) {
        device_pop_and_process_data_buffer();
        if (i == MIN_REQUIRED_FIFO_SIZE / 2)
            enable_irq_at_device_level(my_data);
    }
    spin_lock_irqsave(&my_data->lock, flags);
    my_data->buf_counter -= MIN_REQUIRED_FIFO_SIZE;
    spin_unlock_irqrestore(&my_data->lock, flags);
}
```

In the preceding work structure, we start data processing when enough data has been buffered. We can now provide our IRQ handler, which is responsible for scheduling our work, as follows:

```
/* This is our hard-IRQ handler. */
static irqreturn_t my_interrupt_handler(int irq,
                                        void *dev_id)
{
    u32 status;
    unsigned long flags;
    struct private_struct *my_data = dev_id;
    /* we read the status register to know what to do */
    status = readl(my_data->reg_base + REG_STATUS_OFFSET);
    /*
     * Ack irq at device level. We are safe if another
     * irq pokes since it is disabled at controller
     * level while we are in this handler
     */
    writel(my_data->reg_base + REG_STATUS_OFFSET,
           status | MASK_IRQ_ACK);
    /*
     * Protecting the shared resource, since the worker
     * also accesses this counter
     */
```

```
    spin_lock_irqsave(&my_data->lock, flags);
    my_data->buf_counter++;
    spin_unlock_irqrestore(&my_data->lock, flags);
    /*
     * Our device raised an interrupt to inform it has
     * new data in its fifo. But is it enough for us
     * to be processed ?
     */
    if (my_data->buf_counter != MIN_REQUIRED_FIFO_SIZE)) {
        /* ack and re-enable this irq at controller level */
        return IRQ_HANDLED;
    } else {
        /* Right. prior to scheduling the worker and
         * returning from this handler, we need to
         * disable the irq at device level
         */
        writel(my_data->reg_base + REG_STATUS_OFFSET,
                MASK_IRQ_DISABLE);
        schedule_work(&my_work);
    }
    /* This will re-enable the irq at controller level */
    return IRQ_HANDLED;
};
```

The comments in the IRQ handler code are meaningful enough. `schedule_work()` is the function that schedules our work. Finally, we can write our probe method that will request our IRQ and register the previous handler, as follows:

```
static int foo_probe(struct platform_device *pdev)
{
    struct resource *mem;
    struct private_struct *my_data;
    my_data = alloc_some_memory(
                        sizeof(struct private_struct));
    mem = platform_get_resource(pdev, IORESOURCE_MEM, 0);
    my_data->reg_base = ioremap(ioremap(mem->start,
                            resource_size(mem));););
```

```
    if (IS_ERR(my_data->reg_base))
        return PTR_ERR(my_data->reg_base);
    /*
     * workqueue initialization. "work_handler" is
     * the callback that will be executed when our work
     * is scheduled.
     */
    INIT_WORK(&my_data->my_work, work_handler);
    spin_lock_init(&my_data->lock);
    my_data->irq = platform_get_irq(pdev, 0);
    if (devm_request_irq(&pdev->dev, my_data->irq,
                         my_interrupt_handler, 0,
                         pdev->name, my_data))
        handler_this_error()
    return 0;
}
```

The structure of the preceding probe method shows without a doubt that we are facing a platform device driver. Generic IRQ and workqueue APIs have been used here for initializing our workqueue and registering our handler.

Locking from within an interrupt handler

It is common to use spinlocks on SMP systems, as this guarantees mutual exclusion at the CPU level. Therefore, if a resource is shared only with a threaded bottom half (that is, it is never accessed from the hard IRQ), it is better to use mutexes, as we see in the following example:

```
static int my_probe(struct platform_device *pdev)
{
    int irq;
    int ret;
    irq = platform_get_irq(pdev, i);
    ret = devm_request_threaded_irq(&pdev->dev, irq, NULL,
                my_threaded_irq, IRQF_ONESHOT,
                dev_name(dev), my_data);
[...]
    return 0;
```

```
}

static irqreturn_t my_threaded_irq(int irq, void *dev_id)
{
    struct priv_struct *my_data = dev_id;
    /* Save FIFO Underrun & Transfer Error status */
    mutex_lock(&my_data->fifo_lock);
    /*
     * Accessing the device's buffer through i2c
     */
    device_get_i2c_buffer_and_push_to_fifo();
    mutex_unlock(&ldev->fifo_lock);
    return IRQ_HANDLED;
}
```

However, if the shared resource is accessed from within the hard-interrupt handler, you must use the _irqsave variant of the spinlock, as in the following example, starting with the probe method:

```
static int my_probe(struct platform_device *pdev)
{
    int irq;
    int ret;
    [...]
    irq = platform_get_irq(pdev, 0);
    if (irq < 0)
        goto handle_get_irq_error;
    ret = devm_request_threaded_irq(&pdev->dev, irq,
                    hard_handler, threaded_handler,
                    IRQF_ONESHOT, dev_name(dev), my_data);
    if (ret < 0)
        goto err_cleanup_irq;
     [...]
    return 0;
}
```

Now that the probe method has been implemented, let's implement the top half—that is, the hard-IRQ handler—as follows:

```
static irqreturn_t hard_handler(int irq, void *dev_id)
{
    struct priv_struct *my_data = dev_id;
    u32 status;
    unsigned long flags;
    /* Protecting the shared resource */
    spin_lock_irqsave(&my_data->lock, flags);
    my_data->status = __raw_readl(
            my_data->mmio_base + my_data->foo.reg_offset);
    spin_unlock_irqrestore(&my_data->lock, flags);
    /* Let us schedule the bottom-half */
    return IRQ_WAKE_THREAD;
}
```

The return value of the top half will wake the threaded bottom half, which is implemented as follows:

```
static irqreturn_t threaded_handler(int irq, void *dev_id)
{
    struct priv_struct *my_data = dev_id;
    spin_lock_irqsave(&my_data->lock, flags);
    /* doing sanity depending on the status */
    process_status(my_data->status);
    spin_unlock_irqrestore(&my_data->lock, flags);
    /*
     * content of status not needed anymore, let's do
     * some other work
     */
    [...]
    return IRQ_HANDLED;
}
```

There is a case where protection may not be necessary between the hard IRQ and its threaded counterpart when the `IRQF_ONESHOT` flag is set while requesting the IRQ line. This flag keeps the interrupt disabled after the hard-interrupt handler has finished. With this flag set, the IRQ line is disabled until the threaded handler has been run. This way, the hard handler and its threaded counterpart will never compete, and a lock for a resource shared between the two might not be necessary.

Summary

In this chapter, we discussed the fundamental elements to start driver development, presenting all the mechanisms frequently used in drivers such as work scheduling and time management, interrupt handling, and locking primitives. This chapter is very important since it discusses topics other chapters in this book rely on.

For instance, the next chapter, dealing with character devices, will use some of the elements discussed in this chapter.

4
Writing Character Device Drivers

Unix-based systems expose hardware to user space by means of special files, all created in the /dev directory upon device registration with the system. Programs willing to access a given device must locate its corresponding device file in /dev and perform the appropriate system call on it, which will be redirected to the driver of the underlying device associated with that special file. Though system calls redirection is done by an operating system, what system calls are supported depends on the type of device and the driver implementation.

On the topic of types of devices, there are many of them from a hardware point of view, which are, however, grouped into two families of special device files in /dev – these are **block devices** and **character devices**. They are differentiated by the way they are accessed, their speed, and the way data is transferred between them and the system. Typically, character devices are slow and transfer data to or from user applications sequentially byte by byte (one character after another – hence their name). Such devices include serial ports and input devices (keyboards, mouses, touchpads, video devices, and so on). On the other hand, block devices are fast, since they are accessed quite frequently and transfer data in blocks. Such devices are essentially storage devices (hard drives, CD-ROMs, solid-state drives, and so on).

In this chapter, we will focus on character devices and their drivers, their APIs, and their common data structures. We will introduce most of their concepts and write our first character device driver.

We will cover the following topics in this chapter:

- The concept of major and minor
- Character device data structure introduction
- Creating a device node
- Implementing file operations

The concept of major and minor

Linux has always enforced device file identification by a unique identifier, composed of two parts, a **major** and a **minor**. While other file types (links, directories, and sockets) may exist in /dev, character or block device files are recognizable by their types, which can be seen using the ls -l command:

```
$ ls -la /dev
crw-------  1 root root    254,    0 août  22 20:28 gpiochip0
crw-------  1 root root    240,    0 août  22 20:28 hidraw0
[...]
brw-rw----  1 root disk    259,    0 août  22 20:28 nvme0n1
brw-rw----  1 root disk    259,    1 août  22 20:28 nvme0n1p1
brw-rw----  1 root disk    259,    2 août  22 20:28 nvme0n1p2
[...]
crw-rw----+ 1 root video    81,    0 août  22 20:28 video0
crw-rw----+ 1 root video    81,    1 août  22 20:28 video1
```

From the preceding excerpt, in the first column, c identifies character device files and b identifies block device files. In the fifth and sixth columns, we can see, respectively, major and minor numbers. The major number either identifies the type of device or can be bound to a driver. The minor number either identifies a device locally to the driver or devices of the same type. This explains the fact that some device files have the same major number in the preceding output.

Now that we are done with the basic concepts of character devices in a Linux system, we can start exploring the kernel code, starting from introducing the main data structures.

Character device data structure introduction

A character device driver represents the most basic device driver in the kernel sources. Character devices are represented in the kernel as instances of `struct cdev`, declared in `include/linux/cdev.h`:

```
struct cdev {
    struct kobject kobj;
    struct module *owner;
    const struct file_operations *ops;
    dev_t dev;
[...]
};
```

The preceding excerpt has listed elements of our interest only. The following shows the meaning of these elements in this data structure:

- `kobj`: This is the underlying kernel object for this character device object, used to enforce the Linux device model. We will discuss this in *Chapter 14, Introduction to the Linux Device Model*.

- `owner`: This should be set with the `THIS_MODULE` macro.

- `ops`: This is the set of file operations associated with this character device.

- `dev`: This is the character device identifier.

With this data structure introduced, the next logical one for discussion is the one exposing file operations that system calls will rely on. Let's then introduce the data structure that allows interaction between user space and kernel space through the character device.

An introduction to device file operations

The `cdev->ops` element points to the file operations supported by a given device. Each of these operations is the target of a particular system call, in a manner that, when the system call is invoked by a program in user space on the character device, this system call is redirected in the kernel to its file operation counterpart in `cdev->ops`. `struct file_operations` is the data structure that holds these operations. It looks like the following:

```
struct file_operations {
    struct module *owner;
    loff_t (*llseek) (struct file *, loff_t, int);
```

```
    ssize_t (*read) (struct file *, char __user *,
                    size_t, loff_t *);
    ssize_t (*write) (struct file *, const char __user *,
                    size_t, loff_t *);
    unsigned int (*poll) (struct file *,
                            struct poll_table_struct *);
    int (*mmap) (struct file *, struct vm_area_struct *);
    int (*open) (struct inode *, struct file *);
    int (*flush) (struct file *, fl_owner_t id);
    long (*unlocked_ioctl) (struct file *, unsigned int,
                            unsigned long);
    int (*release) (struct inode *, struct file *);
    int (*fsync) (struct file *, loff_t, loff_t,
                int datasync);
    int (*flock) (struct file *, int, struct file_lock *);
    [...]
};
```

The preceding excerpt lists only the important methods of the structure, especially the ones that are relevant to the needs of this book. The full code is in `include/linux/fs.h` in kernel sources. Each of these callbacks is the backend of a system call, and none of them are mandatory. The following explains the meanings of elements in the structure:

- `struct module *owner`: This is a mandatory field that should point to the module owning this structure. It is used for proper reference counting. Most of the time, it is set to THIS_MODULE, a macro defined in `<linux/module.h>`.

- `loff_t (*llseek) (struct file *, loff_t, int);`: This method is used to move the current cursor position in the file given as the first parameter. On a successful move, the function must return the new position, or else a negative value must be returned. If this method is not implemented, then every seek performed on this file will succeed by modifying the position counter in the `file` structure (`file->f_pos`), except the seek relative to end-of-file, which will fail.

- `ssize_t (*read) (struct file *, char *, size_t, loff_t *);`: The role of this function is to retrieve data from the device. Since the return value is a "signed size" type, this function must return either the number (positive) of bytes successfully read, or else return an appropriate negative code on error. If this function is not implemented, then any `read()` system call on the device file will fail, returning with `-EINVAL` (an "invalid argument").

- `ssize_t (*write) (struct file *, const char *, size_t, loff_t *);`: The role of this function is to send data to the device. Like the `read()` function, it must return a positive number, which, in this case, represents the number of bytes that have been written successfully, or else return an appropriately negative code on error. In the same way, if it is not implemented in the driver, then the `write()` system call attempt will fail with `-EINVAL`.

- `int (*flush) (struct file *, fl_owner_t id);`: This operation is invoked when the file structure is being released. Like `open`, `release` can be NULL.

- `unsigned int (*poll) (struct file *, struct poll_table_struct *);`: This file operation must return a bitmask describing the status of the device. It is the kernel backend for both `poll()` and `select()` system calls, both used to query whether the device is writable, readable, or in some special state. Any caller of this method will block until the device enters the requested state. If this file operation is not implemented, then the device is always assumed to be readable, writable, and in no special state.

- `int (*mmap) (struct file *, struct vm_area_struct *);`: This is used to request part or all of the device memory to be mapped to a process address space. If this file operation is not implemented, then any attempt to invoke the `mmap()` system call on the device file will fail, returning `-ENODEV`.

- `int (*open) (struct inode *, struct file *);` This file operation is the backend of the `open()` system call, which, if not implemented (if NULL), will result in the success of any attempt to open the device and the driver won't be notified of the operation.

- `int (*release) (struct inode *, struct file *);`: This is invoked when the file is being released, in response to the `close()` system call. Like `open`, `release` is not mandatory and can be NULL.

- `int (*fsync) (struct file *, loff_t, loff_t, int datasync);`: This operation is the backend of the `fsync()` system call, whose purpose is to flush any pending data. If it is not implemented, any call to `fsync()` on the device file will fail, returning `-EINVAL`.

- `long (*unlocked_ioctl) (struct file *, unsigned int, unsigned long);`: This is the backend of the `ioctl` system call, whose purpose is to extend the commands that can be sent to the device (such as formatting a track of a floppy disk, which is neither reading nor writing). The commands defined by this function will extend a set of predefined commands that are already recognized by the kernel without referring to this file operation. Thus, for any command that is not defined (either because this function is not implemented or because it does not support the specified command), the system call will return `-ENOTTY`, to say *"No such ioctl for device"*. Any non-negative value returned by this function is passed back to the calling program to indicate successful completion.

Now that we are familiar with the file operation callbacks, let's delve into the kernel insights and learn how files are handled for a better understanding of the mechanisms behind character devices.

File representation in the kernel

Looking at the file operation table, at least one of the parameters of each operation is either the `struct inode` or `struct file` type. `struct inode` refers to a file on the disk. However, to refer an open file (associated with a file descriptor within a process), the `struct file` structure is used.

The following is the declaration of an `inode` structure:

```
struct inode {
    [...]
    union {
        struct pipe_inode_info   *i_pipe;
        struct cdev     *i_cdev;
        char            *i_link;
        unsigned        i_dir_seq;
    };
    [...]
}
```

The most important field in the structure is the `union`, especially the `i_cdev` element, which is set when the underlying file is a character device. This makes it possible to switch back and forth between `struct inode` and `struct cdev`.

On the other hand, `struct file` is a filesystem data structure holding information about a file (its type, character, block, pipe, and so on), most of which is only relevant to the OS. The `struct file` structure (defined in `include/linux/fs.h`) has the following definition:

```
struct file {
[...]
    struct path f_path;
    struct inode *f_inode;
    const struct file_operations *f_op;
    loff_t f_pos;
    void *private_data;
[...]
}
```

In the preceding data structure, `f_path` represents the actual path of the file in the filesystem, and `f_inode` is the underlying `inode` that points to this opened file. This makes it possible to switch back and forth between `struct file` and the underlying `cdev`, through the `f_inode` element. `f_op` represents the file operation table. Because `struct file` represents an open file descriptor, it tracks the current read/write position within its opened instance. This is done through the `f_pos` element, and `f_pos` is the current read/write position.

Creating a device node

The creation of a device node makes it visible to users and allows users to interact with the underlying device. Linux requires intermediate steps before the device node is created and the following section discusses these steps.

Device identification

To precisely identify devices, their identifiers must be unique. Although identifiers can be dynamically allocated, most drivers still use static identifiers for compatibility reasons. Whatever the allocation method, the Linux kernel stores file device numbers in elements of `dev_t` type, which is a 32-bit unsigned integer in which the major is represented by the first 12 bits, and the minor is coded on the 20 remaining bits.

All of this is stated in `include/linux/kdev_t.h`, which contains several macros, including those that, given a `dev_t` type variable, can return either a minor or a major number:

```
#define MINORBITS      20
#define MINORMASK      ((1U << MINORBITS) - 1)


#define MAJOR(dev)     ((unsigned int) ((dev) >> MINORBITS))
#define MINOR(dev)     ((unsigned int) ((dev) & MINORMASK))
#define MKDEV(ma,mi)   (((ma) << MINORBITS) | (mi))
```

The last macro accepts a minor and a major number and returns a `dev_t` type identifier, which the kernel uses to keep identifiers. The preceding excerpt also describes how the `character device` identifier is built using a bit shift. At this point, we can get deep into the code, using the APIs that the kernel provides for code allocation.

Registration and deregistration of character device numbers

There are two ways to deal with device numbers – **registration** (the static method) and **allocation** (the dynamic method). Registration, also called **static allocation**, is only useful if you know in advance which major number you want to start with, after making sure it does not clash with another driver using the same major (though this is not always predictable). Registration is a brute-force method in which you let the kernel know what device numbers you want by providing the starting major/minor pair and the number of minors, and it either grants them to you or not (depending on availability). The function to use for device number registration is the following:

```
int register_chrdev_region(dev_t first, unsigned int count,
                         char *name);
```

This method returns 0 on success, or a negative error code when it fails. The `first` parameter is the identifier you must have built using the major number and the first minor of the desired range. You can use the `MKDEV(maj, min)` macro to achieve that. `count` is the number of consecutive device minors required, and `name` should be the name of the associated device or driver.

However, note that `register_chrdev_region()` works well if you know exactly which device numbers you want, and of course, those numbers must be available on your running system. Because this can be a source of conflict with other device drivers, it is considered preferable to use dynamic allocation, with which the kernel happily allocates a major number for you on the fly. `alloc_chrdev_region()` is the API you must use for dynamic allocation. The following is its prototype:

```
int alloc_chrdev_region(
                    dev_t *dev, unsigned int firstminor,
                    unsigned int count, char *name);
```

This method returns 0 on success, or a negative error code on failure. `dev` is the only output parameter. It represents the first number (built using the allocated major and the first minor requested) that the kernel assigned. `firstminor` is the first of the requested range of minor numbers, `count` is the number of consecutive minors you need, and `name` should be the name of the associated device or driver.

The difference between static allocation and dynamic allocation is that, with the former, you should know in advance which device number is needed. When it comes to loading the driver on another machine, there is no guarantee that the chosen number is free on that machine, and this can lead to conflict and trouble. New drivers are encouraged to use dynamic allocation to obtain a major device number, rather than choosing a number randomly from the ones that are currently free, which will probably lead to a clash. In other words, your drivers are better using `alloc_chrdev_region()` rather than `register_chrdev_region()`.

Initializing and registering a character device on the system

The registration of a character device is made by specifying a device identifier (of `dev_t` type). In this chapter, we will be using dynamic allocation, using `alloc_chrdev_region()`. After the identifier has been allocated, you must initialize the character device and add it to the system using `cdev_init()` and `cdev_add()`, respectively. The following are their prototypes:

```
void cdev_init(struct cdev *cdev,
                const struct file_operations *fops);
int cdev_add (struct cdev * p, dev_t dev, unsigned count);
```

In `cdev_init()`, `cdev` is the structure to initialize, and `fops` the `file_operations` instance for this device, making it ready to add to the system. In `cdev_add()`, `p` is the `cdev` structure for the device, `dev` is the first device number for which this device is responsible (obtained dynamically), and `count` is the number of consecutive minor numbers corresponding to this device. When it succeeds, `cdev_add()` returns 0, or else it returns a negative error code.

The respective reverse operation of `cdev_add()` is `cdev_del()`, which removes the character device from the system and has the following prototype:

```
void cdev_del(struct cdev *);
```

At this step, the device is part of the system but not physically present. In other words, it is not visible in `/dev` yet. For the node to be created, you must use `device_create()`, which has the following prototype:

```
struct device * device_create(struct class *class,
                              struct device *parent,
                              dev_t devt,
                              void *drvdata,
                              const char *fmt, ...)
```

This method creates a device and registers it with Sysfs. In its argument, `class` is a pointer to `struct class` that this device should be registered to, `parent` is a pointer to the `struct device` parent of this new device if there is any, `devt` is the device number for the char device to be added, and `drvdata` is the data to be added to the device for callbacks.

> **Important Note**
>
> In case of multiple minors, the `device_create()` and `device_destroy()` APIs can be put in a `for` loop, and the `<device name format>` string can be appended with the loop counter, as follows:
>
> ```
> device_create(class, NULL, MKDEV(MAJOR(first_devt),
> MINOR(first_devt) + i), NULL, "mynull%d", i);
> ```

Because a device needs an existing class before being created, you must either create a class or use an existing one. For now, we will create a class, and to do that, we need to use the `class_create()` function, declared as the following:

```
struct class * class_create(struct module * owner,
                            const char * name);
```

After then, the class will be visible in /sys/class, and we can create the device using that class. The following is a rough example:

```
#define EEP_NBANK 8
#define EEP_DEVICE_NAME "eep-mem"
#define EEP_CLASS "eep-class"

static struct class *eep_class;
static struct cdev eep_cdev[EEP_NBANK];
static dev_t dev_num;

static int __init my_init(void)
{
    int i;
    dev_t curr_dev;

    /* Request for a major and EEP_NBANK minors */
    alloc_chrdev_region(&dev_num, 0, EEP_NBANK,
                        EEP_DEVICE_NAME);
    /* create our device class, visible in /sys/class */
    eep_class = class_create(THIS_MODULE, EEP_CLASS);

    /* Each bank is represented as a character device (cdev) */
    for (i = 0; i < EEP_NBANK; i++) {
        /* bind file_operations to the cdev */
        cdev_init(&my_cdev[i], &eep_fops);
        eep_cdev[i].owner = THIS_MODULE;

        /* Device number to use to add cdev to the core */
        curr_dev = MKDEV(MAJOR(dev_num),
                         MINOR(dev_num) + i);

        /* Make the device live for the users to access */
        cdev_add(&eep_cdev[i], curr_dev, 1);

        /* create a node for each device */
        device_create(eep_class,
```

```
            NULL,       /* no parent device */
            curr_dev,
            NULL,       /* no additional data */
            EEP_DEVICE_NAME "%d", i); /* eep-mem[0-7] */
    }
    return 0;
}
```

In the preceding code, `device_create()` will create a node for each device, - /dev/ eep-mem0, dev/eep-mem1, and so on, with our class represented by `eep_class`. Additionally, devices can also be viewed under /sys/class/eep-class. In the meantime, the reverse operation is the following:

```
for (i = 0; i < EEP_NBANK; i++) {
    device_destroy(eep_class,
            MKDEV(MAJOR(dev_num), (MINOR(dev_num) +i)));
    cdev_del(&eep_cdev[i]);
}
class_unregister(eep_class);
class_destroy(eep_class);
unregister_chrdev_region(chardev_devt, EEP_NBANK);
```

In the preceding code, `device_destroy()` will remove a device node from / dev, `cdev_del()` will make the system forget about this character device, `class_ unregister()` and `class_destroy()` will deregister and remove the class from the system, and finally, `unregister_chrdev_region()` will release our device number.

Now that we are familiar with all the prerequisites about character devices, we can start implementing a file operation, which allows users to interact with the underlying device.

Implementing file operations

After introducing file operations in the previous section, it is time to implement those to enhance the driver capabilities and expose the device's methods to user space (by means of system calls, of course). Each of these methods has its particularities, which we will highlight in this section.

Exchanging data between the kernel space and user space

As we have seen while introducing the file operation table, the read and write methods are used to exchange data with the underlying device. Both being system calls means that data will originate from or be in destination to user space. While looking at the read and write method prototypes, the first point that catches our attention is the use of __user. This is a cookie used by **Sparse** (a semantic checker used by the kernel to find possible coding faults) to let the developer know they are about to use an untrusted pointer (or a pointer that may be invalid in the current virtual address mapping) improperly, which they should not dereference but, instead, use dedicated kernel functions to access the memory to which this pointer points.

This leads us to two principal functions that allow us to exchange data between a kernel and user space, copy_from_user() and copy_to_user(), which copy a buffer from user space to kernel space and vice versa, respectively:

```
unsigned long copy_from_user(void *to,
               const void __user *from, unsigned long n)
 unsigned long copy_to_user(void __user *to,
               const void *from, unsigned long n)
```

In both cases, pointers prefixed with __user point to the user space (untrusted) memory. n represents the number of bytes to copy, either to or from user space. from represents the source address, and to is the destination address. Each of these returns the number of bytes that could not be copied, if any, while they return 0 on success. Note that these routines may sleep as they run in a user context and do not need to be invoked in an atomic context.

Implementing the open file operation

The open file operation is the backend of the open system call. One usually uses this method to perform device and data structure initializations, after which it should return 0 on success, or a negative error code if something went wrong. The prototype of the open file operation is defined as follows:

```
 int (*open) (struct inode *inode, struct file *filp);
```

If it is not implemented, device opening will always succeed, but the driver won't be aware, which is not necessarily a problem if the device needs no special initialization.

Per-device data

As we have seen in file operation prototypes, there is almost always a `struct file` argument. `struct file` has an element free of use, that is `private_data`. `file->private_data`, if set, will be available to other system calls invoked on the same file descriptor. You can use this field during the lifetime of the file descriptor. It is good practice to set this field in the `open` method as it is always the first system call on any file.

The following is our data structure:

```
struct pcf2127 {
    struct cdev cdev;
    unsigned char *sram_data;
    struct i2c_client *client;
    int sram_size;
    [...]
};
```

Given this data structure, the `open` method would look like the following:

```
static unsigned int sram_major = 0;
static struct class *sram_class = NULL;

static int sram_open(struct inode *inode,
                     struct file *filp)
{
    unsigned int maj = imajor(inode);
    unsigned int min = iminor(inode);

    struct pcf2127 *pcf = NULL;
    pcf = container_of(inode->i_cdev,
                       struct pcf2127, cdev);
    pcf->sram_size = SRAM_SIZE;

    if (maj != sram_major || min < 0 ){
        pr_err ("device not found\n");
        return -ENODEV; /* No such device */
    }
```

```
    /* prepare the buffer if the device is
     * opened for the first time
       */
    if (pcf->sram_data == NULL) {
        pcf->sram_data =
                    kzalloc(pcf->sram_size, GFP_KERNEL);
        if (pcf->sram_data == NULL) {
            pr_err("memory allocation failed\n");
            return -ENOMEM;
        }
    }
    filp->private_data = pcf;
    return 0;
}
```

Most of the time, the open operation does some initialization and requests resources that will be used while a user keeps an open instance of the device node. Everything that has to be done in this operation must be undone and released when the device is closed, as we'll see in the next operation.

Implementing the release file operation

The release method is called when the device gets closed, the reverse of the open method. You must then undo everything you have done in the open operation. It could be literally, freeing any private memory allocated and shutting down the device (if supported), and discarding every buffer on the last closing (if the device supports multi-opening, or if the driver can handle more than one device at a time).

The following is an excerpt of a release function:

```
static int sram_release(struct inode *inode,
                        struct file *filp)
{
    struct pcf2127 *pcf = NULL;
    pcf = container_of(inode->i_cdev,
                        struct pcf2127, cdev);

    mutex_lock(&device_list_lock);
    filp->private_data = NULL;
```

```
    /* last close? */
    pcf2127->users--;
    if (!pcf2127->users) {
        kfree(tx_buffer);
        kfree(rx_buffer);
        tx_buffer = NULL;
        rx_buffer = NULL;


        [...]

        if (any_other_dynamic_struct)
            kfree(any_other_dynamic_struct);
    }
    mutex_unlock(&device_list_lock);
    return 0;
}
```

The preceding code releases all the resources acquired when the device node has been open. This is literally all that needs to be done in this file operation. If the device node is backed by a hardware device, this operation can also put this device in the appropriate state.

At this point, we are able to implement the entry (open) and exit (release) points for the character device. All that remains now is to implement each possible operation that can be done in between.

Implementing the write file operation

The write method is used to send data to the device; whenever a user calls the write() system call on the device's file, the kernel implementation that is called ends up being invoked. Its prototype is as follows:

```
ssize_t (*write) (struct file *filp, const char __user *buf,
                  size_t count, loff_t *pos);
```

This file operation must return the number of bytes (size) written, and the following are the definitions of its arguments:

- `*buf` represents the data buffer coming from the user space.
- `count` is the size of the requested transfer.
- `*pos` indicates the start position from which data should be written in the file (or in the corresponding memory region if the character device file is memory-backed).

Generally, in this file operation, the first thing to do is to check for bad or invalid requests coming from the user space (for example, check for size limitations in case of a memory-backed device and size overflow). The following is an example:

```
/* if trying to Write beyond the end of the file,
 * return error. "filesize" here corresponds to the size
 * of the device memory (if any)
 */
if (*pos >= filesize) return -EINVAL;
```

After the checks, it is common to make some adjustments, especially with `count`, to not go beyond the file size. This step is not mandatory either:

```
/* filesize corresponds to the size of device memory */
if (*pos + count > filesize)
    count = filesize - *pos;
```

The next step is to find the location from which you will start to write. This step is relevant only if the device is backed by physical memory in which the `write()` method is supposed to store given data:

```
/* convert pos into valid address */
void *from = pos_to_address(*pos);
```

Finally, you can copy data from user space into kernel memory, after which you can perform the `write` operation on the backing device and adjust `*pos`, as in the following excerpt:

```
if (copy_from_user(dev->buffer, buf, count) != 0) {
    retval = -EFAULT;
    goto out;
}
/* now move data from dev->buffer to physical device */
```

```
write_error = device_write(dev->buffer, count);
if (write_error)
    return -EFAULT;

/* Increase the current position of the cursor in the file,
 * according to the number of bytes written and finally,
 * return the number of bytes copied
 */
*pos += count;
return count;
```

The following is an example of the `write` method, which summarizes the steps described so far:

```
ssize_t
eeprom_write(struct file *filp, const char __user *buf,
             size_t count, loff_t *f_pos)
{
    struct eeprom_dev *eep = filp->private_data;
    int part_origin = PART_SIZE * eep->part_index;
    int register_address;
    ssize_t retval = 0;

    /* step (1) */
    if (*f_pos >= eep->part_size)
        /* Can't write beyond the end of a partition. */
        return -EINVAL;

    /* step (2) */
    if (*pos + count > eep->part_size)
        count = eep->part_size - *pos;

    /* step (3) */
    register_address = part_origin + *pos;

    /* step(4) */
    /* Copy data from user space to kernel space */
```

```
    if (copy_from_user(eep->data, buf, count) != 0)
        return -EFAULT;
    /* step (5) */
    /* perform the write to the device */
    if (write_to_device(register_address, buff, count)
        < 0){
        pr_err("i2c_transfer failed\n");
        return -EFAULT;
    }

    /* step (6) */
    *f_pos += count;
    return count;
}
```

After the data has been read and processed, it might be necessary to write the processing output back. Since we have started with writing, the next operation we may think of is read, as we will see in the next section.

Implementing the read file operation

The read method has the following prototype:

```
ssize_t (*read) (struct file *filp, char __user *buf,
                 size_t count, loff_t *pos);
```

This operation is the backend of the read() system call. Its arguments are described as follows:

- *buf is the buffer we receive from user space.
- count is the size of the requested transfer (the size of the user buffer).
- *pos indicates the start position from which data should be read in the file.

It must return the size of the data that has been successfully read. This size can be less than count though (for example, when reaching the end of the file before reaching the count requested by the user).

Implementing the read operation looks like the write one, since some sanity checks need to be performed. First, you can prevent reading beyond the file size and return an end-of-file response:

```
if (*pos >= filesize)
    return 0; /* 0 means EOF */
```

Then, you should make sure the number of bytes read can't go beyond the file size and you can adjust count appropriately:

```
if (*pos + count > filesize)
    count = filesize - (*pos);
```

Next, you can find the location from which you will start the read, after which you can copy the data into the user space buffer and return an error on failure, and then advance the file's current position according to the number of bytes read and return the number of bytes copied:

```
/* convert pos into valid address */
void *from = pos_to_address (*pos);
sent = copy_to_user(buf, from, count);
if (sent)
    return -EFAULT;
*pos += count;
return count;
```

The following is an example of a driver read() file operation, intended to give an overview of what can be done here:

```
ssize_t  eep_read(struct file *filp, char __user *buf,
                  size_t count, loff_t *f_pos)
{
    struct eeprom_dev *eep = filp->private_data;

    if (*f_pos >= EEP_SIZE) /* EOF */
        return 0;

    if (*f_pos + count > EEP_SIZE)
        count = EEP_SIZE - *f_pos;
```

```
    /* Find location of next data bytes */
    int part_origin  =  PART_SIZE * eep->part_index;
    int eep_reg_addr_start  =  part_origin + *pos;


    /* perform the read from the device */
    if (read_from_device(eep_reg_addr_start, buff, count)
        < 0){
        pr_err("i2c_transfer failed\n");
        return -EFAULT;
    }


    /* copy from kernel to user space */
    if(copy_to_user(buf, dev->data, count) != 0)
        return -EIO;


    *f_pos += count;
    return count;
}
```

Though reading and writing data moves the cursor position, there is an operation whose main purpose is to move the cursor position without touching data at all. Such an operation helps to start writing or reading data from anywhere by moving the cursor to the desired position.

Implementing the llseek file operation

The `llseek` file operation is the kernel backend for the `lseek()` system call, used to move the cursor position within a file. Its prototype looks as follows:

```
loff_t (*llseek) (struct file *filp, loff_t offset,
                  int whence);
```

This callback must return the new position in the file. The following are the definitions of its parameters:

- loff_t is an offset, relative to the current file position, which defines how much of it will be changed.
- whence defines where to seek from. The possible values are as follows:
  - SEEK_SET: To put the cursor to a position relative from the beginning of the file
  - SEEK_CUR: To put the cursor to a position relative to the current file position
  - SEEK_END: To adjust the cursor to a position relative to the end of the file

When implementing this operation, it is a good practice to use the switch statement to check every possible whence case, since they are limited, and adjust the new position accordingly:

```
switch( whence ){
    case SEEK_SET:/* relative from the beginning of file */
        newpos = offset; /* offset become the new position */
        break;
    case SEEK_CUR: /* relative to current file position */
        /* just add offset to the current position */
        newpos = file->f_pos + offset;
        break;
    case SEEK_END: /* relative to end of file */
        newpos = filesize + offset;
        break;
    default:
        return -EINVAL;
}
/* Check whether newpos is valid **/
if ( newpos < 0 )
    return -EINVAL;
/* Update f_pos with the new position */
filp->f_pos = newpos;
/* Return the new file-pointer position */
return newpos;
```

After the preceding kernel backend excerpt, the following is an example of a user program that will successively read and seek into a file. The underlying driver will then execute the `llseek()` file operation entry:

```
#include <unistd.h>
#include <fcntl.h>
#include <sys/types.h>
#include <stdio.h>

#define CHAR_DEVICE "foo"

int main(int argc, char **argv)
{
    int fd = 0;
    char buf[20];

    if ((fd = open(CHAR_DEVICE, O_RDONLY)) < -1)
        return 1;

    /* Read 20 bytes */
    if (read(fd, buf, 20) != 20)
        return 1;
    printf("%s\n", buf);

    /* Move the cursor to ten time relative to
     * its actual position
     */
    if (lseek(fd, 10, SEEK_CUR) < 0)
        return 1;
    if (read(fd, buf, 20) != 20)
        return 1;
    printf("%s\n",buf);

    /* Move the cursor seven time, relative from
     * the beginning of the file
     */
    if (lseek(fd, 7, SEEK_SET) < 0)
```

```
        return 1;
    if (read(fd, buf, 20) != 20)
        return 1;
    printf("%s\n",buf);

    close(fd);
    return 0;
}
```

The code produces the following output:

```
jma@jma:~/work/tutos/sources$ cat toto
Lorem ipsum dolor sit amet, consectetur adipiscing elit, sed do
eiusmod tempor incididunt ut labore et dolore magna aliqua.
jma@jma:~/work/tutos/sources$ ./seek
Lorem ipsum dolor si
nsectetur adipiscing
psum dolor sit amet,
jma@jma:~/work/tutos/sources$
```

In this section, we have explained the concept of seeking with an example, showing how relative and absolute seeking works. Now that we are done with operations moving data around, we can switch to the next operation, sensing the readability or writability of data in a character device.

The poll method

The poll method is the backend of both the poll() and select() system calls. These system calls are used to passively (by sleeping, without wasting CPU cycles) sense the readability/writability in a file. To support these system calls, the driver must implement poll, which has the following prototype:

```
unsigned int (*poll) (struct file *, struct poll_table_struct
*);
```

The kernel function at the heart of this method implementation is poll_wait(), defined in <linux/poll.h>, which is the header you must include in the driver code. It has the following declaration:

```
void poll_wait(struct file * filp,
            wait_queue_head_t * wait_address, poll_table *p)
```

poll_wait() adds the device associated with a struct file structure (given as the first parameter) to a list of those that can wake up processes (put to sleep in the struct wait_queue_head_t structure given as the second parameter), according to events registered in the struct poll_table structure given as the third parameter. A user process can call the poll(), select(), or epoll() system calls to add a set of files to a list on which it needs to wait, in order to be aware of the associated (if any) device's readiness. The kernel will then call the poll entry of the driver associated with each device file. The poll method of each driver should then call poll_wait() in order to register events for which the process needs to be notified with the kernel, put that process to sleep until one of these events occurs, and register the driver as one of those that can wake the process up. The usual way is to use a wait queue per event type (one for readability, another one for writability, and eventually one for an exception if needed), according to events supported by the select() (or poll()) system call.

The return value of the (*poll) file operation must have POLLIN | POLLRDNORM set if there is data to read, POLLOUT | POLLWRNORM if the device is writable, and 0 if there is no new data and the device is not yet writable. In the following example, we assume the device supports both blocking read and write. Of course, you can implement only one of these. If the driver does not define this method, the device will be considered as always readable and writable, and poll() or select() system calls return immediately.

Implementing the poll operation may require adapting the read or write file operations in a way that, on write, readers are notified of the readability, and on read, writers are notified of the writability:

```
#include <linux/poll.h>


/* declare a wait queue for each event type (read, write ...)
*/
static DECLARE_WAIT_QUEUE_HEAD(my_wq);
static DECLARE_WAIT_QUEUE_HEAD(my_rq);


static unsigned int eep_poll(struct file *file,
                             poll_table *wait)
{
    unsigned int reval_mask = 0;


    poll_wait(file, &my_wq, wait);
    poll_wait(file, &my_rq, wait);
```

```
    if (new_data_is_ready)
        reval_mask |= (POLLIN | POLLRDNORM);
    if (ready_to_be_written)
        reval_mask |= (POLLOUT | POLLWRNORM);
    return reval_mask;
}
```

In the preceding snippet, we have implemented the `poll` operation, which can put processes to sleep if a device is not writable or readable. However, there is no notification mechanism when any of those states change. Therefore, a `write` operation (or any operation making data available, such as an IRQ) must notify processes sleeping in the readability wait queue; the same applies to the `read` operation (or any operation making the device ready to be writable), which must notify processes sleeping in the writability wait queue. The following is an example:

```
wake_up_interruptible(&my_rq); /* Ready to read */
/* set flag accordingly in case poll is called */
new_data_is_ready = true;


wake_up_interruptible(&my_wq); /* Ready to be written to */
ready_to_be_written = true;
```

More precisely, you can notify a readable event either from within the driver's `write()` method, meaning that the written data can be read back, or from within an IRQ handler, meaning that an external device sent some data that can be read back. On the other hand, you can notify a writable event either from within the driver's `read()` method, meaning that the buffer is empty and can be filled again, or from within an IRQ handler, meaning that the device has completed a data-send operation and is ready to accept data again. Do not forget to set flags back to `false` when the state changes.

The following is an excerpt of code that uses `select()` on a given character device in order to sense data availability:

```
#include <unistd.h>
#include <fcntl.h>
#include <stdio.h>
#include <stdlib.h>
#include <sys/select.h>


#define NUMBER_OF_BYTE 100
```

```c
#define CHAR_DEVICE "/dev/packt_char"
char data[NUMBER_OF_BYTE];

int main(int argc, char **argv)
{
    int fd, retval;
    ssize_t read_count;
    fd_set readfds;

    fd = open(CHAR_DEVICE, O_RDONLY);
    if(fd < 0)
        /* Print a message and exit*/
        [...]

    while(1){
        FD_ZERO(&readfds);
        FD_SET(fd, &readfds);

        ret = select(fd + 1, &readfds, NULL, NULL, NULL);
        /* From here, the process is already notified */
        if (ret == -1) {
            fprintf(stderr, "select: an error ocurred");
            break;
        }

        /* we are interested in one file only */
        if (FD_ISSET(fd, &readfds)) {
            read_count = read(fd, data, NUMBER_OF_BYTE);
            if (read_count < 0)
                /* An error occurred. Handle this */
                [...]

            if (read_count != NUMBER_OF_BYTE)
                /* We have read less than needed bytes */
                [...] /* handle this */
            else
```

```
            /* Now we can process the data we have read */
            [...]
        }
    }
    close(fd);
    return EXIT_SUCCESS;
}
```

In the preceding code sample, we used `select()` without timeout, in a way that means we will be notified of "read" events only. From that line, the process is put to sleep until it is notified of the event for which it registered itself.

The ioctl method

A typical Linux system contains around 350 **system calls** (**syscalls**), but only a few of them are linked to file operations. Sometimes, devices may need to implement specific commands that are not provided by system calls, and especially the ones associated with files. In this case, the solution is to use **input/output control** (**ioctl**), which is a method by which you extend a list of commands associated with a device. You can use it to send special commands to devices (reset, shutdown, configure, and so on). If the driver does not define this method, the kernel will return an `-ENOTTY` error to any `ioctl()` system call. The following is its prototype:

```
long ioctl(struct file *f, unsigned int cmd,
           unsigned long arg);
```

In the preceding prototype, `f` is the pointer to the file descriptor representing an opened instance of the device, `cmd` is the ioctl command, and `arg` is a user parameter, which can be the address of any user memory on which the driver can call `copy_to_user()` or `copy_from_user()`. To be concise, and for obvious reasons, an IOCTL command needs to be identified by a number, which should be unique to the system. The unicity of IOCTL numbers across the system will prevent us from sending the right command to the wrong device, or passing the wrong argument to the right command (with a duplicated IOCTL number). Linux provides four helper macros to create an IOCTL identifier, depending on whether there is data transfer or not and the direction of the transfer. Their respective prototypes are as follows:

```
_IO(MAGIC, SEQ_NO)
_IOR(MAGIC, SEQ_NO, TYPE)
_IOW(MAGIC, SEQ_NO, TYPE)
_IORW(MAGIC, SEQ_NO, TYPE)
```

Their descriptions are as follows:

- _IO: The IOCTL command does not need data transfer.

- _IOR: This means that we're creating an IOCTL command number for passing information from the kernel to user space (which is reading data). The driver will be allowed to return the sizeof(TYPE) bytes to the user without this return value being considered as an error.

- _IOW: This is the same as _IOR, but the user sends data to the driver this time.

- _IOWR: The IOCTL command needs both write and read parameters.

What their parameters mean (in the order they are passed) is described here:

- A number coded on 8 bits (0 to 255), called the **magic number**.

- A sequence number or command ID, also on 8 bits.

- A data type, if any, that will inform the kernel about the size to be copied. This could be the name of a structure or a data type.

This is well documented in Documentation/ioctl/ioctl-decoding.txt in the kernel sources, and existing IOCTL commands are listed in Documentation/ioctl/ioctl-number.txt, a good place to start when you need to create your own IOCTL commands.

Generating an IOCTL number (a command)

It is recommended to generate your own IOCTL numbers in a dedicated header file, since this header should be available in user space as well. In other words, you should handle the duplication (by means of symbolic links, for example) of the IOCTL header file so that there is one in the kernel and one in user space, which can be included in user apps. Let's now generate some IOCTL numbers in a real example, and let's call this header eep_ioctl.h:

```
#ifndef PACKT_IOCTL_H
#define PACKT_IOCTL_H
/* We need to choose a magic number for our driver,
 * and sequential numbers for each command:
 */
#define EEP_MAGIC 'E'
#define ERASE_SEQ_NO 0x01
#define RENAME_SEQ_NO 0x02
```

```
#define GET_FOO 0x03
#define GET_SIZE 0x04


/*
 * Partition name must be 32 byte max
 */
#define MAX_PART_NAME 32


/*
 * Now let's define our ioctl numbers:
 */
#define EEP_ERASE _IO(EEP_MAGIC, ERASE_SEQ_NO)
#define EEP_RENAME_PART _IOW(EEP_MAGIC, RENAME_SEQ_NO, \
                            unsigned long)
#define EEP_GET_FOO  _IOR(EEP_MAGIC, GET_FOO, \
                            struct my_struct *)
#define EEP_GET_SIZE _IOR(EEP_MAGIC, GET_SIZE, int *)
#endif
```

After the commands have been defined, the header needs to be included in the final code. Moreover, because they are all unique and limited, it is a good practice to use a `switch` ... `case` statement to handle each command and return a `-ENOTTY` error code when an undefined `ioctl` command is called. The following is an example:

```
#include "eep_ioctl.h"
static long eep_ioctl(struct file *f, unsigned int cmd,
                     unsigned long arg)
{
    int part;
    char *buf = NULL;
    int size = 2048;

    switch(cmd){
        case EEP_ERASE:
            erase_eepreom();
            break;
        case EEP_RENAME_PART:
```

```
            buf = kmalloc(MAX_PART_NAME, GFP_KERNEL);
            copy_from_user(buf, (char *)arg,
                            MAX_PART_NAME);
            rename_part(buf);
            break;
        case EEP_GET_SIZE:
            if (copy_to_user((int*)arg,
                            &size, sizeof(int)))
                return -EFAULT;
            break;
        default:
            return -ENOTTY;
    }
    return 0;
}
```

Both the kernel and user space must include the header files that contain the IOCTL commands. Therefore, in the first line of the preceding excerpt, we have included eep_ioctl.h, which is the header file where our IOCTL commands are defined.

If you think your IOCTL command will need more than one argument, you should gather those arguments in a structure and just pass a pointer to the structure to ioctl.

Now, from the user space, you must use the same ioctl header as in the driver's code:

```
#include <stdio.h>
#include <stdlib.h>
#include <fcntl.h>
#include <unistd.h>
#include "eep_ioctl.h" /* our ioctl header file */

int main()
{
    int size = 0;
    int fd;
    char *new_name = "lorem_ipsum";

    fd = open("/dev/eep-mem1", O_RDWR);
    if (fd < 0){
```

```
        printf("Error while opening the eeprom\n");
        return 1;
    }

    /* ioctl to erase partition */
    ioctl(fd, EEP_ERASE);
    /* call to get partition size */
    ioctl(fd, EEP_GET_SIZE, &size);
    /* rename partition */
    ioctl(fd, EEP_RENAME_PART, new_name);

    close(fd);
    return 0;
}
```

In the preceding code, we have demonstrated the use of kernel IOCTL commands from user space. That said, all throughout this section, we have learned how to implement the character device's `ioctl` callback and how to exchange data between the kernel and user space.

Summary

In this chapter, we have demystified character devices, and we have seen how to let users interact with our driver through device files. We learned how to expose file operations to user space and control their behavior from within the kernel. We went so far that you are even able to implement multi-device support.

The next chapter is a bit more hardware-oriented, as it deals with the device tree, a mechanism that allows hardware devices present on the system to be declared to the kernel. See you in the next chapter.

Section 2 - Linux Kernel Platform Abstraction and Device Drivers

In this section, we'll first introduce the concept of a device tree, allowing us to declare and describe the non-discoverable devices present on the system, and then we'll learn how to deal with such devices. While dealing with such devices, we will introduce the concept of platform devices and their drivers, and we will learn how to write I2C and SPI device drivers.

The following chapters will be covered in this section:

5
Understanding and Leveraging the Device Tree

The device tree is an easy-to-read hardware description file, with a JSON-like formatting style. It is a simple tree structure where devices are represented by nodes and their properties. These properties can either be empty (that is, just the key to describe Boolean values) or key-value pairs, where the value can contain an arbitrary byte stream. This chapter is a simple introduction to device trees. Every kernel subsystem or framework has its own device tree binding, and we will talk about those specific bindings when we deal with the relevant topics.

The device tree originated from **Open Firmware (OF)**, which is a standard endorsed by computer companies, and whose main purpose is to define interfaces for computer firmware systems. That said, you can find out more about device tree specification at `http://www.devicetree.org/`. Therefore, this chapter will cover the basics of the device tree, including the following:

- Understanding the basic concept of the device tree mechanism
- Describing data types and their APIs
- Representing and addressing devices
- Handling resources

Understanding the basic concept of the device tree mechanism

The support of the device tree is enabled in the kernel by setting the CONFIG_OF option to Y. To pull the device tree API from your driver, you must add the following headers:

```
#include <linux/of.h>
#include <linux/of_device.h>
```

The device tree supports a few data types and writing conventions that we can summarize with a sample node description:

```
/* This is a comment */
// This is another comment
node_label: nodename@reg{
    string-property = "a string";
    string-list = "red fish", "blue fish";
    one-int-property = <197>; /* One cell in the property */
    int-list-property = <0xbeef 123 0xabcd4>;
    mixed-list-property = "a string", <35>,[0x01 0x23 0x45];
    byte-array-property = [0x01 0x23 0x45 0x67];
    boolean-property;
};
```

In the preceding example, `int-list-property` is a property where each number (or cell) is a 32-bit integer (`uint32`), and there are three cells in this property. Here, `mixed-list-property` is, as its name suggests, a property with mixed element types.

The following are some definitions of the data types used in the device tree:

- Text strings are represented with double quotes. You can use commas to create a list of the strings.

- Cells are 32-bit unsigned integers delimited by angle brackets.

- Boolean data is nothing more than an empty property. The true or false value depends on the property being there or not.

We have easily enumerated the types of data that can be found in the device tree. Before we start learning about the APIs that can be used to parse this data, first, let's understand how the device tree naming convention works.

The device tree naming convention

Every node must have a name in the form of <name>[@<address>], where <name> is a string that can be up to 31 characters in length, and [@<address>] is optional, depending on whether the node represents an addressable device or not. That said, <address> should be the primary address used to access the device. For example, for a memory-mapped device, it must correspond to the starting address of its memory region, the bus device address for an I2C device, and the chip-select index (relative to the controller) for an SPI device node.

The following presents some examples of device naming:

```
i2c@021a0000 {
    compatible = "fsl,imx6q-i2c", "fsl,imx21-i2c";
    reg = <0x021a0000 0x4000>;
    [...]


    expander@20 {
        compatible = "microchip,mcp23017";
        reg = <20>;
        [...]
    };
};
```

In the preceding device tree excerpt, the I2C controller is a memory-mapped device. Therefore, the address part of the node name corresponds to the beginning of its memory region, relative to the **System on Chip (SoC)** memory map. However, the expander is an I2C device. Thus, the address part of its node name corresponds to its I2C address.

An introduction to the concept of aliases, labels, phandles, and paths

Aliases, labels, phandles, and paths are keywords that you need to be familiar with when dealing with the device tree. It is likely that you will face at least one, if not all, of these terms as and when you deal with device drivers. To describe these terms, let's take the following device tree excerpt as an example:

```
aliases {
    ethernet0 = &fec;
    gpio0 = &gpio1;
    [...];
};
bus@2000000 { /* AIPS1 */
    gpio1: gpio@209c000 {
        compatible = "fsl,imx6q-gpio", "fsl,imx35-gpio";
        reg = <0x0209c000 0x4000>;
        interrupts = <0 66 IRQ_TYPE_LEVEL_HIGH>,
                     <0 67 IRQ_TYPE_LEVEL_HIGH>;
        gpio-controller;
        #gpio-cells = <2>;
        interrupt-controller;
        #interrupt-cells = <2>;
    };
    [...];
};
bus@2100000 { /* AIPS2 */
    [...]
    i2c1: i2c@21a0000 {
        compatible = "fsl,imx6q-i2c", "fsl,imx21-i2c";
        reg = <0x021a0000 0x4000>;
        interrupts = <0 36 IRQ_TYPE_LEVEL_HIGH>;
        clocks = <&clks IMX6QDL_CLK_I2C1>;
    };
};

&i2c1 {
```

```
    eeprom-24c512@55 {
        compatible = "atmel,24c512";
        reg = <0x55>;
    };
    accelerometer@1d {
        compatible = "adi,adxl345";
        reg = <0x1d>;
        interrupt-parent = <&gpio1>;
        interrupts = <24 IRQ_TYPE_LEVEL_HIGH>,
        <25 IRQ_TYPE_LEVEL_HIGH>;
        [...]
    };
 [...]
};
```

In the device tree, there are two ways in which a node can be referenced: by a path or by a **phandle**. By referencing a node using its path, you must explicitly specify the full path of this node in the device tree source, which might be complicated for a deeply nested node. On the other hand, a phandle is a unique 32-bit value associated with a node that is used to uniquely identify that node. It is assigned to the node through a `phandle` property and, sometimes, duplicated in a `linux,phandle` property for historical reasons.

However, the device tree source format allows labels to be attached to any node or property value. Given that a label must be unique all over the device tree source for a given board, it became obvious that it could be used to identify a node as well, and a decision was made to do so. Thus, logic has been added to the **device tree compiler** (**DTC**) so that, whenever a label name is prefixed with an ampersand (&) in a cell property, it is replaced with the phandle of the node to which this label is attached. Moreover, using the same logic, whenever a label is prefixed with an ampersand outside of a cell (a simple value assignment), it is replaced by the full path of the node to which the label is attached. This way, the phandle and path references can be automatically generated by referencing a label instead of explicitly specifying a phandle value or the full path to a node.

> **Note**
>
> Labels are only used in the device tree source format and are not encoded within the **device tree blob** (**DTB**). Whenever a node is labeled and referenced somewhere else using this label, at compile time, the dtc tool will remove that label from the node and add a phandle property to that node, generating and assigning a unique 32-bit value. The dtc tool will then use this phandle in every cell where the node has been referenced by the label (prefixed with an ampersand).

Back to our preceding excerpt, the gpio@0209c000 node is labeled gpio1, and this label is also used as a reference. This will instruct the DTC to generate a phandle for this node. Therefore, in the accelerometer@1d node, inside the interrupt-parent property, the cell value (&gpio1) will be replaced by the phandle of the node to which gpio1 is attached (the assignment inside a cell). In the same way, inside the aliases node, &gpio1 will be replaced with the full path of the node to which gpio1 is attached (the assignment outside of a cell).

After compiling and decompiling our original device tree excerpt, we obtain the following, where labels no longer exist and label references have been replaced either by a phandle or by the full node path:

```
aliases {
    gpio0 = "/soc/aips-bus@2000000/gpio@209c000";
    ethernet0 = "/soc/aips-bus@2100000/ethernet@2188000";
    [...]
};
aips-bus@2000000 {
    gpio@209c000 {
        compatible = "fsl,imx6q-gpio", "fsl,imx35-gpio";
        gpio-controller;
        #interrupt-cells = <0x2>;
        interrupts = <0x0 0x42 0x4 0x0 0x43 0x4>;
        phandle = <0x40>;
        reg = <0x209c000 0x4000>;
        #gpio-cells = <0x2>;
        interrupt-controller;
    };
};
aips-bus@2100000 {
```

```
i2c@21a8000 {
    compatible = "fsl,imx6q-i2c", "fsl,imx21-i2c";
    clocks = <0x4 0x7f>;
    interrupts = <0x0 0x26 0x4>;
    reg = <0x21a8000 0x4000>;

    eeprom-24c512@55 {
        compatible = "atmel,24c512";
        reg = <0x55>;
    };

    accelerometer@1d {
        compatible = "adi,adxl345";
        interrupt-parent = <0x40>;
        interrupts = <0x18 0x4 0x19 0x4>;
        reg = <0x1d>;
    };
};
};
```

In the preceding snippet, the accelerometer node has its interrupt-parent property cell assigned the value of 0x40. Looking at the gpio@209c000 node, we can see that this value corresponds to the value of its phandle property, which has been generated at compile time by the DTC. It's the same for the aliases node, where the node references have been replaced by their full paths.

This leads us to the definition of aliases; aliases are simply nodes that are referenced via their absolute paths for a quick lookup. The aliases node can be seen as a fast lookup table. Unlike labels, aliases do appear in the output device tree, although paths are generated by referencing labels. With an alias, a handle to the node it is referring to is obtained by simply searching for it in the aliases section rather than searching for it in the entire device tree as it is done while looking up by phandle. Aliases can be seen as a shortcut or similar to the aliases we set in our Unix shell to refer to a complete/long/ repetitive path/command.

The Linux kernel dereferences the aliases rather than using them directly in the device tree source. When using of_find_node_by_path() or of_find_node_opts_by_path() to find a node given its path, if the supplied path does not start with /, then the first element of the path must be a property name in the /aliases node. That element is replaced with the full path from the alias.

> **Note**
>
> Labeling a node is only useful if the node is intended to be referenced from the property of another node. You can consider a label as a pointer to a node, either by the path or by the reference.

Understanding overwriting nodes and properties

After taking a closer look at the result of the decompiled excerpt, you should notice one more thing: in the original sources, the i2c@21a0000 node has been referenced through its label as an external node (&i2c1 { [...] }) with some content inside. However, oddly, after decompiling, the final content of the i2c@21a0000 node has been merged with the content of the external reference, and the external reference node no longer exists.

This is the third usage of labels: allowing you to overwrite nodes and properties. In the external reference, any new content (such as nodes or properties) will be appended to the original node content at compile time. However, in the case of duplication (either nodes or properties), the content of the external reference will take precedence over the original content.

Consider the following example:

```
bus@2100000 { /* AIPS2 */
    [...]
    i2c1: i2c@21a0000 {
        [...]
        status = "disabled";
    };
};
```

Let's reference the i2c@21a0000 node through its i2c1 label, as follows:

```
&i2c1 {
    [...]
    status = "okay";
};
```

We will see that the result from the compilation will be the status property having the value of "okay". This is because the content of the external reference will have taken precedence over the original content.

To summarize, latter definitions always overwrite earlier definitions. For entire nodes to be overwritten, you simply have to redefine them as you would do so for properties.

Device tree sources and compilers

The **device tree** (also referred to as **DT**) comes in two forms. The first is the textual form, which represents the sources (also referred to as **DTS**). And the second is the binary blob form, which represents the compiled device tree, also referred to as **DTB** (for **device tree blob**) or **FDT** (for **flattened device tree**). Source files have a .dts extension, while the binary forms have either a .dtb or .dtbo extension. .dtbo is a particular extension that is used for compiled device tree overlays (**DTBO** means **device tree blob for overlay**), as we will see in the next section. There are also .dtsi text files (where the i at the end means "include"). These host SoC-level definitions and are intended to be included in .dts files, hosting the board-level definitions.

The syntax of the device tree allows you to use /include/ or #include to include other files. This inclusion mechanism makes it possible to use #define, but above all, it allows you to factorize the common aspects of several platforms in the shared files.

This factorization allows you to split the source files into tree levels, with the most common being the SoC level, which is provided by the SoC vendor (for example, NXP), the **System on Module (SoM)** level (for example, Engicam), and, finally, the carrier board or customer board level.

Therefore, all electronic boards using the same SoC do not redefine all of the peripherals of the SoC from scratch: this description is factored into a common file. By convention, such *common* files use the .dtsi extension, while final device trees use the .dts extension.

In the Linux kernel sources, ARM device tree source files can be found under the `arch/arm/boot/dts/` and `arch/arm64/boot/dts/<vendor>/` directories for the 32-bit and 64-bit ARM SoCs/boards, respectively. In either directory, there is a `Makefile` file that lists the device tree source files that can be compiled.

The utility used to compile the DTS files into DTB files is called DTC. The DTC sources exist in two places:

- **As a standalone upstream project**: The DTC upstream project is maintained in `https://git.kernel.org/cgit/utils/dtc/dtc.git`. It is pulled into the Linux kernel source tree on a regular basis.

- **In-kernel**: The Linux version of the DTC can be found in the kernel source directory under `scripts/dtc/`. New versions are pulled from the upstream project on a regular basis. The DTC is built by the Linux kernel build process as a dependency when needed (for example, before compiling the device tree). You can use the `make scripts` command if you wish to build it explicitly in the Linux kernel source tree.

From the main directory in the kernel sources, you can either compile a specific device tree or all device trees for a specific SoC. In either case, the appropriate config option to enable this or these device tree files must be enabled. For a single device tree compilation, the make target is the name of the `.dts` file with `.dts` changed to `.dtb`. For all of the enabled device trees to be compiled, the make target that you should use is `dtbs`. In both cases, you should make sure that the config option that enables the `dtb` has been set.

Consider the following excerpt from `arch/arm/boot/dts/Makefile`:

```
dtb-$(CONFIG_SOC_IMX6Q) += \
    imx6dl-alti6p.dtb \
    imx6dl-aristainetos_7.dtb \
[...]
    imx6q-hummingboard.dtb \
    imx6q-hummingboard2.dtb \
    imx6q-hummingboard2-emmc-som-v15.dtb \
    imx6q-hummingboard2-som-v15.dtb \
    imx6q-icore.dtb \
[...]
```

By enabling `CONFIG_SOC_IMX6Q`, you can either compile all of the device tree files listed in there or target a specific device tree. By running `make dtbs`, the kernel DTC will compile all the device tree files listed in the enabled config options.

First, let's make sure the appropriate config option has been set:

```
$ grep CONFIG_SOC_IMX6Q .config
CONFIG_SOC_IMX6Q =y
```

Then, let's compile all of the device tree files:

```
ARCH=arm CROSS_COMPILE=arm-linux-gnueabihf- make dtbs
```

Assuming our platform is an ARM64 platform, we would use the following:

```
ARCH=arm64 CROSS_COMPILE=aarch64-linux-gnu- make dtbs
```

Once again, the right kernel config options must be set.

You could target a particular device tree build (let's say imx6q-hummingboard2.dts), using a command such as the following:

```
ARCH=arm CROSS_COMPILE=arm-linux-gnueabihf- make imx6q-
hummingboard2.dtb
```

It must be noted that, given a compiled device tree (.dtb) file, you can do the reverse operation and extract the source (.dts) file:

```
$ dtc -I dtb -O dts arch/arm/boot/dts/imx6q-hummingboard2.dtb >
path/to/my_devicetree.dts
```

For debug purposes, it might be useful to expose to user space the current device tree of a running system, that is, the so-called live device tree. To do so, the kernel CONFIG_PROC_DEVICETREE config option must be enabled. Then, you can explore and walk through the device tree in the /proc/device-tree directory.

If installed on the running system, the DTC can be used to convert the filesystem tree into a more readable form using the following command:

```
# dtc -I fs -O dts /sys/firmware/devicetree/base > MySBC.dts
```

After this command returns, the MySBC.dts file will contain the sources corresponding to the current device tree.

The device tree overlay

Device tree overlaying is a mechanism that allows you to patch a live device tree, that is, modify the current device tree at runtime. It allows you to update the current device tree at runtime by updating existing nodes and properties or creating new ones. However, it does not allow you to delete a node or a property.

A device tree overlay has the following format:

```
/dts-v1/;
/plugin/; /* allow undefined label references and record them
*/


/{
    fragment@0 { /* first child node */
        target=<phandle>; /* phandle of the target node to
extend */
    or
        target-path="/path"; /* full path of the node to extend
*/

        __overlay__ {
            property-a;  /* add property-a to the target */
            property-b = <0x80>; /* update property-b value */
            node-a { /* add to an existing, or create a node-a
*/

                ...

            };
        };
    };

    fragment@1 {
        /* second fragment overlay ... */
    };

    /* more fragments follow */
}
```

From the preceding excerpt, we can note that each node from the base device tree that needs to be overlayed must be enclosed inside a `fragment` node in the overlay device tree.

Then, each fragment has two elements:

- One of these two properties could be as follows:

    - `target-path`: This specifies the absolute path to the node that the fragment will modify.

    - `target`: This specifies the relative path to the node alias (prefixed with an ampersand symbol) that the fragment will modify.

- A node, named `__overlay__`, that contains the changes that should be applied to the referred node. Such changes can be new nodes (which are added), new properties (which are added), or existing properties (which are overridden with the new value). There is no removal operation possible since a property or a node cannot be removed.

Now that we are comfortable with the basics of device tree overlaying, we can learn how they are compiled and turned into a binary blob that can be loaded on demand.

Building device tree overlays

Unless a device tree overlay adds new nodes under the root node only (in which case, it could specify / in the `target-path` property in the fragment), it would be much easier to specify the target node via its phandle (`<&label_name>`) as it would save us from manually computing the node's full path (especially if it is nested).

The thing is, there is no direct correlation or link between the base device tree and the overlay. They are each built, standalone, on their sides. Therefore, referencing a remote node (that is, a node in the base device tree) from the device tree overlay will raise errors, and the build of the overlay will fail because of undefined references or labels. It would be like building a dynamically linked application without room for symbol resolution.

To address this issue, -@ command-line flag support has been added to the DTC. This flag must be specified for both the base device tree and all of the overlays to be compiled. It will instruct the DTC to generate extra nodes in the root (such as __symbols__, __fixups__, and __local_fixups__) that contain resolution data for the translation of phandle names. These extra nodes are spread as follows:

- When the -@ option is added to build an overlay, it recognizes the /plugin/; line that marks a device tree fragment/object. That line controls the generation of __fixups__ and __local_fixups__ nodes.

- When the -@ option is added to build the base device tree, /plugin/; is not present, so the source is recognized as being the base device tree, which causes the generation of __symbols__ nodes only.

These extra nodes add room for symbol resolution.

> **Note**
>
> Support for the -@ option can only be found in dtc version 1.4.4 or later. Only Linux kernel versions v4.14 or higher includes a built-in version of dtc that meets this requirement. This option is not needed if only target-path properties (that is, non-phandle-based) are used in the device tree overlay.

Building a binary device tree overlay follows the same process as building a traditional binary device tree. For example, let's consider the following base device tree. Let's call it base.dts:

```
/dts-v1/;
/ {
        foo: foonode {
                foo-bool-property;
                foo-int-property = <0x80>;
                status = "disabled";
        };
};
```

Then, let's build this base device tree with the following command:

```
dtc -@ -I dts -O dtb -o base.dtb base.dts
```

In the next step, let's consider the following device tree overlay. Let's call it `foo-verlay.dts`:

```
/dts-v1/;
/plugin/;
/ {
        fragment@1 {
                target = <&foo>;
                __overlay__ {
                        overlay-1-property;
                        status = "okay";
                        bar: barnode {
                                bar-property;
                        };
                };
        };
};
```

In the preceding device tree overlay, the `status` property of the `foo` node in the base device tree has been modified from `disabled` to `okay`, which will activate this node. Following this, the `overlay-1-property` Boolean property has been added, and finally, a `bar` sub-node has been added with a single Boolean property. This device tree overlay can be compiled with the following command:

```
dtc -@ -I dts -O dtb -o foo-overlay.dtbo foo-overlay.dts
```

As you can see, the `-@` flag has been added on both sides, enabling room for symbol resolution.

> **Note**
> In the Yocto build system, you could add this flag to the machine configuration or, during development, to the `local.conf` file, as follows: `DEVICETREE_FLAGS += "-@"`.

For the Linux kernel to build the device tree overlay, you should add it to the `Makefile` device tree of your SoC architecture, for example, `arch/arm64/boot/dts/freescale/Makefile` or `arm/arm/boot/dts/Makefile`, along with the `dtbo` extension, as follows:

```
dtb-y += foo-overlay.dtbo
```

Now that we can manage to build our own device tree overlays, let's consider the logical next step, which consists of loading these overlays into the system.

Loading device tree overlays via configfs

This section is named such that it mentions the way the device tree overlay is going to be loaded (configfs) since there is not only one way to load device tree overlays. In this section, we will focus on doing this on a running system whose kernel has already booted and the root filesystem is already mounted.

In order to do this, your kernel must have been compiled with CONFIG_OF_OVERLAY and CONFIG_CONFIGFS for the following steps to work. The following is a check, assuming the kernel config is available in the target:

```
~# zcat /proc/config.gz | grep CONFIGFS
CONFIG_CONFIGFS_FS=y
~# zcat /proc/config.gz | grep OF_OVERLAY
CONFIG_OF_OVERLAY=y
~#
```

Now it's time to insert the DTBs into a running kernel using configfs. First, we mount the configfs filesystem if it has not already been mounted on your system:

```
# mount -t configfs none /sys/kernel/config
```

When configfs has been mounted properly, the directory should be populated with base subdirectories (device-tree/overlays), which, according to our mount path, will result in /sys/kernel/config/device-tree/overlays, as demonstrated in the following:

```
# mkdir -p /sys/kernel/config/device-tree/overlays/
```

Then, each overlay entry must be added from within the overlays directory. It has to be noted that overlay entries are created and manipulated using a standard filesystem I/O.

To load an overlay, a directory corresponding to this overlay must be created under the `overlays` directory. For our example, let's use the name `foo` :

```
# mkdir /sys/kernel/config/device-tree/overlays/foo
```

Next, to effectively load the overlay, you can `echo` the overlay firmware file path to the `path` property file, as follows:

```
# echo /path/to/foo-overlay.dtbo > /sys/kernel/config/device-tree/overlays/foo/path
```

Alternatively, you can `cat` the contents of the overlay to the `dtbo` file:

```
# cat foo.dtbo > /sys/kernel/config/device-tree/overlays/foo/dtbo
```

After that, the overlay file will be applied, and devices will be created/destroyed as required.

To remove the overlay and undo its changes, you should simply `rmdir` the corresponding overlay directory. In our example, it should be as follows:

```
# rmdir /sys/kernel/config/device-tree/overlays/foo
```

Although you have loaded the device tree overlay dynamically, it won't be sufficient; the device driver for the added device node needs to be loaded for the device to work unless this driver is built-in and enabled (that is, selected with `y` during `make menuconfig`).

At this stage, we are done with our device tree compilation-related stuff. Now, we can learn how to write our own device trees, starting with device addressing and representation.

Representing and addressing devices

In the device tree, a node is the representational unit of a device. In other words, a device is represented by at least one node. Following this, device nodes can either be populated with other nodes (therefore, creating a parent-child relationship) or with properties (which would describe the device corresponding to the node they populate).

While each device can operate standalone, there are situations where a device might want to be accessed by its parent or where a parent might want to access one of its children. For example, such situations occur when a bus controller (the parent node) wants to access one or more of the devices (declared as a sub-node) sitting on its bus. Typical examples include I2C controllers and I2C devices, SPI controllers and SPI devices, CPUs and memory-mapped devices, and more. Thus, the concept of device addressing has emerged. Device addressing has been introduced with a `reg` property, which is used in each addressable device but whose meaning or interpretation depends on the parent (most of the time, they are bus controllers). The meaning and interpretation of `reg` in a child device depends on the `#address-cells` and `#size-cells` properties of its parent. The # (sharp) character that prefixes `size-cells` and `address-cells` can be considered to mean "length of."

Each addressable device gets a `reg` property that is a list of tuples in the form of `reg = <address0 size0 [address1 size1] [address2 size2] ... >`, where each tuple represents an address range used by the device. `#size-cells` indicates how many 32-bit cells are used to represent the size, which might be 0 if the size is not relevant. On the other hand, `#address-cells` indicates how many 32-bit cells are used to represent the address. In other words, the address element of each tuple is interpreted according to `#address-cells`; this uses the same size as the size element, which is interpreted according to `#size-cells`.

To sum up, addressable devices inherit from the `#size-cell` and `#address-cell` properties of their parent, which is the node that represents the bus controller most of the time. The presence of `#size-cell` and `#address-cell` in a given device does not affect the device itself but its children if they are addressable.

Now that we have seen how addressing works in a general manner, let's address specific addressing for non-discoverable devices, starting with SPI and I2C.

Handling SPI and I2C device addressing

SPI and I2C devices both belong to non-memory-mapped devices because their addresses are not accessible to the CPU. Instead, the parent device's driver (the bus controller driver) will perform indirect access on behalf of the CPU. Each I2C/SPI device node is always represented as a sub-node of the I2C/SPI controller node that the device sits on. For a non-memory-mapped device, the `#size-cells` property is 0, and the size element in the addressing tuple is empty. This means that the `reg` property for this kind of device is always one cell. The following is an example:

```
&i2c3 {
    [...]
```

```
        status = "okay";

    temperature-sensor@49 {
        compatible = "national,lm73";
        reg = <0x49>;
    };

    pcf8523: rtc@68 {
        compatible = "nxp,pcf8523";
        reg = <0x68>;
    };
};

&ecspi1 {
    fsl,spi-num-chipselects = <3>;
    cs-gpios = <&gpio5 17 0>, <&gpio5 17 0>, <&gpio5 17 0>;
    status = "okay";
    [...]

    ad7606r8_0: ad7606r8@1 {
        compatible = "ad7606-8";
        reg = <1>;
        spi-max-frequency = <1000000>;
        interrupt-parent = <&gpio4>;
        interrupts = <30 0x0>;
        convst-gpio = <&gpio6 18 0>;
    };
};
```

If you look at the SoC-level file in arch/arm/boot/dts/imx6qdl.dtsi, you will notice that #size-cells and #address-cells are set to 0, for the former, and 1, for the latter, respectively, in both the I2C and SPI controller nodes (labeled i2c3 and ecspi1). This helps you to understand their reg property, which is only one cell for the address value and none for the size value.

The I2C device's `reg` property is used to specify the device's address on the bus. For SPI devices, `reg` represents the index of the chip-select line assigned to the device among the list of chip selects the controller node has. For example, for the `ad7606r8` ADC, the chip-select index is 1, which corresponds to `<&gpio5 17 0>` in `cs-gpios`. This is the list of chip selects of the controller node. The binding of other controllers might differ, and you should refer to their documentation in `Documentation/devicetree/bindings/spi`.

Memory-mapped devices and device addressing

This section addresses simple memory-mapped devices where the memory region is accessible by the CPU. With such device nodes, the `reg` property still defines the device's address, and the `reg = <address0 size0 [address1 size1] [address2 size2] ... >` pattern takes place. Each region is represented with a tuple of cells, where the first cell is the base address of the memory region, and the second cell is the size of the region. It could be translated into the following pattern: `reg = <base0 length0 [base1 length1] [address2 length2] ... >`. Here, each tuple represents an address range used by the device.

Let's consider the following example:

```
soc {
    #address-cells = <1>;
    #size-cells = <1>;
    compatible = "simple-bus";
    aips-bus@02000000 { /* AIPS1 */
        compatible = "fsl,aips-bus", "simple-bus";
        #address-cells = <1>;
        #size-cells = <1>;
        reg = <0x02000000 0x100000>;
        [...];

    spba-bus@02000000 {
        compatible = "fsl,spba-bus", "simple-bus";
        #address-cells = <1>;
        #size-cells = <1>;
        reg = <0x02000000 0x40000>;
        [...]
```

```
        ecspi1: ecspi@02008000 {
            #address-cells = <1>;
            #size-cells = <0>;
            compatible = "fsl,imx6q-ecspi", "fsl,imx51-
ecspi";
            reg = <0x02008000 0x4000>;
            [...]
        };

        i2c1: i2c@021a0000 {
            #address-cells = <1>;
            #size-cells = <0>;
            compatible = "fsl,imx6q-i2c", "fsl,imx21-i2c";
            reg = <0x021a0000 0x4000>;
          [...]
        };
      };
    };
  };
```

In the preceding excerpt, device nodes that have simple-bus in their compatible properties are SoC internal memory-mapped bus controllers connecting IP cores (such as I2C, SPI, USB, Ethernet, and other internal SoC IPs) to the CPU. Their sub-nodes, which are either IP core or other internal buses, inherit from the #address-cells and #size-cells properties. We can see that the i2c@021a0000 I2C controller (labeled i2c1) is connected to the spba-bus@02000000 bus. All of them will appear as platform devices on the system at runtime. On the other hand, this I2C controller changes its addressing scheme by defining its own #address-cells and #size-cells properties, which the I2C devices connected to it will inherit. This is the same for the SPI controller.

To summarize, in the real world, you should not interpret a reg property in a node without knowing the #size-cells and #address-cells properties of its parent. Memory-mapped devices must have the size field of their reg property set with the size of the memory regions of the devices, but also the address field, which must be defined such that it corresponds to the beginning of the device memory regions in the SoC memory map, as shown in the SoC datasheet.

> **Note**
>
> The simple-bus-compatible string also indicates that the bus has no special driver, that there's no way to dynamically probe the bus, and that direct child nodes (exclusively, level-1 children) will be registered as platform devices. Sometimes, this is used in board-level device trees to instantiate GPIO-based fixed regulators.

Handling resources

The main purpose of a device driver is to provide a set of driving functions for a given device and expose its capabilities to users. Here, the objective is to gather the device's configuration parameters, especially resources (such as the memory region, interrupt line, DMA channel, and more) that will help the driver to perform its job.

The struct resource

Once probed, device resources assigned to the device (either in the device or the board/ machine file) are gathered and allocated either by of_platform or by the platform cores using struct resource, as follows:

```
struct resource {
    resource_size_t start;
    resource_size_t end;
    const char *name;
    unsigned long flags;
[...]
};
```

The following lists the meanings of the elements in the data structure:

- start: Depending on the resource flag, this can be the starting address of a memory region, an IRQ line number, a DMA channel number, or a register offset.

- end: This is the end of the memory region or the end of the register offset. In the case of an IRQ or DMA channel, most of the time, the start and end values are the same.

- name: This is the name of the resource if any. We discuss it in more detail in the next section.

- `flags`: This indicates the type of resource. Possible values include the following:

  - `IORESOURCE_IO`: This indicates the PCI/ISA I/O port region. `start` is the first port of the region, and `end` is the last one.

  - `IORESOURCE_MEM`: This is used for the I/O memory regions. `start` indicates the starting address of the region, and `end` indicates where it ends.

  - `IORESOURCE_REG`: This refers to the register offsets. It is mostly used with MFD devices. `start` indicates the offset relative to a parent device register, and `end` indicates where the register section ends.

  - `IORESOURCE_IRQ`: The resource is an IRQ line number. In this case, either both `start` and `end` have the same value or `end` is irrelevant.

  - `IORESOURCE_DMA`: This indicates that the resource is a DMA channel number. You should consider `end` in the same way as an IRQ. However, what happens when you have more than one cell for the DMA channel identifier or when you have multiple DMA controllers is not very well defined. `IORESOURCE_DMA` is not scalable for multiple controller systems.

There is one instance of this data structure allocated per resource. That means, for a device that is assigned two memory regions and one IRQ line, there will be three data structures allocated. Moreover, resources of the same type will be allocated and indexed (starting from 0) in the order they are declared in the device tree (or the board file). This means that the first memory region assigned will have index 0, and so on.

To get the appropriate resource, we will use a generic API, `platform_get_resource()`, given the resource type and its index in this type. This function is defined as follows:

```
struct resource *platform_get_resource(
                    struct platform_device *dev,
                    unsigned int type, unsigned int num)
```

In the preceding prototype, `dev` is the platform device that we write the driver for, `type` is the resource type, and `num` is the index of this resource in the same type. On success, the function returns a valid pointer to `struct resource`, or NULL otherwise.

When there is more than one resource in the same type, using indexes could be misleading. You might decide to rely on the resource name as an alternative, and the named variant of `platform_get_resource()` will be introduced, which is `platform_get_resource_byname()`. This function is given a resource flag (or type) and its name and returns the appropriate resource, whatever the order they are declared in.

It is defined as follows:

```
struct resource *platform_get_resource_byname(
                          struct platform_device *dev,
                          unsigned int type,
                          const char *name)
```

To understand how to use this function, first, let's introduce the concept of named resources. We will discuss this next.

The concept of named resources

When the driver expects a list of resources of a certain type (let's say two IRQ lines, where the first one is for Tx and the second for Rx), there is no guarantee of the way the list will be ordered, and the driver must make no assumption. What happens if the driver logic is hardcoded so that it expects Rx IRQ first, but the device tree has been populated with Tx first? To avoid such mismatches, the concept of named resources (such as clocks, IRQs, DMA channels, and memory regions) has been introduced. This consists of defining the resource list and naming them. This is so that, whatever their indexes are, a given name will always match the resource. The concept of a named resource also makes it easy to read and understand the device tree resource assignment.

The corresponding properties to name the resources are as follows:

- reg-names: This is the list of names for memory regions in the reg property.
- interrupt-names: This gives a name to each interrupt line in the interrupts property.
- dma-names: This is for the dma property.
- clock-names: This is to name the clocks inside the clocks property. Note that clocks won't be discussed in this book.

To demonstrate the concept, let's consider the following fake device node entry:

```
fake_device {
    compatible = "packt,fake-device";
    reg = <0x4a064000 0x800>,
          <0x4a064800 0x200>,
          <0x4a064c00 0x200>;
    reg-names = "ohci", "ehci", "config";
    interrupts = <0 66 IRQ_TYPE_LEVEL_HIGH>,
                 <0 67 IRQ_TYPE_LEVEL_HIGH>;
    interrupt-names = "ohci", "ehci";
};
```

In the preceding example, the device is assigned three memory regions and two interrupt lines. The resource name list and resources respect a one-to-one mapping. This means, for example, that the name at index 0 will be assigned to the resource at the same index. The code in the driver to extract each named resource is as follows:

```
struct resource *res_mem_config, resirq, *res_mem1;
int txirq, rxirq;


/*let's grab region <0x4a064000 0x800>*/
res_mem1 = platform_get_resource_byname(pdev,
                    IORESOURCE_MEM, "ohci");
/*let's grab region <0x4a064c00 0x200>*/
res_mem_config = platform_get_resource_byname(pdev,
                    IORESOURCE_MEM, "config");


txirq = platform_get_resource_byname(pdev,
                        IORESOURCE_IRQ, "ohci");
rxirq = platform_get_resource_byname(pdev,
                        IORESOURCE_MEM, "ehci");
```

As you can see, requesting resources in the named manner is less error-prone. That said, both `platform_get_resource()` and `platform_get_resource_byname()` are generic APIs used to deal with resources. However, there are dedicated APIs that allow you to reduce the development effort (such as `platform_get_irq_byname()`, `platform_get_irq()`, or `platform_get_and_ioremap_resource()`), as we will learn in later chapters.

Extracting application-specific data

Application-specific data is data that is beyond the common resources (neither IRQ numbers, memory regions, regulators, nor clocks). These are arbitrary properties and child nodes that can be assigned to a device. Usually, such properties use a manufacture prefix. These can be any kind of string, Boolean, or integer value, along with their API defined in `drivers/of/base.c` in the Linux sources. Note that the examples we discuss here are not exhaustive. Now, let's reuse the node that was defined earlier in this chapter:

```
node_label: nodename@reg{
    string-property = "a string";
    string-list = "red fish", "blue fish";
    one-int-property = <197>; /* One cell property */
    /* in the following line, each number (cell) is a
     * 32-bit integer(uint32). There are 3 cells in
     * this property */
    int-list-property = <0xbeef 123 0xabcd4>;
    mixed-list-property = "a string", <0xadbcd45>,
                          <35>, [0x01 0x23 0x45];
    byte-array-property = [0x01 0x23 0x45 0x67];
    one-cell-property = <197>;
    boolean-property;
};
```

In the sections that follow, we will learn how to obtain each property in the preceding device tree node excerpt.

Extracting string properties

The following is an excerpt of the previous example, showing a single string and multiple string properties:

```
string-property = "a string";
string-list = "red fish", "blue fish";
```

Back in the driver, there are different APIs that can be used depending on the need. You should use of_property_read_string() to read a single string value property. Its prototype is defined as follows:

```
int of_property_read_string(const struct device_node *np,
                            const char *propname,
                            const char **out_string)
int of_property_read_string_index(const struct
                                  device_node *np,
                                  const char *propname,
                                  int index,
                                  const char **output)
int of_property_read_string_array(
                    const struct device_node *np,
                    const char *propname,
                    const char **out_strs,
                    size_t sz)
```

In the preceding functions, np is the node from which the string property needs to be read. propname is the name of the property hosting the string or string list, and sz is the number of array elements to read. out_string is an output parameter whose pointer value will be changed to point to the string value(s).

of_property_read_string() and of_property_read_string_index() return -ENODATA if the property does not have a value. Alternatively, they return -EINVAL if the property does not exist at all. Additionally, -EILSEQ is returned if the string is not NULL-terminated within the length of the property data. Finally, on success, they return 0.

`of_property_read_string_array()` searches for the specified property in the given device tree node, retrieves a list of NULL-terminated string values (actually a pointer to these strings, not a copy) inside that property, and assigns it to `out_strs`. In addition to the values returned by `of_property_read_string()` and `of_property_read_string_index()`, this function is special in the way that it returns the number of strings that have been read if the target array of pointers is not NULL. The `out_strs` parameter can be omitted (NULL) if you simply wish to count the number of strings in the property.

The following code shows how you can use them:

```
size_t count;
const char **res;
const char *my_string = NULL;
const char *blue_fish = NULL;
of_property_read_string(pdev->dev.of_node,
                    "string-property", &my_string);
of_property_read_string_index(pdev->dev.of_node,
                    "string-list", 1, &blue_fish);
count = of_property_read_string_array(dp,
                    "string-list", res, count);
```

In the preceding example, we learned how to extract string properties, either from a single-value property or from a list. It must be noted that a pointer to the string (or string list) is returned, not a copy. Additionally, in the last line, we see how to extract a given number of NULL-terminated string elements in the array. Here again, a pointer to these elements are returned, not copies of these.

Reading cells and unsigned 32-bit integers

Here are our `int` properties:

```
one-int-property = <197>;
int-list-property = <1350000 0x54dae47 1250000 1200000>;
```

Back in the driver, as with string properties, there is a set of APIs that you can choose according to your need. You should use `of_property_read_u32()` to read a cell value. Its prototype is defined as follows:

```
int of_property_read_u32(const struct device_node *np,
                         const char *propname,
                         u32 *out_value)
int of_property_read_u32_index(
                         const struct device_node *np,
                         const char *propname,
                         u32 index, u32 *out_value)
int of_property_read_u32_array(
                         const struct device_node *np,
                         const char *propname,
                         u32 *out_values, size_t sz)
```

The preceding APIs behave in the same way as their `_string`, `_string_index`, and `_string_array` counterparts. This is because they all return 0 on success, `-EINVAL` if the property does not exist at all, or `-ENODATA` when the property does not have a value. The differences are that the type of value to read, in this case, is u32, that we have an `-EOVERFLOW` error if the property data isn't large enough, and that `out_values` must be allocated first because the values to be returned have been copied instead.

The following is the usage of these APIs in our example, with `int` and our list of `int` properties:

```
unsigned int number;
of_property_read_u32(pdev->dev.of_node,
                     "one-cell-property", &number);
```

You can use `of_property_read_u32_array` to read a list of cells. Its prototype is as follows:

```
int of_property_read_u32_array(
                         const struct device_node *np,
                         const char *propname,
                         u32 *out_values, size_t sz);
```

Here, `sz` is the number of array elements to read. Take a look at `drivers/of/base.c` to see how to interpret its return value:

```
unsigned int cells_array[4]; /* return value by copy */
if (!of_property_read_u32_array(pdev->dev.of_node,
    "int-list-property", cells_array, 4))
    dev_info(&pdev->dev, "u32 list read successfully\n");
/* can now process values in cells_array */
[...]
```

Here, we have demonstrated how easy it would be to deal with an array of cells, that is, arrays of 32-bit integer values. In the next section, we will see how to do that with Boolean properties.

Handling Boolean properties

You should use `of_property_read_bool()` to read the Boolean property whose name is given in the second argument of the function:

```
bool my_bool = of_property_read_bool(pdev->dev.of_node,
                                     "boolean-property");
if(my_bool){
    /* boolean is true */
} else
    /* boolean is false */
}
```

The preceding example demonstrates how to deal with Boolean properties. Now we can learn about far more complex APIs, starting with extracting and parsing sub-nodes.

Extracting and parsing sub-nodes

Note that you are allowed to add whatever sub-node you want to a device node. This usage is common in numerous use cases, such as population partitions in an MTD device node or describing regulator constraints in a power management chip node. For example, given a node representing a flash memory device, partitions can be represented as nested sub-nodes. The following excerpt shows how this is achieved:

```
eeprom: ee24lc512@55 {
    compatible = "microchip,24xx512";
    reg = <0x55>;
```

```
    partition@1 {
        read-only;
        part-name = "private";
        offset = <0>;
        size = <1024>;
    };

    config@2 {
        part-name = "data";
        offset = <1024>;
        size = <64512>;
    };
};
```

You can use `for_each_child_of_node()` to walk through sub-nodes of the given node:

```
structdevice_node *np = pdev->dev.of_node;
structdevice_node *sub_np;
for_each_child_of_node(np, sub_np) {
    /* sub_np will point successively to each sub-node */
    [...]
    int size;
    of_property_read_u32(client->dev.of_node,
                    "size", &size);

    ...
}
```

In the preceding excerpt, we learned how to iterate over the sub-nodes of a given device tree node.

Summary

The time to switch from hardcoded device configurations to device trees has come. This chapter gave you all the bases to handle device trees. Now you have the necessary skills to customize or add whatever node and property you want to the device tree and extract them from your driver.

In the next chapter, we will discuss the I2C driver, and use the device tree API to enumerate and configure our I2C devices.

6
Introduction to Devices, Drivers, and Platform Abstraction

The **Linux Device Model** (**LDM**) is a concept that was introduced in the Linux kernel to describe and manage kernel objects (those requiring reference counting, for example, such as files, devices, buses, and even drivers), as well as their hierarchies and how they are bound to others. LDM introduced object life cycle management, reference counting, an **object-oriented** (**OO**) programming style in the kernel, and other advantages (such as code reusability and refactoring, automatic resource releasing, and more), which will not be discussed here.

Since reference counting and life cycle management are at the lowest level of LDM, we will discuss higher representations, such as dealing with common kernel data objects and structures, including **devices**, **drivers**, and **buses**.

In this chapter, we will cover the following topics:

- Linux kernel platform abstraction and data structures
- Device and driver matching mechanism explained

Linux kernel platform abstraction and data structures

The Linux device model is built on top of some fundamental data structures, including `struct device`, `struct device_driver`, and `struct bus_type`. The first data structure represents the device to be driven, the second is the data structure of each software entity intended to drive the device, and the latter represents the channel between the device and the CPU.

Device base structure

Devices help extract either physical or virtual devices. They are built on top of the `struct device` structure, which is worth introducing first, as described in `include/linux/device.h`:

```
struct device {
    struct device           *parent;
    struct kobject          kobj;
    struct bus_type         *bus;
    struct device_driver    *driver;
    void *platform_data;
    void *driver_data;
    struct dev_pm_domain    *pm_domain;
    struct device_node      *of_node;
    struct fwnode_handle    *fwnode;
    dev_t           devt;
    u32             id;
    [...]
};
```

Let's look at each element in this structure:

- `parent`: This is the device's "parent" device, the device that this device is attached to. In most cases, a parent device is some sort of bus or host controller. If `parent` is `NULL`, then the device is a top-level device. This is the case for bus controller devices for example.

- `kobj`: This is the lowest-level data structure and is used to track a kernel object (bus, driver, device, and so on). This is the centerpiece of LDM. We will discuss this in *Chapter 14*, *Introduction to the Linux Device Model*.

- `bus`: This specifies the type of bus the device is on. It is the channel between the device and the CPU.

- `driver`: This specifies which driver has allocated this device.

- `platform_data`: This provides platform data that's specific to the device. This field is automatically set when the device is declared from within the board file. In other words, it points to board-specific structures from within the board setup file that describe the device and how it is wired. It helps minimize the use of `#ifdefs` inside the device driver code. It contains resources such as chip variants, GPIO pin roles, and interrupt lines.

- `driver_data`: This is a private pointer for driver-specific information. The bus controller driver is responsible for providing helper functions, which are accessors that are used to get/set this field.

- `pm_domain`: This specifies power management-specific callbacks that are executed during system power state changes: suspend, hibernation, system resume, and during runtime PM transitions, along with subsystem-level and driver-level callbacks.

- `of_node`: This is the device tree node that's associated with this device. This field is automatically filled by the **Open Firmware** (**OF**) core when the device is declared from within the device tree. You can check whether `platform_data` or `of_node` is set to determine where exactly the device has been declared.

- `id`: This is the device instance.

Devices are rarely represented by bare device structures since most subsystems track extra information about the devices they host; instead, the structure is frequently embedded within a higher-level representation of the device. This is the case for the `struct i2c_client`, `struct spi_device`, `struct usb_device`, and `struct platform_device` structures, which all embed a `struct device` element in their members (`spi_device->dev`, `i2c_client->dev`, `usb_device->dev`, and `platform_device->dev`).

Device driver base structure

The next structure we need to introduce is the `struct device_driver` structure. This structure is the base element of any device driver. In object-oriented languages, this structure would be the base class, which would be inherited by each device driver.

This data structure is defined in `include/linux/device/driver.h` like so:

```
struct device_driver {
    const char        *name;
    struct bus_type   *bus;
    struct module     *owner;
    const struct of_device_id   *of_match_table;
    const struct acpi_device_id *acpi_match_table;
    int (*probe) (struct device *dev);
    int (*remove) (struct device *dev);
    void (*shutdown) (struct device *dev);
    int (*suspend) (struct device *dev,
                        pm_message_t state);
    int (*resume) (struct device *dev);
    const struct dev_pm_ops *pm;
};
```

Let's look at each element in this structure:

- name: This is the name of the device driver. It's used as a fallback (that is, it matches this name with the device name) when no matching method succeeds.

- bus: This field is mandatory. It represents the bus that the devices of this driver belong to. Driver registration will fail if this field is not set because it is its probe method that is responsible for matching the driver with devices.

- owner: This field specifies the module owner.

- of_match_table: This is the open firmware table. It represents the array of `struct of_device_id` elements that are used for device tree matching.

- acpi_match_table: This is the ACPI match table. This is the same as `of_match_table` but for ACPI matching, which will not be discussed in this tutorial.

- probe: This function is called to query the existence of a specific device, whether this driver can work with it, and then bind the driver to a specific device. The bus driver is responsible for calling this function at given moments. We will discuss this shortly.

- `remove`: When a device is removed from the system, this method is called to unbind it from this driver.

- `shutdown`: This command is issued when the device is about to be turned off.

- `suspend`: This is a callback that allows you to put the device into sleep mode, mostly in a low-power state.

- `resume`: This is invoked by the driver core to wake up a device that has been in sleep mode.

- pm: This represents a set of power management callbacks for devices that matched this driver.

In the preceding data structure, the `shutdown`, `suspend`, `resume`, and pm elements are optional as they are used for power management purposes. Providing these elements depends on the capability of the underlying device (whether it can be shut down, suspended, or perform other power management-related capabilities).

Driver registration

First, you should keep in mind that registering a device consists of inserting that device into the list of devices that are maintained by its bus driver. In the same way, registering a device driver consists of pushing this driver into the list of drivers that's maintained by the driver of the bus that it sits on top of. For example, registering a USB device driver will result in inserting that driver into the list of drivers that are maintained by the USB controller driver. The same goes for registering an SPI device driver, which will queue the driver into the list of drivers that are maintained by the SPI controller driver. `driver_register()` is a low-level function that's used to register a device driver with the bus. It adds the driver to the bus's list of drivers. When a device driver is registered with the bus, the core walks through the bus's list of devices and calls the bus's `match()` callback for each device that does not have a driver associated with it to find out whether there are any devices that the driver can handle. When a match occurs, the device and the device driver are bound together. The process of associating a device with a device driver is called binding.

You probably never want to use `driver_register()` as-is; it is up to the bus driver to provide a bus-specific registration function, which will be a wrapper based on `driver_register()`. So far, bus-specific registration functions have always matched the `{bus_name}_register_driver()` pattern. For example, the registration functions for the USB, SPI, I2C, and PCI drivers would be `usb_register_driver()`, `spi_register_driver()`, `i2c_register_driver()`, and `pci_register_driver()`, respectively.

The recommended place to register/unregister the driver is within the `init/exit` functions of the module, which are executed at the module loading/unloading stages, respectively. In lots of cases, registering/unregistering the driver is the only action you will want to execute within those `init/exit` functions. In such cases, each bus core provides a specific helper macro, which will be expanded as the `init/exit` functions of the module and internally call the bus-specific registering/unregistering function. Those bus macros follow the `module_{bus_name}_driver(__{bus_name}_driver);` pattern, where `__{bus_name}_driver` is the driver structure of the corresponding bus. The following table shows a non-exhaustive list of buses that are supported in Linux, along with their macros:

Bus Name	(Un)Register Macro
I2C	`module_i2c_driver(__i2c_driver);`
SPI	`module_spi_driver(__spi_driver);`
Platform (that is, pseudo-platform)	`module_platform_driver(__platform_driver);`
USB	`module_usb_driver(__usb_driver);`
PCI	`module_pci_driver(__pci_driver);`

Table 6.1 – Some buses, along with their (un)registration macros

The bus controller code is responsible for providing such macros, but this is not always the case. For example, the MDIO bus driver (a 2-wire serial bus that's used to control network devices) does not provide a `module_mdio_driver()` macro. You should check whether this macro exists for the bus that the device sits on top of to write the driver before using it. The following code blocks show two examples of different buses – one using the bus-provided registering/unregistering macro, and another not using it. Let's see what the code looks like when we don't use the macro:

```
static struct platform_driver mypdrv = {
    .probe = my_pdrv_probe,
    .remove = my_pdrv_remove,
    .driver = {
        .name = KBUILD_MODNAME,
        .owner = THIS_MODULE,
    },
};

static int __init my_drv_init(void)
{
    /* Registering with Kernel */
```

```
    platform_driver_register(&mypdrv);
    return 0;
}

static void __exit my_pdrv_remove (void)
{
    /* Unregistering from Kernel */
    platform_driver_unregister(&my_driver);
}

module_init(my_drv_init);
module_exit(my_pdrv_remove);
```

The preceding example does not use the macro at all. Now, let's look at an example that uses the macro:

```
static struct platform_driver mypdrv = {
    .probe = my_pdrv_probe,
    .remove = my_pdrv_remove,
    .driver = {
        .name = KBUILD_MODNAME,
        .owner = THIS_MODULE,
    },
};
module_platform_driver(my_driver);
```

Here, you can see how the code is factorized, which is a serious plus when you're writing a driver.

Exposing the supported devices in the driver

The kernel must be aware of the devices that are supported by a given driver and whether they are present on the system so that whenever one of them appears on the system (the bus), the kernel knows which driver is in charge of it and runs its probe function. That said, the probe() function of the driver will only be run if this driver is loaded (which is a userspace operation); otherwise, nothing will happen. The next section will explain how to manage driver auto-loading so that when the device appears, its driver is automatically loaded, and its probe function is called.

If we have a look at each bus-specific device driver structure (`struct platform_driver`, `struct i2c_driver`, `struct spi_driver`, `struct pci_driver`, and `struct usb_driver`), we will see that there is an `id_table` field whose type depends on the bus type. This field should be given an array of device IDs that correspond to those supported by the driver. The following table shows the common buses, along with their device ID structures:

Bus type	Device ID structure
I2C	`struct i2c_device_id`
SPI	`struct spi_device_id`
platform (that is, pseudo-platform)	`struct platform_device_id`
PCI	`struct pci_device_id`
USB	`struct usb_device_id`

Table 6.2 – Some buses, along with their device identification data structures

I intentionally omitted two special cases: the device tree and ACPI. They can expose devices so that they can be declared either from within the device tree or ACPI using the `driver.of_match_table` or `driver.acpi_match_table` fields, which are not direct elements of the bus-specific driver structure:

OF	`struct of_device_id`
ACPI	`struct acpi_device_id`

Table 6.3 – Pseudo buses, along with their device identification data structures

These structures are all defined in `include/linux/mod_devicetable.h` in the kernel sources, and their names match the `{bus_name}_device_id` pattern. We have already discussed each structure in the appropriate chapters. So, let's look at an example that exposes SPI devices using both `struct spi_device_id` and `struct of_device_id` for declaring the device tree (new and recommended) of this driver (`http://elixir.free-electrons.com/linux/v4.10/source/drivers/gpio/gpio-mcp23s08.c`):

```
static const struct spi_device_id mcp23s08_ids[] = {
    { "mcp23s08", MCP_TYPE_S08 },
    { "mcp23s17", MCP_TYPE_S17 },
    { "mcp23s18", MCP_TYPE_S18 },
    { },
};
```

```
static const struct of_device_id mcp23s08_spi_of_match[] = {
    {
        .compatible = "microchip,mcp23s08",
        .data = (void *) MCP_TYPE_S08,
    },
    {
        .compatible = "microchip,mcp23s17",
        .data = (void *) MCP_TYPE_S17,
    },
    {
        .compatible = "microchip,mcp23s18",
        .data = (void *) MCP_TYPE_S18,
    },
    { },
};

static struct spi_driver mcp23s08_driver = {
    .probe  = mcp23s08_probe, /* don't care about this */
    .remove = mcp23s08_remove, /* don't care about this */
    .id_table = mcp23s08_ids,
    .driver = {
        .name    = "mcp23s08",
        .of_match_table =
                of_match_ptr(mcp23s08_spi_of_match),
    },
};
```

The preceding excerpt shows how a driver can declare the devices it supports. Since our example is an SPI driver, the data structure that is involved is struct spi_device_id, in addition to struct of_device_id, which is used in any driver that needs to match a device according to their compatible string in the driver.

Now that we are done learning the way a driver can expose the device it supports, let's get deeper in the device and driver binding mechanism to understand what happens under the hood when there is a match between a device and a driver.

Device/driver matching and module (auto) loading

Please pay attention to this section, even though we will partially repeat what we discussed previously. The **bus** is the fundamental element that device drivers and devices rely on. From a hardware point of view, the bus is the link between devices and the CPU, while from a software point of view, the bus driver is the link between devices and their drivers. Whenever a device or driver is added/registered with the system, it is automatically added to a list that's maintained by the driver of the bus that it sits on top of. For example, registering a list of I2C devices that can be managed by a given driver (i2c, of course) will result in queueing those devices into a global list that maintains the I2C adapter driver, as well as providing a USB device table that will insert those devices into the list of devices that's maintained by the USB controller driver. Another example involves registering a new SPI driver, which will insert this driver into the list of drivers that's maintained by the SPI controller driver. Without this, there would be no way for the kernel to know which driver should handle which device.

Every device driver should expose the list of devices it supports and should make that list accessible to the driver core (especially to the bus driver). This list of devices is called `id_table` and is declared and filled from within the driver code. This table is an array of device IDs, where each ID's type depends on the device's type (I2C, SPI, USB, and so on). In this manner, whenever a device appears on the bus, the bus driver will walk through its device driver's list and look into each ID table for the entry that corresponds to this new device. Every driver that contains the device ID in their table will have their `probe()` function run, with the new device given as a parameter. This process is called the matching loop. It works similarly for drivers. Whenever a new driver is registered with the bus, the bus driver will walk through the list of its devices and look for the device IDs that appear in the registered driver's `id_table`. For each hit, the corresponding device will be given as a parameter to the `probe()` function of the driver, which will be run as many times as there are hits.

The problem with the matching loop is that only loaded modules will have their probe functions invoked. In other words, the matching loop will be useless if the corresponding module is not loaded (`insmod`, `modprobe`) or built-in. You'll have to manually load the module before the device appears on the bus. The solution to this issue is module auto-loading. Since, most of the time, module loading is a userspace action (when the kernel does not request the module itself using the `request_module()` function), the kernel must find a way to expose drivers, along with their device tables, to the userspace. Thus came a macro called `MODULE_DEVICE_TABLE()`:

```
MODULE_DEVICE_TABLE(<bus_type_name>,  <array_of_ids>)
```

This macro is used to support hot-plugging, which describes which devices each specific driver can support. At compilation time, the build process extracts this information out of the driver and builds a human-readable table called `modules.alias`, which is located in the `/lib/modules/kernel_version/` directory.

The `<bus_type_name>` parameter should be the generic name of the bus that you need to add module auto-loading support to. It should be `spi` for an SPI bus, `of` for a device tree, `i2c` for I2C, and so on. In other words, it should be one of the elements of the first column (of the **bus type**) of the previous table (knowing that not all the buses are listed). Let's add module auto-loading support to the same driver we used previously (gpio-mcp23s08):

```
MODULE_DEVICE_TABLE(spi, mcp23s08_ids);
MODULE_DEVICE_TABLE(of, mcp23s08_spi_of_match);
```

Now, let's see what these two lines do when they're added to the `modules.alias` file on an i.MX6-based board running a Yocto-based image:

```
root:/lib/modules/5.10.10+fslc+g8dc0fcb# cat modules.alias
# Aliases extracted from modules themselves.
alias fs-msdos msdos
alias fs-binfmt_misc binfmt_misc
alias fs-configfs configfs
alias iso9660 isofs
alias fs-iso9660 isofs
alias fs-udf udf
alias of:N*T*Cmicrochip,mcp23s17* gpio_mcp23s08
alias of:N*T*Cmicrochip,mcp23s18* gpio_mcp23s08
alias of:N*T*Cmicrochip,mcp23s08* gpio_mcp23s08
alias spi:mcp23s17 gpio_mcp23s08
alias spi:mcp23s18 gpio_mcp23s08
alias spi:mcp23s08 gpio_mcp23s08
alias usb:v0C72p0011d*dc*dsc*dp*ic*isc*ip*in* peak_usb
alias usb:v0C72p0012d*dc*dsc*dp*ic*isc*ip*in* peak_usb
alias usb:v0C72p000Dd*dc*dsc*dp*ic*isc*ip*in* peak_usb
alias usb:v0C72p000Cd*dc*dsc*dp*ic*isc*ip*in* peak_usb
alias pci:v00008086d000015B8sv*sd*bc*sc*i* e1000e
alias pci:v00008086d000015B7sv*sd*bc*sc*i* e1000e
[...]
```

```
alias usb:v0416pA91Ad*dc*dsc*dp*ic0Eisc01ip00in* uvcvideo
alias of:N*T*Ciio-hwmon* iio_hwmon
alias i2c:lm73 lm73
alias spi:ad7606-4 ad7606_spi
alias spi:ad7606-6 ad7606_spi
alias spi:ad7606-8 ad7606_spi
```

The second part of the solution is the kernel informing the userspace about some events (called **uevents**) through *netlink sockets*. Right after a device appears on a bus, this bus code will create and emit an event that contains the corresponding module alias (for example, `pci:v00008086d000015B8sv*sd*bc*sc*i*`). This event will be caught by your system hotplug manager (**udev** on most machines), which will parse the `module.alias` file while looking for an entry with the same alias and load the corresponding module (for example, e1000). As soon as the module is loaded, the device will be probed. This is how the simple `MODULE_DEVICE_TABLE()` macro can change your life.

Device declaration – populating devices

Device declaration is not part of the LDM. It consists of declaring devices that are present (or not) on the system, while the module device table involves feeding the drivers with devices they support. There are three places you can declare/populate devices:

- From the board file or in a separate module (older and now deprecated)
- From the **device tree** (the new and recommended method)
- From the **Advanced Configuration and Power Interface** (**ACPI**), which will not be discussed here

To be handled by a driver, any declared device should exist at least in one module device table; otherwise, the device will simply be ignored, unless a driver with this device ID in its module device table gets loaded or has already been loaded.

Bus structure

Finally, there's the `struct bus_type` structure, which is the structure that the kernel internally represents a bus with (whether it is physical or virtual). The **bus controller** is the root element of any hierarchy. Physically speaking, a bus is a channel between the processor and one or more devices. From a software point of view, the bus (`struct bus_type`) is the link between devices (`struct device`) and drivers (`struct device_driver`). Without this, nothing would be appended to the system, since the bus (`bus_type`) is responsible for matching the devices and drivers:

```
struct bus_type {
    const char     *name;
    struct device     *dev_root;
    int (*match)(struct device *dev,
                    struct device_driver *drv);
    int (*probe)(struct device *dev);
    int (*remove)(struct device *dev);
    /* [...] */
};
```

Let's look at the elements in this structure:

- name: This is the bus's name as it will appear in /sys/bus/.

- match: This is a callback that's called whenever a new device or driver is added to the bus. The callback must be smart enough and should return a nonzero value when there is a match between a device and a driver. Both are given as parameters. The main purpose of a match callback is to allow a bus to determine whether a particular device can be handled by a given driver or the other logic if the given driver supports a given device. Most of the time, the verification process is done with a simple string comparison (the device and driver name, or a table and **device tree** (**DT**)-compatible property). For enumerated devices (such as PCI and USB), the verification process is done by comparing the device IDs that are supported by the driver with the device ID of the given device, without sacrificing bus-specific functionality.

- probe: This is a callback that's called when a new device or driver is added to the bus *and* once a match has occurred. This function is responsible for allocating the specific bus device structure and calling the given driver's probe function, which is supposed to manage the device (we allocated this earlier).

- remove: This is called when a device is removed from the bus.

When the device that you wrote the driver for sits on a physical bus called the **bus controller**, it must rely on the driver of that bus, called the **controller driver**, which is responsible for sharing bus access between devices. The controller driver offers an abstraction layer between your device and the bus. Whenever you perform a transaction (read or write) on an I2C or USB bus, for example, the I2C/USB bus controller transparently takes care of that in the background (managing the clock, shifting data, and so on). Every bus controller driver exports a set of functions to ease the development of drivers for the devices sitting on that bus. This works for every bus (I2C, SPI, USB, PCI, SDIO, and so on).

Now that we have looked at the bus driver and how modules are loaded, we will discuss the matching mechanism, which tries to bind a particular device to its drivers.

Device and driver matching mechanism explained

Device drivers and devices are always registered with the bus. When it comes to exporting the devices that are supported by the driver, you can use `driver.of_match_table`, `driver.of_match_table`, or `<bus>_driver.id_table` (which is specific to the device type; for example, `i2c_device.id_table` or `platform_device.id_table`).

Each bus driver has the responsibility of providing its match function, which is run by the kernel whenever a new device or device driver is registered with this bus. That said, there are three matching mechanisms for platform devices, all of which consist of string comparison. Those matching mechanisms are based on the DT table, ACPI table, device, and driver name. Let's see how the pseudo-platform and i2c buses implement their matching functions using those mechanisms:

```
static int platform_match(struct device *dev,
                          struct device_driver *drv)
{
    struct platform_device *pdev =
                        to_platform_device(dev);
    struct platform_driver *pdrv =
                        to_platform_driver(drv);

    /* Only bind to the matching driver when
     * driver_override is set
     */
```

```
    if (pdev->driver_override)
        return !strcmp(pdev->driver_override, drv->name);

    /* Attempt an OF style match first */
    if (of_driver_match_device(dev, drv))
        return 1;

    /* Then try ACPI style match */
    if (acpi_driver_match_device(dev, drv))
    return 1;

    /* Then try to match against the id table */
    if (pdrv->id_table)
        return platform_match_id(pdrv->id_table,
                                    pdev) != NULL;

    /* fall-back to driver name match */
    return (strcmp(pdev->name, drv->name) == 0);
}
```

The preceding code shows the **pseudo-platform** bus matching function, which is defined in drivers/base/platform.c. The following code shows the I2C bus matching function, which is defined in drivers/i2c/i2c-core.c:

```
static const struct i2c_device_id *i2c_match_id(
            const struct i2c_device_id *id,
            const struct i2c_client *client)
{
    while (id->name[0]) {
        if (strcmp(client->name, id->name) == 0)
            return id;
        id++;
    }
    return NULL;
}

static int i2c_device_match(struct device *dev, struct
```

```
                    device_driver *drv)
{
    struct i2c_client *client = i2c_verify_client(dev);
    struct i2c_driver *driver;

    if (!client)
        return 0;

    /* Attempt an OF style match */
    if (of_driver_match_device(dev, drv))
        return 1;

    /* Then ACPI style match */
    if (acpi_driver_match_device(dev, drv))
        return 1;

    driver = to_i2c_driver(drv);
    /* match on an id table if there is one */
    if (driver->id_table)
        return i2c_match_id(driver->id_table,
                            client) != NULL;
    return 0;
}
```

Case study – the OF matching mechanism

In the device tree, each device is represented by a node and declared as a child of its bus node. At boot time, the kernel (the **OF** core) parses every bus node (as well as their sub-nodes, which are the devices that are sitting on it) in the device tree. For each device node, the kernel will do the following:

- Identify the bus that this node belongs to.

- Allocate a platform device and initialize it according to the properties contained in the node using the of_device_alloc() function. built_pdev->dev.of_node will be set with the current device tree node.

- Walk through the list of device drivers associated with (maintained by) the previously identified bus using the bus_for_each_drv() function.

- For each driver in the list, the core will do the following:

  A.   Call the bus match function, given as the parameter that the driver found and the previously built device structure; that is, `bus_found->match(cur_drv, cur_dev);`.

  B.   If the DT matching mechanism is supported by this bus driver, the bus match function will then call `of_driver_match_device()`, given the same parameters that were mentioned previously; that is, `of_driver_match_device(ur_drv, cur_dev)`.

  C.   `of_driver_match_device` will walk through the `of_match_table` table (which is an array of struct `of_device_id` elements) that's associated with the current driver. For each `of_device_id` in the array, the kernel will compare the compatible property of both the current `of_device_id` element and `built_pdev->dev.of_node`. If they are the same (let's say that there's a match), the probe function of the current driver will be run.

- If no driver that supports this device is found, this device will be registered with the bus anyway. Then, the probing mechanism will be deferred to a later date so that whenever a new driver is registered with this bus, the core will walk through the list of devices that are maintained by the bus; any devices without any drivers associated with them will be probed again. For each, the compatible property of associated `of_node` will be compared to the compatible property of each `of_device_id` in the `of_match_table` array that's associated with the freshly registered driver.

This is how drivers are matched with devices that are declared from within the device tree. This works in the same manner for each type of device declaration (board file, ACPI, and so on).

Summary

In this chapter, you learned how to deal with devices and drivers, as well as how they are tied to each other. We have also demystified the matching mechanism. Make sure you understand this before moving on to *Chapter 7, Understanding the Concept of Platform Devices and Drivers, Chapter 8, Writing I2C Device Drivers*, and *Chapter 9, Writing SPI Device Drivers* , which will deal with device driver development. This will involve working with devices, drivers, and bus structures.

In the next chapter, we will delve into *platform driver development* in detail.

7
Understanding the Concept of Platform Devices and Drivers

The Linux kernel handles devices by using the concept of buses, that is, the links between the CPU and these devices. Some buses are smart enough and embed a discoverability logic to enumerate devices sitting on them. With such buses, early in the bootup phase, the Linux kernel requests these buses for the devices they have enumerated as well as the resources (such as interrupt lines and memory regions) they need to work correctly. PCI, USB, and SATA buses all come under this family of discoverable buses.

Unfortunately, the reality is not always so beautiful. There are a number of devices that the CPU is still unable to detect. Most of these non-discoverable devices are on-chip, although some of them sit on slow or dumb buses that do not support device discoverability.

As a result, the kernel must provide mechanisms for receiving information about the hardware and users must inform the kernel where these devices can be found. In the Linux kernel, these non-discoverable devices are known as **platform devices**. Because they do not sit on known buses such as I2C, SPI, or any non-discoverable bus, the Linux kernel has implemented the concept of a **platform bus** (also referred to as a **pseudo platform bus**) in order to maintain the paradigm according to which devices are always connected to CPUs through buses.

In this chapter, we will learn how and where to instantiate platform devices as well as their resources, and learn how to write their drivers, that is, platform drivers. To achieve that, the chapter will be split into the following topics:

- Understanding the platform core abstraction in the Linux kernel
- Dealing with platform devices
- Platform driver abstraction and architecture
- Example of writing a platform driver from scratch

Understanding the platform core abstraction in the Linux kernel

To cover the long list of non-discoverable devices that are being increasingly used as **System on Chips (SoCs)** are becoming more and more popular, the platform core has been introduced. Within this framework, the three most important data structures are as follows: the one representing the platform device, another representing its resource, and the final data structure, representing the platform driver.

A platform device is represented in the kernel as an instance of `struct platform_device`, defined in `<linux/platform_device.h>` as follows:

```
struct platform_device {
    const char        *name;
    u32               id;
    struct device     dev;
    u32               num_resources;
    struct resource *resource;
    const struct platform_device_id  *id_entry;
    struct mfd_cell *mfd_cell;
};
```

In the previous data structure, `name` is the platform device name. The name assigned to the platform device must be chosen with care. Platform devices are matched against their respective drivers in the pseudo platform bus matching function, that is, `platform_match()`. In this function, under certain circumstances (no device tree or ACPI support and no `id` table match), matching falls back to name matching, comparing the driver's name and the platform device's name.

dev is the underlying device structure for the Linux device model, and id is used to extend this device name. The following describes how id is used:

- Where id is -1 (which corresponds to the PLATFORM_DEVID_NONE macro), the underlying device name will be the same as the platform device name. The platform core will do the following:

```
dev_set_name(&pdev->dev, "%s", pdev->name);
```

- Where id is -2 (which corresponds to the PLATFORM_DEVID_AUTO macro), the kernel will automatically generate a valid ID and will name the underlying device as follows:

```
dev_set_name(&pdev->dev, "%s.%d.auto", pdev->name,
             <auto_id>);
```

- In any other case, id will be used as follows:

```
dev_set_name(&pdev->dev, "%s.%d", pdev->name,
             pdev->id);
```

resource is an array of resources assigned to the platform device, and num_resources is the number of elements in that array.

In the case of a platform device and driver matching by ID table, pdev->id_entry, which is of the struct platform_device_id type, will point to the matching ID table entry that caused the platform driver to match with this platform device.

Regardless of how platform devices have been registered, they need to be driven by appropriate drivers, that is, platform drivers. Such drivers must implement a set of callbacks that are used by the platform core as and when devices appear/disappear on the platform bus.

Platform drivers are represented in the Linux kernel as an instance of struct platform_driver, defined as follows:

```
struct platform_driver {
    int (*probe)(struct platform_device *);
    int (*remove)(struct platform_device *);
    void (*shutdown)(struct platform_device *);
    int (*suspend)(struct platform_device *,
                   pm_message_t state);
    int (*resume)(struct platform_device *);
```

```
        struct device_driver driver;
        const struct platform_device_id *id_table;
        bool prevent_deferred_probe;
};
```

The following describes the elements used in this data structure:

- `probe()`: This is the function that gets called when a device claims your driver after a match occurs. Later, we will see how `probe` is called by the core. Its declaration is as follows:

  ```
  int my_pdrv_probe(struct platform_device *pdev)
  ```

  The kernel is responsible for providing the `platform_device` parameter. `probe` is called by the bus driver when the device driver is registered in the kernel.

- `remove()`: This is called to get rid of the driver when it is no longer needed by devices and its declaration looks like the following:

  ```
  static void my_pdrv_remove(struct platform_device *pdev)
  ```

- `driver` is the underlying driver structure for the device model and must be provided with a name (which must be chosen with care as well), an owner, and some other fields (such as a device tree matching table), which we will see later. When it comes to the platform driver, before the driver and device match, the `platform_device.name` and `platform_driver.driver.name` fields must be identical.

- `id_table` is one of the ways provided by the platform driver to the bus code to bind actual devices to the driver. The other way is via the device tree, which will be discussed in the *Provisioning supported devices in the driver* section.

Now that we have introduced both platform device and platform driver data structures, let's move beyond this and try to understand how they are created and registered with the system.

Dealing with platform devices

Before we start writing platform drivers, this section will teach you how and where to instantiate platform devices. Only after that will we delve into the platform driver implementation. Platform device instantiation in the Linux kernel has existed for a long time and has been improved all over the kernel versions, and we will discuss the specificities of each instantiation method in this section.

Allocating and registering platform devices

Since there is no way for platform devices to make themselves known to the system, they must be populated manually and registered with the system along with their resources and private data. In the early platform core days, platform devices were declared in the board file, `arch/arm/mach-*` (which is `arch/arm/mach-imx/mach-imx6q.c` for i.MX6), and made known to the kernel with `platform_device_add()` or `platform_device_register()`, depending on how each platform device has been allocated.

This leads us to conclude that there are two methods for allocating and registering platform devices.

A static method, which entails enumerating platform devices in the code and calling `platform_device_register()` on each of them for their registration. This API is defined as follows:

```
int platform_device_register(struct platform_device *pdev)
```

One example of such usage is the following:

```
static struct platform_device my_pdev = {
    .name             = "my_drv_name",
    .id               = 0,
    .ressource        = jz4740_udc_resources,
    .num_ressources   = ARRY_SIZE(jz4740_udc_resources),
};

int foo()
{
    [...]
    return platform_device_register(&my_pdev);
}
```

With static registration, platform devices are statically initialized and passed to `platform_device_register()`. This method also allows for the addition of bulk platform devices. The corresponding code can use `platform_add_devices()`, which accepts an array of pointers to platform devices and the number of elements in the array. This function is defined as follows:

```
int platform_add_devices(struct platform_device **devs,
                    int num);
```

On the other hand, dynamic allocation requires an initial call to `platform_device_alloc()` to dynamically allocate and initialize the platform device, defined as follows:

```
struct platform_device *platform_device_alloc(
                            const char *name, int id);
```

In parameters, `name` is the base name of the device we're adding, and `id` is the instance ID of the platform device. In the event of success, this function returns a valid platform device object, or `NULL` on error.

Platform devices allocated in this way are registered with the system using the `platform_device_add()` API exclusively, defined as follows:

```
int platform_device_add(struct platform_device *pdev);
```

As an only parameter, the platform device has been allocated using `platform_device_alloc()`. In the event of success, this function returns `0`, or a negative error code in the event of failure.

If platform device registration with `platform_device_add()` fails, you should release the memory occupied by the platform device structure using `platform_device_put()`, defined as follows:

```
void platform_device_put(struct platform_device *pdev);
```

The following is an example of how it is used:

```
status = platform_device_add(evm_led_dev);
if (status < 0) {
     platform_device_put(evm_led_dev);
 [...]
}
```

That said, whatever way the platform device has been allocated and registered, it must be unregistered with `platform_device_unregister()`, defined as follows:

```
void platform_device_unregister(struct platform_device *pdev)
```

Do note that `platform_device_unregister()` internally calls `platform_device_put()`.

Now that we know how to instantiate platform devices in the traditional way, let's see how minimalistic it could be to achieve the same goal by adopting another approach, the device tree.

How not to allocate platform devices to your code

As we said earlier, platform devices used to be populated either in board files or from other drivers. There was no flexibility in using this method. Adding/removing platform devices would require, in the best case, recompiling a module, and, in the worst case, recompiling the whole kernel. This deprecated method is not portable either.

Nowadays, there is the device tree, which has existed for quite a long time now. This mechanism is used to declare non-discoverable devices present in the system. As a standalone entity, it can be built independently from the kernel or other modules. It turns out that this could be the best alternative for populating platform devices.

This is achieved by declaring these platform devices as nodes in the device tree, under a node whose compatible string property is simple-bus. This compatible string means the node is regarded as a bus that requires no specific handling or driver. Moreover, it indicates that there is no way to dynamically probe the bus and that the only way to locate the child devices of this bus is by using the address information in the device tree. Then Linux, in its implementation, ends up creating a platform device for each first-level sub-node under the node that has simple-bus in the compatible property.

The following is a demonstration:

```
foo {
    compatible = "simple-bus";

    bar: bar@0 {
        compatible = "labcsmart,something";
        [...]

        baz: baz@0 {
            compatible = "labcsmart,anotherthing";
            [...]
        }
    }

    foz: foz@1 {
        compatible = "company,product";
        [...]
    };
}
```

In the preceding example, only the `bar` and `foz` nodes will be registered as platform devices. `baz` will not (its direct parent does not have `simple-bus` in a compatible string). Thus, if there is any platform driver with `company,  product` and/or `labcsmart,something` in their compatible matching table, then these platform devices will be probed.

The common use is under the SoC node or on-chip memory mapped buses to declare on-chip devices. Another frequent usage is to declare regulator devices, as follows:

```
regulators {
    compatible = "simple-bus";
    #address-cells = <1>;
    #size-cells = <0>;

    reg_usb_h1_vbus: regulator@0 {
        compatible = "regulator-fixed";
        reg = <0>;
        regulator-name = "usb_h1_vbus";
        regulator-min-microvolt = <5000000>;
        regulator-max-microvolt = <5000000>;
        enable-active-high;
        startup-delay-us = <2>;
        gpio = <&gpio7 12 0>;
    };

    reg_panel: regulator@1 {
        compatible = "regulator-fixed";
        reg = <1>;
        regulator-name = "lcd_panel";
        enable-active-high;
        gpio = <&gpio1 2 0>;
    };
};
```

In the preceding example, two platform devices will be registered, each corresponding to a fixed regulator.

Working with platform resources

At the opposite end of hot-pluggable devices that are enumerated, and which advertise resources they need, the kernel has no idea of what platform devices are present on your system, what they are capable of, or what they need in order to work properly. There is no auto-negotiation process, so any information provided to the kernel about the resources required by a given platform device would be welcome. Such resources could be IRQ lines, DMA channels, memory regions, I/O ports, and so on.

From within the platform code, a resource is represented as an instance of a struct resource, defined in `include/linux/ioport.h` as follows:

```
struct resource {
    resource_size_t start;
    resource_size_t end;
    const char *name;
    unsigned long flags;
};
```

In the preceding data structure, start/end point to the beginning/end of the resource. They indicate where I/O or memory regions begin and terminate. Because IRQ lines, buses, and DMA channels do not have ranges, it is usual to assign the same value to start/end.

flags is a mask that characterizes the type of resource, for example, IORESOURCE_BUS. Possible values are as follows:

- IORESOURCE_IO for PCI/ISA I/O ports
- IORESOURCE_MEM for memory regions
- IORESOURCE_REG for register offsets
- IORESOURCE_IRQ for IRQ lines
- IORESOURCE_DMA for DMA channels
- IORESOURCE_BUS for a bus

Finally, the name element identifies or describes the resource, since the resource can be extracted by name.

Assigning such resources to a platform device can be done in two ways, first, in the same compilation unit where the platform device has been declared and registered, and second, from the device tree.

In this section, we have described the resource and shown their usage. Let's now see how they can be fed to the platform device in the next section.

Platform resource provisioning – the old and deprecated way

This method is to be used with kernels that do not support the device tree or when such a choice does not exist. It is mainly used with a **multi-function device (MFD)**, where a master (the enclosing) chip shares its resources with subdevices.

With this method, resources are provided in the same way as platform devices. The following is an example:

```
static struct resource foo_resources[] = {
    [0] = { /* The first memory region */
        .start = 0x10000000,
        .end   = 0x10001000,
        .flags = IORESOURCE_MEM,
        .name  = "mem1",
    },
    [1] = {
        .start = JZ4740_UDC_BASE_ADDR2,
        .end   = JZ4740_UDC_BASE_ADDR2 + 0x10000 -1,
        .flags = IORESOURCE_MEM,
        .name  = "mem2",
    },
    [2] = {
        .start = 90,
        .end   = 90,
        .flags = IORESOURCE_IRQ,
        .name  = "mc-irq",
    },
};
```

The preceding excerpt shows three resources (two of the memory type and one of the IRQ type) whose types can be identified with `IORESOURCE_IRQ` and `IORESOURCE_MEM`. The first one is a 4 KB memory region, while the second is also a memory region whose range is defined by a macro, and finally, the IRQ 90.

Assigning these resources to a platform device is a straightforward operation. If the platform device has been allocated statically, the resources should be assigned in the following manner:

```
static struct platform_device foo_pdev = {
        .name = "foo-device",
        .resource               = foo_resources,
        .num_ressources = ARRY_SIZE(foo_resources),
[...]
};
```

For a dynamically allocated platform device, this is done from within a function, and it should be something like the following:

```
struct platform_device *foo_pdev;
[...]
my_pdev = platform_device_alloc("foo-device", ...);
if (!my_pdev)
            return -ENOMEM;

my_pdev->resource = foo_resources;
my_pdev->num_ressources = ARRY_SIZE(foo_resources);
```

There are several helper functions for getting data out of the resource array; these include the following:

```
struct resource *platform_get_resource(
                    struct platform_device *pdev,
                    unsigned int type, unsigned int n);
struct resource *platform_get_resource_byname(
                    struct platform_device *pdev,
                    unsigned int type, const char *name);
int platform_get_irq(struct platform_device *pdev,
                    unsigned int n);
```

The n parameter says which resource of that type is desired, with zero indicating the first one. Thus, for example, a driver could find its second MMIO region with the following:

```
r = platform_get_resource(pdev, IORESOURCE_MEM, 1);
```

The use of this function is explained in the *Handling resources* section of *Chapter 5, Understanding and Leveraging the Device Tree.*

Understanding the concept of platform data

The struct resource that we have used so far is adequate to instantiate resources for a simple platform device, but many devices are more complex than that. This data structure can only encode a limited number of types of information. As an extension to that, `platform_device.device.platform_data` is used to assign any other extra information to the platform device.

Data can be enclosed in a bigger structure and assigned to this extra field. This data can be of whatever type the driver can understand, but most of the time, they are other data types that are not a part of the resource types we enumerated in the preceding section. These can be regulator constraints or even pointers to per-device functions, for example.

Let's describe, in the following code block, extra platform data that corresponds to a set of pointers to functions that are private to the platform device type:

```
static struct foo_low_level foo_ops = {
        .owner            = THIS_MODULE,
        .hw_init          = trizeps_pcmcia_hw_init,
        .socket_state     = trizeps_pcmcia_socket_state,
        .configure_socket = trizeps_pcmcia_configure_socket,
        .socket_init      = trizeps_pcmcia_socket_init,
        .socket_suspend   = trizeps_pcmcia_socket_suspend,
[...]
};
```

For a statically allocated platform device, we would execute the following command:

```
static struct platform_device my_device = {
        .name = "foo-pdev",
        .dev  = {
                .platform_data    = (void*)&trizeps_pcmcia_ops,
        },
        [...]
};
```

If the platform device was found to be allocated dynamically, it would be assigned using the `platform_device_add_data()` helper, defined as follows:

```
int platform_device_add_data(struct platform_device *pdev,
                        const void *data, size_t size);
```

The preceding function returns 0 in the event of success. `data` is the data to use as platform data, and `size` is the size of this data.

Back to our example of a set of functions, we would do the following:

```
int ret;
[...]

ret = platform_device_add_data(foo_pdev,
                    &foo_ops, sizeof(foo_ops));

if (ret == 0)
    ret = platform_device_add(trizeps_pcmcia_device);

if (ret)
    platform_device_put(trizeps_pcmcia_device);
```

From within the platform driver, the platform device will have its `pdev->dev. platform_dat` element pointing to the platform data. Although we would dereference this field, it is recommended to use the kernel-provided function, `dev_get_platdata()`, defined as follows:

```
void *dev_get_platdata(const struct device *dev)
```

Then, to get back the enclosing set of function structures, the driver could do the following:

```
struct foo_low_level *my_ops =
        dev_get_platdata(&pdev->dev);
```

There is no way for the driver to check the type of data that is passed. Drivers must simply assume that they have been supplied a structure of the expected type because the platform data interface lacks any sort of type checking.

Platform resource provisioning – the new and recommended way

The first resource provisioning method has a few drawbacks, including the fact that any change would require a rebuilding of either the kernel or the module that has changed and this could increase the kernel size.

With the arrival of the device tree, things have become simpler. To keep things compatible, memory regions, interrupts, and DMA resources specified in the device tree are converted into an instance of struct resources by the platform core so that `platform_get_resource()`, `platform_get_resource_by_name()`, or even `platform_get_irq()` can return the appropriate resource, irrespective of whether this resource has been populated the traditional way or from the device tree. This can be verified in the *Handling resources* section of *Chapter 5, Understanding and Leveraging the Device Tree*.

Needless to say, the device tree allows passing any type of information that the driver may need to know. Such data types can be used to pass any device-/driver-specific data. This can be accomplished by reading the *Extracting application-specific data* section, still in *Chapter 5, Understanding and Leveraging the Device Tree*.

However, the device tree code is unaware of the specific structure used by a given driver for its platform data, so it will be unable to provide that information in that form. To pass such extra data, drivers can use the `.data` field in the `of_device_id` entry that caused the platform device and the driver to match. This field could then point to the platform data. See the *Provisioning supported devices in the driver* section in this chapter.

Drivers expecting platform data in the traditional way should check the `platform_device->dev.platform_data` pointer. If there is a non-null value there, this means that the device was instantiated in the traditional way along with the platform data and the device tree was not used; the platform data should be used as usual. If, however, the device has been instantiated from the device tree code, the `platform_data` pointer will be NULL, indicating that the information must be acquired from the device tree directly. In this case, the driver will find a `device_node` pointer in the platform device's `dev.of_node` field. The various device tree access routines (most notably `of_get_property()`) can then be used to pull the required data from the device tree.

Now that we are familiar with platform resource provisioning, let's learn how to design a platform device driver.

Platform driver abstraction and architecture

Let's get warned before going further. Not all platform devices are handled by platform drivers (or, should I say, pseudo platform drivers). Platform drivers are dedicated to devices not based on conventional buses. I2C devices or SPI devices are platform devices, but rely on I2C or SPI buses, respectively, and not on platform buses. Everything needs to be done manually with the platform driver.

Probing and releasing the platform devices

The platform driver entry point is the probe method, invoked after a match with a platform device has occurred. This probe method has the following prototype:

```
int pdrv_probe(struct platform_device *pdev)
```

`pdev` corresponds to the platform device that has been instantiated in the traditional way or a fresh one allocated by the platform core because of the associated device tree node having a direct parent with `simple-bus` in its compatible property. Platform data and resources, if any, will also be set accordingly. If the platform device in the parameter is the one expected by the driver, this method must return 0. Otherwise, an appropriate negative error code must be returned.

Whether the platform device has been instantiated the traditional way or from the device tree, its resources can be extracted using conventional platform core APIs, `platform_get_resource()`, `platform_get_resource_by_name()`, or even `platform_get_irq()`, and similar APIs.

The probe method must request any resource (GPIOs, clocks, IIO channels, and so on) and data required by the driver. If mappings need to be done, the probe method is the appropriate place to do that.

Everything that has been done in the `probe` method must be undone when the device leaves the system or when the platform driver is unregistered. The `remove` method is the appropriate place to achieve that. It must have the following prototype:

```
int pdrv_remove(struct platform_device *dev)
```

This function must return 0 only if everything has been undone and cleaned up. Otherwise, an appropriate error code must be returned so that users can be informed. In the parameter, the same platform device that has been passed to the `probe` function.

In the *Example of writing a platform driver from scratch* section, we discuss all the tips and particularities of implementing the `probe` and `remove` methods.

Now that the driver callbacks are ready, we can populate the devices that are going to be handled by this driver.

Provisioning supported devices in the driver

Our platform driver is useless on its own. To make it useful to devices, it must inform the kernel what devices it can manage. To achieve that, an ID table must be provided, assigned to the `platform_driver.id_table` field. This will allow platform device matching. However, to extend this feature to module autoloading, the same table must be given to `MODULE_DEVICE_TABLE` so that module aliases are generated.

Back to the code, each entry in that table is of the `struct platform_device_id` type, defined as follows:

```
struct platform_device_id {
    char name[PLATFORM_NAME_SIZE];
    kernel_ulong_t driver_data;
};
```

In the preceding data structure, `name` is a descriptive name for the device, and `driver_data` is the driver state value. It can be set with a pointer to a per-device data structure. The following is an example, an excerpt from `drivers/mmc/host/mxs-mmc.c`:

```
static const struct platform_device_id mxs_ssp_ids[] = {
    {
            .name = "imx23-mmc",
            .driver_data = IMX23_SSP,
    }, {
            .name = "imx28-mmc",
            .driver_data = IMX28_SSP,
    }, {
            /* sentinel */
    }
};
MODULE_DEVICE_TABLE(platform, mxs_ssp_ids);
```

When `mxs_ssp_ids` is assigned to the `platform_driver.id_table` field, platform devices will be able to match this driver based on their name matching any `platform_device_id.name` entry. `platform_device.id_entry` will point to the entry in this table that triggered the match.

To allow matching platform devices declared in the device tree using their compatible string, the platform driver must set `platform_driver.driver.of_match_table` with a list of elements of `struct of_device_id`; then, to allow module autoloading from device tree matching as well, this device tree match table must be given to `MODULE_DEVICE_TABLE`. The following is an example:

```
static const struct of_device_id mxs_mmc_dt_ids[] = {
    {
        .compatible = "fsl,imx23-mmc",
        .data = (void *) IMX23_SSP,
    },{
        .compatible = "fsl,imx28-mmc",
        .data = (void *) IMX28_SSP,
    }, {
        /* sentinel */
    }
};
MODULE_DEVICE_TABLE(of, mxs_mmc_dt_ids);
```

If `of_device_id` is set in the driver, the matching is judged by any `of_device_id.compatible` element matching the value of the compatible property in the device node. To get the `of_device_id` entry that caused the match, the driver should call `of_match_device()`, passing as parameters the device tree match table and the underlying device structure, `platform_device.dev`.

The following is an example:

```
static int mxs_mmc_probe(struct platform_device *pdev)
{
    const struct of_device_id *of_id =
        of_match_device(mxs_mmc_dt_ids, &pdev->dev);
    struct device_node *np = pdev->dev.of_node;
    [...]
}
```

After these match tables have been defined, they can be assigned to the platform driver structure as follows:

```
static struct platform_driver imx_uart_platform_driver = {
    .probe = imx_uart_probe,
    .remove = imx_uart_remove,

    .id_table = imx_uart_devtype,
    .driver = {
        .name = "imx-uart",
        .of_match_table = imx_uart_dt_ids,
    },
};
```

With the preceding final platform driver data structure, we give devices a chance to match the platform driver based on either the device tree match table, or on the ID table, and finally, based on the driver's name.

Driver initialization and registration

Registering a platform driver with the kernel is as simple as calling `platform_driver_register()` or `platform_driver_probe()` in the module initialization function. Then, to get rid of a platform driver that has been registered, the module must call `platform_driver_unregister()` to unregister this driver.

The following are the respective prototypes of these functions:

```
int platform_driver_register(struct platform_driver *drv);
void platform_driver_unregister(struct platform_driver *);
int platform_driver_probe(struct platform_driver *drv,
                int (*probe)(struct platform_device *))
```

The difference between the two probing functions is as follows:

- `platform_driver_register()` registers and puts the driver into a list of drivers maintained by the kernel, meaning that its `probe()` function can be called on demand whenever a new match with a platform device occurs. To prevent the driver from being inserted and registered in that list, the next function can be used. That said, any platform driver registered with `platform_driver_register()` must be unregistered with `platform_driver_unregister()`.

- With `platform_driver_probe()`, the kernel immediately runs the matching loop to check whether there are platform devices that can be matched against this platform driver and will call the `probe` method for each device with a match. If no device is found at that time, the platform driver is simply ignored. This method prevents a deferred probe since it does not register the driver on the system. Here, the `probe` function is placed in an `__init` section, which is freed (provided the driver is compiled statically, that is, built-in) when the kernel boot has completed, thereby preventing a deferred probe, and reducing the driver's memory footprint. Use this method if you are 100% sure the device is present in the system:

  ```
  ret = platform_driver_probe(&mypdrv, my_pdrv_probe);
  ```

  The `probe()` procedure can be placed in an `__init` section if the device is known not to be hot-pluggable.

The appropriate time to register the platform driver is right after the module gets loaded, which allows us to say that the appropriate place to do that is from the module initialization function. The same goes for unregistering the driver, which must be done on the unloading path of the module. Thanks to that, we can say that the `module_exit` function is the appropriate place to unregister a platform driver.

The following is a typical demonstration of simple platform driver registering with the platform core:

```
#include <linux/module.h>
#include <linux/kernel.h>
#include <linux/init.h>
#include <linux/platform_device.h>

static int my_pdrv_probe (struct platform_device *pdev)
{
    pr_info("Hello! device probed!\n");
    return 0;
}

static void my_pdrv_remove(struct platform_device *pdev)
{
    pr_info("good bye reader!\n");
}

static struct platform_driver mypdrv = {
```

```
        .probe      = my_pdrv_probe,
        .remove     = my_pdrv_remove,
[...]
};

static int __init foo_init(void)
{
    [...] /* My init code */
    return platform_driver_register(&mypdrv);
}
module_init(foo_init);

static void __exit foo_cleanup(void)
{
    [...] /* My clean up code */
    platform_driver_unregister(&my_driver);
}
module_exit(foo_cleanup);
```

Our module does nothing else in the initialization/exit function apart from registering/ unregistering with the platform driver with the platform bus core. This is a frequent situation, where the module init/exit methods are lite and dummy. In this case, we can get rid of module_init() and module_exit(), and use the module_platform_ driver() macro, as follows:

```
[...]
static int my_pdrv_probe(struct platform_device *pdev)
{
    [...]
}

static void my_pdrv_remove(struct platform_device *pdev)
{
    [...]
}

static struct platform_driver mypdrv = {
```

```
        [...]
};
```

```
module_platform_driver(mypdrv);
```

Although we have only inserted snapshot code in the preceding example, this is enough to cover the demonstration of reducing the code and making it cleaner.

To make sure we are on the same knowledge base, let's summarize all the concepts we have learned in this chapter into a real platform driver.

Example of writing a platform driver from scratch

This section will try to summarize as far as possible the knowledge that has been acquired so far throughout the chapter. Now, let's imagine a platform device that is memory-mapped and that the memory range through which it can be controlled starts at 0x02008000 and is 0x4000 in size. Then, let's say this platform device can interrupt the CPU upon completion of its jobs and that this interrupt line number is 31. To keep things simple, let's not require any other resource for this device (we could have imagined clocks, DMAs, regulators, and so on).

To start with, let's instantiate this platform device from the device tree. If you remember, for a node to be registered as a platform device by the platform core, the direct parent of this node must have simple-bus in its compatible string list, and this is what is implemented here:

```
demo {
    compatible = "simple-bus";

    demo_pdev: demo_pdev@0 {
        compatible = "labcsmart,demo-pdev";
        reg = <0x02008000 0x4000>;
        interrupts = <0 31 IRQ_TYPE_LEVEL_HIGH>;
    };
};
```

If we had to instantiate this platform device along with its resources in the traditional way, we would have to do something like the following, in a file that we will call `demo-pdev-init.c`:

```
#include <linux/module.h>
#include <linux/init.h>
#include <linux/platform_device.h>

#define DEV_BASE 0x02008000
#define PDEV_IRQ 31

static struct resource pdev_resource[] = {
    [0] = {
        .start = DEV_BASE,
        .end   = DEV_BASE + 0x4000,
        .flags = IORESOURCE_MEM,
    },
    [1] = {
        .start = PDEV_IRQ,
        .end   = PDEV_IRQ,
        .flags = IORESOURCE_IRQ,
    },
};
```

In the preceding, we first defined the platform resources. Now we can instantiate the platform device and assign these resources as follows, still in `demo-pdev-init.c`:

```
struct platform_device demo_pdev = {
    .name         = "demo_pdev",
    .id           = 0,
    .num_resources = ARRAY_SIZE(pdev_resource),
    .resource     = pdev_resource,
    /*.dev  = {
    *    .platform_data  = (void*)&big_struct_1,
    *}*/,
};
```

In the preceding, we commented out the dev-related assignment just to show that we could have defined a bigger structure with any other information expected by the driver and passed this data structure as platform data.

Now that all the data structures are set up, we can register the platform device by adding the following content at the bottom of demo-pdev-init.c:

```
static int demo_pdev_init()
{
    return platform_device_register(&demo_pdev);
}
module_init(demo_pdev_init);

static void demo_pdev_exit()
{
    platform_device_unregister(&demo_pdev);
}
module_exit(demo_pdev_exit);
```

The preceding code does nothing other than register the platform device with the system. In this example, we have used static initialization. That said, things won't be that much different with dynamic initialization.

Now that we are done with platform device instantiation, let's focus on the platform driver itself, whose compilation unit is named demo-pdriver.c, and add the following header files to it:

```
#include <linux/module.h>
#include <linux/init.h>
#include <linux/interrupt.h>
#include <linux/of_platform.h>
#include <linux/platform_device.h>
```

Now that the headers that will allow us to support the necessary APIs and to use the appropriate data structures are in place, let's start by enumerating the devices that can be matched with this platform driver from the device tree:

```
static const struct of_device_id labcsmart_dt_ids[] = {
    {
        .compatible = " labcsmart,demo-pdev",
        /* .data = (void *)&big_struct_1, */
```

```
    }, {
            .compatible = " labcsmart,other-pdev ",
            /* .data = (void *) &big_struct_2, */
    }, {
            /* sentinel */
    }
};
MODULE_DEVICE_TABLE(of, labcsmart_dt_ids);
```

In the preceding, the device tree match table enumerates the supported devices that are differentiated by their compatible property. Here, again, the .data assignment is commented out just to show how we could have passed platform-specific data according to the entry that caused the match with the platform device. This platform-specific structure could have been big_struct_1 or big_struct_2. This is the way I recommend when you need to pass platform data and still want to use the device tree.

To provide an opportunity for matching using an ID table, we populate such a table as follows:

```
static const struct platform_device_id labcsmart_ids[] = {
    {
            .name = "demo-pdev",
            /*.driver_data = &big_struct_1,*/
    }, {
            .name = "other-pdev",
            /*.driver_data = &big_struct_2,*/
    }, {
            /* sentinel */
    }
};
MODULE_DEVICE_TABLE(platform, labcsmart_ids);
```

The preceding requires no special comments, except the commented-out part, which shows how extra data could be used depending on the device being handled.

At this stage, we are good to go with the implementation of the probe method. Still in our demo platform driver compilation unit, we add the following:

```
static u32 *reg_base;
static struct resource *res_irq;

static int demo_pdrv_probe(struct platform_device *pdev)
{
    struct resource *regs;
    regs = platform_get_resource(pdev, IORESOURCE_MEM, 0);
    if (!regs) {
        dev_err(&pdev->dev, "could not get IO memory\n");
        return -ENXIO;
    }

    /* map the base register */
    reg_base = devm_ioremap(&pdev->dev, regs->start,
                            resource_size(regs));
    if (!reg_base) {
        dev_err(&pdev->dev, "could not remap memory\n");
        return 0;
    }

    res_irq = platform_get_resource(pdev,
                            IORESOURCE_IRQ, 0);
    /* we could have used
     * irqnum = platform_get_irq(pdev, 0); */
    devm_request_irq(&pdev->dev, res_irq->start,
                    top_half, IRQF_TRIGGER_FALLING,
                    "demo-pdev", NULL);
[...]
    return 0;
}
```

In the preceding probe method, a lot of things could have been added or done differently. The first case is if our driver was device tree-compliant only, and that the associated platform device was instantiated from the device tree as well. We could have done something like this:

```
struct device_node *np = pdev->dev.of_node;

const struct of_device_id *of_id =
        of_match_device(labcsmart_dt_ids, &pdev->dev);
struct big_struct *pdata_struct = (big_struct*)of_id->data;
```

In the preceding, we grab a reference to the device tree note associated with the platform device, on which we could use whatever device tree-related APIs (such as of_get_property()) to extract data from there. Next, in the case where a platform-specific data structure was passed when provisioning the supported device in the match table, we use of_match_device() to point to the entry that corresponds to the hit and we extract the platform-specific data.

If the match occurred thanks to the ID table, pdev->id_entry would point to the entry thanks to which the match occurred, and pdev->id_entry->driver_data would point to the appropriate bigger structure.

However, if the platform data were declared in the traditional method, we would use something as follows:

```
struct big_struct *pdata_struct =
                    (big_struct*)dev_get_platdata(&pdev->dev);
```

Now that we are done with the probe method, I would like to draw your attention to a particular point: in that method, we exclusively used the devm_ prefixed function to deal with resources. These are functions that take care of releasing the resource at the appropriate time.

What this means is that we don't require a remove method. However, for the sake of pedagogy, the following is what our remove method would look like if we did not use resource-managed helpers:

```
int demo_pdrv_remove(struct platform_device *dev)
{
    free_irq(res_irq->start, NULL);
    iounmap(reg_base);
    return 0;
}
```

In the preceding, the IRQ resource is released, and the memory mapping is destroyed.

Well now, everything is in place. We can initialize and register the platform driver using the following:

```
static struct platform_driver demo_driver = {
    .probe     = demo_pdrv_probe,
    /* .remove = demo_pdrv_remove, */
    .driver    = {
        .owner   = THIS_MODULE,
        .name    = "demo_pdev",
    },
};
module_platform_driver(demo_driver);
```

There is another important point you need to consider: the names of the platform device, platform driver, and even the device node name in the device tree. They are all the same, that is, demo_pdev. This is a way to provide an opportunity to match even by platform device and platform driver names since matching by name is used as a fallback when all of the device tree, ID table, and ACPI matchings fail.

Summary

The kernel pseudo platform bus no longer holds any secrets for you. With a bus matching mechanism, you can understand how, when, and why your driver has been loaded, as well as which device it was written for. We can implement any probe function, based on the matching mechanism we want. Since the main purpose of a device driver is to handle devices, we are now able to populate devices in the system, either in the traditional way or from the device tree. To finish in style, we implemented a functional platform driver example from scratch.

In the next chapter, we will continue learning how to write drivers for non-discoverable devices, but sitting on I2C buses this time.

8
Writing I2C Device Drivers

I2C stands for **Inter-Integrated Circuit**. It is a serial, multi-master, and asynchronous bus invented by Philips (now NXP), though multi-master mode is not widely used. I2C is a two-wire bus, respectively named **Serial Data** (**SDA**) and **Serial Clock** (**SCL**, or **SCK**). An I2C device is a chip that interacts with another device via an I2C bus. On this bus, both SDA and SCL are open-drain/open collectors, meaning that each can drive its output low, but neither can drive its output high without having pull-up resistors. SCL is generated by the master to synchronize data (carried by SDA) transfer over the bus. Both the slave and master can send data (not at the same time, of course), thus making SDA a bidirectional line. That said, the SCL signal is also bidirectional since the slave can *stretch* the clock by keeping the SCL line low. The bus is controlled by the master, which in our case is part of the **System on Chip** (**SoC**). This bus is frequently used in embedded systems to connect serial EEPROM, RTC chips, GPIO expanders, temperature sensors, and so on.

The following diagram shows various devices (also known as slaves) connected to an I2C bus:

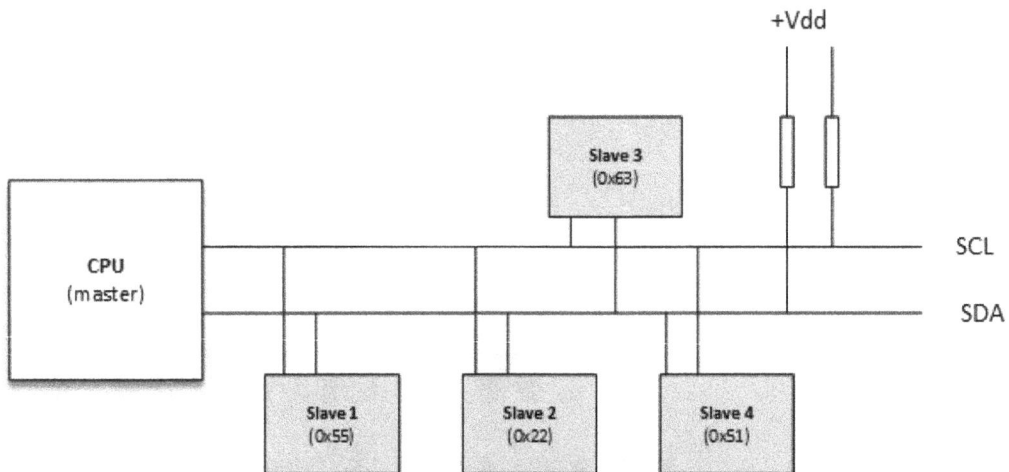

Figure 8.1 – I2C bus and device representation

From the preceding diagram, we can represent the Linux kernel I2C framework as follows:

```
CPU <--platform bus-->i2c adapter<---i2c bus---> i2c slave
```

The CPU is the master hosting the I2C controller, also known as the I2C adapter, which implements the I2C protocol and manages the bus segment that's hosting the I2C devices. In the kernel I2C framework, the adapter is managed by a platform driver while the slave is driven by an I2C driver. However, both drivers use APIs provided by the I2C core. In this chapter, we will be focusing on I2C (slave) device drivers, though references to the adapter will be mentioned if necessary.

Back to the hardware, I2C clock speed varies from 10 kHz to 100 kHz, and from 400 kHz to 2 MHz. There are no strict data transfer speed requirements for I2C, and all the slaves sitting on a given bus will use the same clock speed that the bus has been configured with. This is unlike **Serial Peripheral Interfaces (SPIs)**, where the clock speed is applied on a per-device basis. An example of an I2C controller drive (for example, an i.MX6 chip) can be found at `drivers/i2c/busses/i2c-imx.c` in the kernel source, and the I2C specifications can be found at `https://www.nxp.com/docs/en/user-guide/UM10204.pdf`.

Now that we know that we will be dealing with an I2C device driver, in this chapter, we will cover the following topics:

- I2c framework abstraction in the Linux kernel

- The I2C driver abstraction and architecture

- How not to write I2C device drivers

I2C framework abstractions in the Linux kernel

The Linux kernel I2C framework is made up of a few data structures, with the most important being as follows:

- `i2c_adapter`: Used to abstract the I2C master device. It is used to identify a physical I2C bus.

- `i2c_algorithm`: This abstracts the I2C bus transaction interface. Here, transaction means to transfer, such as read or write operations.

- `i2c_client`: Used to abstract a slave device sitting on the I2C bus.

- `i2c_driver`: The driver of the slave device. It contains a set of specific driving functions to deal with the device.

- `i2c_msg`: This is the low-level representation of one segment of an I2C transaction. This data structure defines the device address, the transaction flags (if it's a transmit or receive, for example), a pointer to the data to send/receive, and the size of the data.

Since the scope of this chapter is limited to slave device drivers, we will focus on the last three data structures. However, to help you understand this, we need to introduce both the adapter and algorithm data structures.

A brief introduction to struct i2c_adapter

The kernel uses `struct i2c_adapter` to represent a physical I2C bus, along with the algorithms that are necessary to access it. It is defined as follows:

```
struct i2c_adapter {
    struct module *owner;
    const struct i2c_algorithm *algo;
    [...]
};
```

In the preceding data structure, we have the following:

- `owner`: Most of the time, this is set with `THIS_MODULE`. This is the owner, and it's used for reference counting.

- `algo`: This is a set of callbacks that are used by the controller (the master) driver to drive the I2C line. These callbacks allow you to generate the signal that's needed for the I2C access cycle.

The algorithm's data structure has the following definition:

```
struct i2c_algorithm {
    int (*master_xfer)(struct i2c_adapter *adap,
                        struct i2c_msg *msgs, int num);
    int (*smbus_xfer)(struct i2c_adapter *adap, u16 addr,
            unsigned short flags, char read_write,
            u8 command, int size,
            union i2c_smbus_data *data);
    /* To determine what the adapter supports */
    u32 (*functionality)(struct i2c_adapter *adap);
[...]
};
```

In the preceding data structure, the unimportant fields have been omitted. Let's look at each element in the excerpt:

- `master_xfer`: This is the core transfer function. It must be provided for this algorithm driver to have plain I2C access. It is invoked when an I2C device driver needs to communicate with the underlying I2C device. However, if it is not implemented (if `NULL`), the `smbus_xfer` function is called instead.

- `smbus_xfer`: This is a function pointer that is set by the I2C controller driver if its algorithm driver can perform SMBus accesses. It is used whenever an I2C chip driver wants to communicate with the chip device using the SMBus protocol. If it is `NULL`, the `master_xfer` function is used instead and the SMBus is emulated.

- `functionality`: This is a function pointer that's called by the I2C core to determine the capabilities of the adapter. It informs you about what kind of reads and writes the I2C adapter driver can do.

In the preceding code, `functionality` is a sanity callback. Either the core or device drivers can invoke it (through `i2c_check_functionality()`) to check whether the given adapter can provide the I2C access we need before we initiate this access. For example, 10-bit addressing mode is not supported by all adapters. Thus, it is safe to call `i2c_check_functionality(client->adapter, I2C_FUNC_10BIT_ADDR)` in the chip driver to check whether it is supported by the adapter. All the flags are in the form of `I2C_FUNC_XXX`. Though each can be checked individually, the I2C core has split them into logical functions, as follows:

```
#define I2C_FUNC_I2C                  0x00000001
#define I2C_FUNC_10BIT_ADDR           0x00000002
#define I2C_FUNC_SMBUS_BYTE    (I2C_FUNC_SMBUS_READ_BYTE | \
                    I2C_FUNC_SMBUS_WRITE_BYTE)
#define I2C_FUNC_SMBUS_BYTE_DATA \
                    (I2C_FUNC_SMBUS_READ_BYTE_DATA | \
                    I2C_FUNC_SMBUS_WRITE_BYTE_DATA)
#define I2C_FUNC_SMBUS_WORD_DATA \
                    (I2C_FUNC_SMBUS_READ_WORD_DATA | \
                    I2C_FUNC_SMBUS_WRITE_WORD_DATA)
#define I2C_FUNC_SMBUS_BLOCK_DATA \
                    (I2C_FUNC_SMBUS_READ_BLOCK_DATA | \
                    I2C_FUNC_SMBUS_WRITE_BLOCK_DATA)
#define I2C_FUNC_SMBUS_I2C_BLOCK \
                    (I2C_FUNC_SMBUS_READ_I2C_BLOCK | \
                    I2C_FUNC_SMBUS_WRITE_I2C_BLOCK)
```

With the preceding code, you can check the `I2C_FUNC_SMBUS_BYTE` flag to make sure that the adapter supports SMBus byte-oriented commands.

This introduction to I2C controllers will be referenced in other sections as needed. It may make sense to understand this first, even though the main purpose of this chapter is to discuss the I2C client driver, which we will do in the next section.

I2C client and driver data structures

The first and most evident data structure is the `struct i2c_client` structure, which is declared like so:

```
struct i2c_client {
    unsigned short flags;
```

```
    unsigned short addr;
    char name[I2C_NAME_SIZE];
    struct i2c_adapter *adapter;
    struct device dev;
    int irq;
};
```

In the preceding data structure, which holds the properties of the I2C device, flags represents the device flags, the most important of which is the one telling us whether this is a 10-bit chip address. addr contains the chip address. In the case of a 7-bit address chip, it will be stored in the lower 7 bits. name contains the device name, which is limited to I2C_NAME_SIZE (set to 20 in include/linux/mod_devicetable.h) characters. adapter is the adapter (remember, it is the I2C bus) that this device sits on. dev is the underlying device structure for the device model, and irq is the interrupt line that's been assigned to the device.

Now that we are familiar with the I2C device data structure, let's focus on its driver, which is abstracted by struct i2c_driver. It can be declared like so:

```
struct i2c_driver {
    unsigned int class;
    /* Standard driver model interfaces */
    int (*probe)(struct i2c_client *client,
                 const struct i2c_device_id *id);
    int (*remove)(struct i2c_client *client);
    int (*probe_new)(struct i2c_client *client);
    void (*shutdown)(struct i2c_client *client);
    struct device_driver driver;
    const struct i2c_device_id *id_table;
};
```

Let's look at each element in the data structure:

- probe: A callback for device binding that should return 0 on success or the appropriate error code on failure.

- remove: A callback for device unbinding. It must undo what has been done in probe.

- shutdown: A callback for device shutdown.

- `probe_new`: The new driver model interface. This will deprecate the legacy `probe` method to get rid of its commonly unused second parameter (that is, the `struct i2c_device_id` parameter).

- `driver`: The underlying device driver model's driver structure.

- `id_table`: The list of I2C devices that are supported by this driver.

The third and last data structure of this series is `struct i2c_msg`, which represents one operation of an I2C transaction. It is declared like so:

```
struct i2c_msg {
    __u16 addr;
    __u16 flags;
#define I2C_M_TEN 0x0010
#define I2C_M_RD 0x0001
    __u16 len;
    __u8 * buf;
};
```

Each element in this data structure is self-explanatory. Let's look at them in more detail:

- `addr`: This is always the slave address.

- `flags`: Because a transaction may be made of several operations, this element represents the flags of this operation. It should be set to 0 in the case of a write operation (that the master sends to the slave). However, it can be ORed with either `I2C_M_RD` if it's a read operation (the master reads from the slave) or `I2C_M_TEN` if the device is a 10-bit chip address.

- `length`: This is the size of data in the buffer. In a read operation, it corresponds to the number of bytes to be read from the device and is stored in `buf`. In the case of a write operation, it represents the number bytes in `buf` to write to the device.

- `buf`: This is the read/write buffer, which must be allocated as per `length`.

> **Note**
>
> Since `i2c_msg.len` is `u16`, you must ensure you are always less than 2^{16} (64k) away with your read/write buffer.

Now that we have discussed the most important I2C data structures, let's look at the APIs (most of which involve the I2C adapter under the hood) that are exposed by the I2C core to get the most out of our devices.

I2C communication APIs

Once the driver and the data structure have been initialized, communication between the slave and the master can take place. Serial bus transactions are just simple matters of register access, either to get or set their content. I2C devices respect this principle.

Plain I2C communication

We will start from the lowest level – i2c_transfer() is the core function that's used to transfer I2C messages. Other APIs wrap this function, which is backed by algo->master_xfer of the adapter. The following is its prototype:

```
int i2c_transfer(struct i2c_adapter *adap,
                 struct i2c_msg *msg, int num);
```

With i2c_transfer(), no stop bit is sent between bytes in the same read/write operation of the same transaction. This is useful for devices that require no stop bit between address write and data read, for example. The following code shows how it can be used:

```
static int i2c_read_bytes(struct i2c_client *client,
                          u8 cmd, u8 *data, u8 data_len)
{
    struct i2c_msg msgs[2];
    int ret;
    u8 *buffer;
    buffer = kzalloc(data_len, GFP_KERNEL);
    if (!buffer)
        return -ENOMEM;;
    msgs[0].addr = client->addr;
    msgs[0].flags = client->flags;
    msgs[0].len = 1;
    msgs[0].buf = &cmd;

    msgs[1].addr = client->addr;
    msgs[1].flags = client->flags | I2C_M_RD;
    msgs[1].len = data_len;
    msgs[1].buf = buffer;
    ret = i2c_transfer(client->adapter, msgs, 2);
    if (ret < 0)
```

```
            dev_err(&client->adapter->dev,
                    "i2c read failed\n");
        else
            memcpy(data, buffer, data_len);
        kfree(buffer);
        return ret;
}
```

If the device requires a stop bit in the middle of a read sequence, you should split your transaction into two parts (two operations) – `i2c_transfer` for address write (a transaction with a single write operation), and another `i2c_transfer` for data read (a transaction with a single read operation), as shown here:

```
static int i2c_read_bytes(struct i2c_client *client,
                          u8 cmd, u8 *data, u8 data_len)
{
    struct i2c_msg msgs[2];
    int ret;
    u8 *buffer;
    buffer = kzalloc(data_len, GFP_KERNEL);
    if (!buffer)
        return -ENOMEM;;
    msgs[0].addr = client->addr;
    msgs[0].flags = client->flags;
    msgs[0].len = 1;
    msgs[0].buf = &cmd;
    ret = i2c_transfer(client->adapter, msgs, 1);
    if (ret < 0) {
        dev_err(&client->adapter->dev,
                "i2c read failed\n");
        kfree(buffer);
        return ret;
    }
    msgs[1].addr = client->addr;
    msgs[1].flags = client->flags | I2C_M_RD;
    msgs[1].len = data_len;
    msgs[1].buf = buffer;
```

```
    ret = i2c_transfer(client->adapter, &msgs[1], 1);
    if (ret < 0)
        dev_err(&client->adapter->dev,
                "i2c read failed\n");
    else
        memcpy(data, buffer, data_len);
    kfree(buffer);
    return ret;
}
```

Otherwise, you can use alternative APIs, such as `i2c_master_send` and `i2c_master_recv`, respectively:

```
int i2c_master_send(struct i2c_client *client,
            const char *buf, int count);
int i2c_master_recv(struct i2c_client *client,
            char *buf, int count);
```

These APIs are both implemented on top of `i2c_transfer()`. `i2c_master_send()` actually implements an I2C transaction with a single write operation, while `i2c_master_recv()` does the same with a single read operation.

The first argument is the I2C device to be accessed. The second parameter is the read/write buffer, while the third represents the number of bytes to read or write. The returned value is the number of bytes being read/written. The following code is a simplified version of our previous excerpt:

```
static int i2c_read_bytes(struct i2c_client *client,
                    u8 cmd, u8 *data, u8 data_len)
{
    struct i2c_msg msgs[2];
    int ret;
    u8 *buffer;
    buffer = kzalloc(data_len, GFP_KERNEL);
    if (!buffer)
        return -ENOMEM;;
    ret = i2c_master_send(client, &cmd, 1);
    if (ret < 0) {
        dev_err(&client->adapter->dev,
```

```
                "i2c read failed\n");
        kfree(buffer);
        return ret;
    }
    ret = i2c_master_recv(client, buffer, data_len);
    if (ret < 0)
        dev_err(&client->adapter->dev,
                "i2c read failed\n");
    else
        memcpy(data, buffer, data_len);
    kfree(buffer);
    return ret;
}
```

With that, we are familiar with how plain I2C APIs are implemented in the kernel. However, there is a category of devices we need to address – SMBus-compatible devices – which are not to be confused with I2C ones, even though they sit on the same physical bus.

System Management Bus (SMBus)-compatible functions

SMBus is a two-wire bus developed by Intel and is very similar to I2C. Moreover, it is a subset of I2C, which means I2C devices are SMBus-compatible, but not the reverse. SMBus is a subset of I2C, which means I2C controllers support most SMBus operations. However, this is not true for SMBus controllers as they may not support all the protocol options that an I2C controller will. Therefore, it is better to use SMBus methods in case you have doubts about the chip you are writing the driver for.

The following are some examples of SMBus APIs:

```
s32 i2c_smbus_read_byte_data(struct i2c_client *client,
                            u8 command);
s32 i2c_smbus_write_byte_data(struct i2c_client *client,
                            u8 command, u8 value);
s32 i2c_smbus_read_word_data(struct i2c_client *client,
                            u8 command);
s32 i2c_smbus_write_word_data(struct i2c_client *client,
                            u8 command, u16 value);
s32 i2c_smbus_read_block_data(struct i2c_client *client,
```

```
                               u8 command, u8 *values);
s32 i2c_smbus_write_block_data(struct i2c_client *client,
                               u8 command, u8 length,
                               const u8 *values);
```

A complete list of SMBus APIs is available in `include/linux/i2c.h` in the kernel sources. Each function is self-explanatory. The following example shows a simple read/write operation that's using SMBus-compatible APIs to access an I2C GPIO expander:

```
struct mcp23016 {
    struct i2c_client    *client;
    struct gpio_chip     chip;
    struct mutex         lock;
};
[...]
static int mcp23016_set(struct mcp23016 *mcp,
            unsigned offset, intval)
{
    s32 value;
    unsigned bank = offset / 8;
    u8 reg_gpio = (bank == 0) ? GP0 : GP1;
    unsigned bit = offset % 8;

    value = i2c_smbus_read_byte_data(mcp->client,
                                     reg_gpio);
    if (value >= 0) {
        if (val)
            value |= 1 << bit;
        else
            value &= ~(1 << bit);
        return i2c_smbus_write_byte_data(mcp->client,
                                         reg_gpio, value);
    } else
        return value;
}
```

The SMBus part is as simple as the list of APIs that are available. Now that we can access the device either using plain I2C functions or SMBus functions, we can start implementing the body of our I2C driver.

The I2C driver abstraction and architecture

The struct i2c_driver structure, as we saw in the previous section, contains the driving methods that are needed to handle the I2C devices it is responsible for. Once added to the bus, the device will need to be probed, which makes the i2c_driver. probe_new method the entry point of the driver.

Probing the I2C device

The probe() callback in the struct i2c_driver structure is invoked any time an I2C device is instantiated on the bus and claims this driver. It is responsible for the following tasks:

- Checking whether the I2C bus controller (the I2C adapter) supports the functionalities needed by the device using the i2c_check_functionality() function

- Checking whether the device is the one we expected

- Initializing the device

- Setting up device-specific data if necessary

- Registering with the appropriate kernel frameworks

Formerly, the probing callback was assigned to the probe element of struct i2c_driver and had the following prototype:

```
int foo_probe(struct i2c_client *client,
              const struct i2c_device_id *id)
```

Because the second argument is rarely used, this callback has been deprecated in favor of probe_new, which has the following prototype:

```
int probe(struct i2c_client *client)
```

In the preceding prototype, the struct i2c_client pointer represents the I2C device itself. This parameter is prebuilt and initialized by the core according to the device's description, which is done either in the device tree or in the board file.

It is not recommended to access the device early in the probe method. Because each I2C adapter has different capabilities, it is better to request for it to know what capabilities it supports and adapt the driver's behavior accordingly:

```
#define CHIP_ID 0x13
#define DA311_REG_CHIP_ID  0x000f

static int fake_i2c_probe(struct i2c_client *client)
{
    int err;
    int ret;
    if (!i2c_check_functionality(client->adapter,
            I2C_FUNC_SMBUS_BYTE_DATA))
        return -EIO;

    /* read family id */
    ret = i2c_smbus_read_byte_data(client, REG_CHIP_ID);
    if (ret != CHIP_ID)
        return (ret < 0) ? ret : -ENODEV;

    /* register with other frameworks */
    [...]
    return 0;
}
```

In the preceding example, we checked whether the underlying adapter supports the type/command the device needs. Only after a successful sanity check, we can safely access the device and go further by allocating resources of any kind and registering with other frameworks if necessary.

Implementing the i2c_driver.remove method

The i2c_driver.remove callback must undo what has been done in the probe function. It must unregister from each framework that was registered with probe and release each resource that was requested. This callback has the following prototype:

```
static int remove(struct i2c_device *client)
```

In the preceding line of code, `client` is the same I2C device data structure that was passed by the core to the `probe` method. This means that whatever data you have stored at probing time can be retrieved here. For example, you may need to process some cleaning or any other stuff based on the private data you set up in the `probe` function:

```
static int mc9s08dz60_remove(struct i2c_client *client)
{
    struct mc9s08dz60 *mc9s;
    /* We retrieve our private data */
    mc9s = i2c_get_clientdata(client);
    /* Which hold gpiochip we want to work on */
    return gpiochip_remove(&mc9s->chip);
}
```

The preceding example is simple and may represent most cases that you will see in drivers. Since this callback is supposed to return zero on success, the failure reasons may include that the device can't power down, is still in use, and so on. This means that there may be situations where you will want to consult the device and perform some additional operations in this callback function.

At this stage in the development process, all the callbacks have been prepared. Now, it is time for the driver to register with the I2C core, as we'll see in the next section.

Driver initialization and registration

I2C drivers register and unregister themselves with the core using the `i2c_add_driver()` and `i2c_del_driver()` APIs, respectively. The former is a macro that's backed by the `i2c_register_driver()` function. The following code shows their respective prototypes:

```
int i2c_add_driver(struct i2c_driver *drv);
void i2c_del_driver(struct i2c_driver *drv);
```

In both functions, `drv` is the I2C driver structure that had been set up previously. The registration API returns zero on success or a negative error code on failure.

Driver registration mostly takes place in the module initialization while unregistering this driver is usually done in module exit method. The following is a typical demonstration of I2C driver registering:

```
static int __init foo_init(void)
{
    [...] /*My init code */
        return i2c_add_driver(&foo_driver);
}
module_init(foo_init);

static void __exit foo_cleanup(void)
{
    [...] /* My clean up code */
        i2c_del_driver(&foo_driver);
}
module_exit(foo_cleanup);
```

If the driver needs to do nothing else but register/unregister the driver during module initialization/cleanup, then the module_i2c_driver() macro can be used to reduce the preceding code, like so:

```
module_i2c_driver(foo_driver);
```

This macro will be expanded at build time to the appropriate initialization/exit methods in the module, which will take care of registering/unregistering the I2C driver.

Provisioning devices in the driver

For the matching loop to be invoked with our I2C driver, the i2c_driver.id_table field must be set with a list of I2C device IDs, each described by an instance of the struct i2c_device_id data structure, which has the following definition:

```
struct i2c_device_id {
    char name[I2C_NAME_SIZE];
    kernel_ulong_t driver_data;
};
```

In the preceding data structure, `name` is a descriptive name for the device, while `driver_data` is the driver state data, which is private to the driver. It can be set with a pointer to a per-device data structure, for example. Additionally, for device matching and module (auto)loading purposes, this same device ID array needs to be given to the `MODULE_DEVICE_TABLE` macro.

However, this does not concern device tree matching. For a device node in the device tree to match our driver, the `i2c_driver.device.of_match_table` element of our driver must be set with a list of elements of the `struct of_device_id` type. Each entry in that list will describe an I2C device that can be matched from the device tree. The following is the definition of this data structure:

```
struct of_device_id {
[...]
    char   compatible[128];
    const void *data;
};
```

In the preceding data structure, `compatible` is a quite descriptive string that can be used in the device tree to match this driver, while `data` can point to anything, such as a per-device resource. In the same way, for module (auto)loading due to a device tree match, this list must be given to the `MODULE_DEVICE_TABLE` macro.

The following is an example:

```
#define ID_FOR_FOO_DEVICE   0
#define ID_FOR_BAR_DEVICE   1

static struct i2c_device_id foo_idtable[] = {
    { "foo", ID_FOR_FOO_DEVICE },
    { "bar", ID_FOR_BAR_DEVICE },
    { },
};
MODULE_DEVICE_TABLE(i2c, foo_idtable);
```

Now, for module loading after a device tree match, we need to do the following:

```
static const struct of_device_id foobar_of_match[] = {
        { .compatible = "packtpub,foobar-device" },
        { .compatible = "packtpub,barfoo-device" },
        {},
};
MODULE_DEVICE_TABLE(of, foobar_of_match);
```

The excerpt shows the final content for i2c_driver, with the respective device table pointers set:

```
static struct i2c_driver foo_driver = {
    .driver        = {
        .name    = "foo",
        /* The below line adds Device Tree support */
        .of_match_table = of_match_ptr(foobar_of_match),
    },
    .probe         = fake_i2c_probe,
    .remove        = fake_i2c_remove,
    .id_table      = foo_idtable,
};
```

In the preceding code, we can see what an I2C driver structure would look like once it has been set up.

Instantiating I2C devices

We will be using the device tree declaration because the board file, though used in old drivers so far, is an era far behind us. I2C devices must be declared as children (sub-nodes) of the bus node they sit on. The following are the properties that are required for their binding:

- reg: This represents the address of the device on the bus.
- compatible: This is a string that is used to match the device with a driver. It must match an entry in the driver's of_match_table.

The following is an example declaration of two I2C devices on the same adapter:

```
&i2c2 { /* Phandle of the bus node */
    pcf8523: rtc@68 {
        compatible = "nxp,pcf8523";
        reg = <0x68>;
    };
    eeprom: ee24lc512@55 { /* eeprom device */
        compatible = "labcsmart,ee24lc512";
        reg = <0x55>;
    };
};
```

The preceding sample declares an RTC chip at address `0x68` and an EEPROM at address `0x55` on the same bus, which is the SoC's I2C bus number 2. The I2C core will rely on the `compatible` string property and the `i2c_device_id` table to bind devices and drivers. A first attempt is made to match the device by compatible string (the `OF` style, which is the device tree); if it fails, then the I2C core tries to match the device by the `id` table.

How not to write I2C device drivers

Deciding not to write the device driver consists of writing the appropriate user code to deal with the underlying hardware. Though it is user code, the kernel always intervenes to ease the development process. I2C adapters are exposed by the kernel in the user space as character devices in the form of `/dev/i2c-<X>`, where `<X>` is the bus number. Once you have opened the character device file that corresponds to the adapter your device sits on, there is a series of commands you can execute.

First, the required headers for dealing with I2C devices from the user space are as follows:

```
#include <linux/i2c-dev.h>
#include <i2c/smbus.h>
#include <linux/i2c.h>
```

The following are the possible commands:

- `ioctl(file, I2C_FUNCS, unsigned long *funcs)`: This command is probably the first command you should issue. It is the equivalent of `i2c_check_functionality()` in the kernel, which returns the necessary adapter functionality (in the `*funcs` argument). The returned flags are also in the form of `I2C_FUNC_*`:

```
unsigned long funcs;
if (ioctl(file, I2C_FUNCS, &funcs) < 0)
        return -errno;
if (!(funcs & I2C_FUNC_SMBUS_QUICK)) {
        /* Oops, SMBus write_quick) not available! */
        exit(1);
}
/* Now it is safe to use SMBus write_quick command */
```

- `ioctl(file, I2C_TENBIT, long select)`: Here, you can select whether the slave you need to talk to is a 10-bit address chip (`select = 1`) or not (`select = 0`).

- `ioctl(file, I2C_SLAVE, long addr)`: This is used to set the chip address you need to talk to on this adapter. The address is stored in the 7 lower bits of `addr` (except for 10-bit addresses, which are passed in the 10 lower bits in this case). This chip may already be in use, in which case you can use `I2C_SLAVE_FORCE` to force usage.

- `ioctl(file, I2C_RDWR, struct i2c_rdwr_ioctl_data *msgset)`: You can use this to perform combined plain I2C read/write transactions without stop bit in between. The structure of interest is `struct i2c_rdwr_ioctl_data`, which has the following definition:

```
struct i2c_rdwr_ioctl_data {
    struct i2c_msg *msgs; /* ptr to array of messages */
    int nmsgs; /* number of messages to exchange */
}
```

The following is an example of using this IOCTL:

```
int ret;
uint8_t buf [5] = {regaddr, '0x55', '0x65',
                        '0x88', '0x14'};

struct i2c_msg messages[] = {
    {
            .addr = dev,
            .buf = buf,
            .len = 5, /* buf size is 5 */
    },
};

struct i2c_rdwr_ioctl_data payload = {
    .msgs = messages,
    .nmsgs = sizeof(messages)
              /sizeof(messages[0]),
};

ret = ioctl(file, I2C_RDWR, &payload);
```

You can also use the `read()` and `write()` calls to make plain I2C transactions (once the address has been set with `I2C_SLAVE`).

*   `ioctl(file, I2C_SMBUS, struct i2c_smbus_ioctl_data *args)`: This command is used to issue an SMBus transfer. The main structure argument has the following prototype:

```
struct i2c_smbus_ioctl_data {
    __u8 read_write;
    __u8 command;
    __u32 size;
    union i2c_smbus_data __user *data;
};
```

In the preceding data structure, `read_write` determines the direction of the transfer – `I2C_SMBUS_READ` to read and `I2C_SMBUS_WRITE` to write. `command` is a command that can be interpreted by the chip. It may be a register address, for example. `size` is the message's length, while `buf` is the message buffer. Note that standardized sizes are already exposed by the I2C core. These are `I2C_SMBUS_BYTE`, `I2C_SMBUS_BYTE_DATA`, `I2C_SMBUS_WORD_DATA`, `I2C_SMBUS_BLOCK_DATA`, and `I2C_SMBUS_I2C_BLOCK_DATA`, for 1, 2, 3, 5, and 8 bytes, respectively. The full list is available at `include/uapi/linux/i2c.h`. The following is an example that shows how to do an SMBus transfer in the user space:

```
uint8_t buf [5] = {'0x55', '0x65', '0x88'};
struct i2c_smbus_ioctl_data payload = {
    .read_write = I2C_SMBUS_WRITE,
    .size = I2C_SMBUS_WORD_DATA,
    .command = regaddr,
    .data = (void *) buf,
};

ret = ioctl (fd, I2C_SLAVE_FORCE, dev);
if (ret < 0)
    /* handle errors */

ret = ioctl (fd, I2C_SMBUS, &payload);
if (ret < 0)
    /* handle errors */
```

Since you can use a simple `read()`/`write()` system call to do a plain I2C transfer (even though a stop bit is sent after each transfer), the I2C core provides the following APIs to perform an SMBus transfer:

```
__s32 i2c_smbus_write_quick(int file, __u8 value);
__s32 i2c_smbus_read_byte(int file);
__s32 i2c_smbus_write_byte(int file, __u8 value);
__s32 i2c_smbus_read_byte_data(int file, __u8 command);
__s32 i2c_smbus_write_byte_data(int file, __u8 command,
                                __u8 value);
__s32 i2c_smbus_read_word_data(int file, __u8 command);
__s32 i2c_smbus_write_word_data(int file, __u8 command,
                                __u16 value);
```

```
__s32 i2c_smbus_read_block_data(int file, __u8 command,
                          __u8 *values);
__s32 i2c_smbus_write_block_data(int file, __u8 command,
                          __u8 length, __u8 *values);
```

You are encouraged to use these functions instead of IOCTLs. If a failure occurs, all of these transactions will return -1; you can check errno for a better understanding of what went wrong. On success, the *_write_* transactions will return 0, while the *_read_* transactions will return the read value, except for *_read_block_*, which will return the number of values that have been read. In block-oriented operations, buffers don't need to be longer than 32 bytes.

Apart from the APIs, which would require that you write some code, you can use a CLI package, i2ctools, which is shipped with the following tools:

- i2cdetect: A command that enumerates the I2C devices sitting on a given adapter.
- i2cget: This is used to dump the content of device registers.
- i2cset: This is used to set the content of device registers.

In this section, we learned how to use user space APIs and commands or command-line tools to communicate with I2C devices. While all these may be useful for prototyping, it could be difficult to handle devices that support interrupts or other kernel-based resources, such as clocks.

Summary

In this chapter, we looked at I2C device drivers. Now, it's time for you to pick any I2C device on the market and write the corresponding driver, along with the necessary device tree support. This chapter talked about the kernel I2C core and its associated API, including device tree support, to give you the necessary skills to talk with I2C devices. You should now be able to write efficient probe functions and register them with the kernel I2C core.

In the next chapter, we will use the skills we have learned in this chapter to develop an SPI device driver.

9
Writing SPI Device Drivers

The **Serial Peripheral Interface** (**SPI**) is (at least) a 4-wire bus – **Master Input Slave Output** (**MISO**), **Master Output Slave Input** (**MOSI**), **Serial Clock** (**SCK**), and **Chip Select** (**CS**) – which is used to connect serial flash and analog-to-digital/digital-to-analog converters. The master always generates the clock. Its speed can reach up to 80 MHz, though there is no real speed limitation (this is much faster than I2C as well). The same applies to the CS line, which is always managed by the master.

Each of these signal names has a synonym:

- Whenever you see **Slave Input Master Output** (**SIMO**), **Slave Data Input** (**SDI**), or **Data Input** (**DI**), they refer to MOSI.

- **Slave Output Master Input** (**SOMI**), **Slave Data Output** (**SDO**), and **Data Output** (**DO**) refer to MISO.

- **Serial Clock** (**SCK**), **Clock** (**CLK**), and **Serial Clock** (**SCL**) refer to SCK.

- $\overline{S}\,\overline{S}$ is the Slave Select line, also called CS. CSx can be used (where x is an index such as CS0, CS1), EN and ENB too, meaning enable. CS is usually an active low signal.

The following diagram shows how SPI devices are connected to the controller via the bus it exposes:

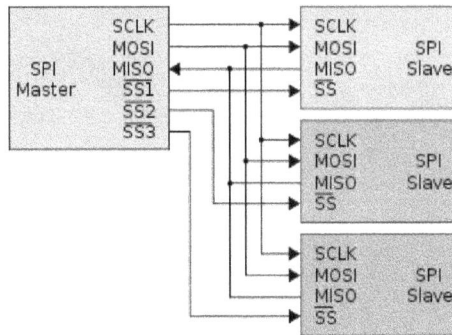

Figure 9.1 – SPI slave devices and master interconnection

From the preceding diagram, we can represent the Linux kernel SPI framework as follows:

```
CPU <--platform bus--> SPI master <---SPI bus---> SPI slave
```

The CPU is the master hosting the SPI controller, also known as the SPI master, which manages the bus segment hosting the SPI slave devices. In the kernel SPI framework, the bus is managed by a platform driver while the slave is driven by an SPI device driver. However, both drivers use APIs provided by the SPI core. In this chapter, we will be focusing on SPI (slave) device drivers, though references to the controller will be mentioned if necessary.

This chapter will walk through SPI driver concepts such as the following:

- Understanding the SPI framework abstraction in the Linux kernel
- Dealing with the SPI driver abstraction and architecture
- Learning how not to write SPI device drivers

Understanding the SPI framework abstractions in the Linux kernel

The Linux kernel SPI framework is made up of a few data structures, the most important of which are the following:

- `spi_controller`, used to abstract the SPI master device.
- `spi_device`, used to abstract a slave device sitting on the SPI bus.
- `spi_driver`, the driver of the slave device.

- `spi_transfer`, which is the low-level representation of one segment of a protocol. It represents a single operation between the master and slave. It expects Tx and/or Rx buffers as well as the length of the data to be exchanged and an optional CS behavior.

- `spi_message`, which is an atomic sequence of transfers.

Let's now introduce each of these data structures, one after the other, starting with the most complex, which represents the SPI controller's data structure.

Brief introduction to struct spi_controller

Throughout this chapter, we will reference the controller because it is deeply coupled with the slaves and other data structures that the SPI framework is made up of. It is necessary therefore to introduce its data structure, represented by `struct spi_controller` and defined as follows:

```
struct spi_controller {
    struct device       dev;
    u16                 num_chipselect;
    u32                 min_speed_hz;
    u32                 max_speed_hz;
    int                 (*setup)(struct spi_device *spi);
    int (*set_cs_timing)(struct spi_device *spi,
                        struct spi_delay *setup,
                        struct spi_delay *hold,
                        struct spi_delay *inactive);
    int     (*transfer)(struct spi_device *spi,
                        struct spi_message *mesg);

    bool    (*can_dma)(struct spi_controller *ctlr,
                        struct spi_device *spi,
                        struct spi_transfer *xfer);

    struct kthread_worker   *kworker;
    struct kthread_work     pump_messages;
    spinlock_t              queue_lock;
    struct list_head        queue;
    struct spi_message      *cur_msg;
```

```
    bool                       busy;
    bool                       running;
    bool                       rt;

    int (*transfer_one_message)(
                    struct spi_controller *ctlr,
                    struct spi_message *mesg);
[...]
    int (*transfer_one_message)(
            struct spi_controller *ctlr,
            struct spi_message *mesg);
    void (*set_cs)(struct spi_device *spi, bool enable);
    int (*transfer_one)(struct spi_controller *ctlr,
                    struct spi_device *spi,
                    struct spi_transfer *transfer);
[...]
    /* DMA channels for use with core dmaengine helpers */
    struct dma_chan    *dma_tx;
    struct dma_chan    *dma_rx;

    /* dummy data for full duplex devices */
    Void               *dummy_rx;
    Void               *dummy_tx;
};
```

Only the important elements required for a better understanding of the data structure used in this chapter are listed in the preceding code. The following list explains their use:

- num_chipselect indicates the number of CSs assigned to this controller. CSs are used to distinguish individual SPI slaves and are numbered from 0.

- min_speed_hz and max_speed_hz are the lowest and the highest transfer speeds supported by this controller, respectively.

- set_cs_timing is a method provided if the SPI controller supports CS timing configuration, in which case the client drivers would call spi_set_cs_timing() with the requested timings. It has been deprecated in recent kernel versions by this patch: https://lore.kernel.org/lkml/20210609071918.2852069-1-gregkh@linuxfoundation.org/).

- `transfer` adds a message to the transfer queue of the controller. On the controller registration path (thanks to `spi_register_controller()`), the SPI core checks whether this field is NULL or not:

  - If NULL, the SPI core will check if either `transfer_one` or `transfer_one_message` is set, in which case it assumes this controller supports message queuing and invokes `spi_controller_initialize_queue()`, which will set this field with `spi_queued_transfer` (which is the SPI core helper to queue SPI messages to the controller's queue and to schedule the message pump `kworker` if it is not already running or busy).

    - Moreover, `spi_controller_initialize_queue()` will create both a dedicated kthread worker (`kworker` element) and a work struct (`pump_messages` element) for this controller. This worker will be scheduled quite often in order to process the message queue in a FIFO order.

    - Next, the controller's `queued` element is set to true by the SPI core.

    - Finally, if the controller's `rt` element has been set to true by the driver prior to calling the registration API, the SPI core will set the scheduling policy of the worker thread to the real-time FIFO policy, with a priority of 50.

  - If NULL, and both `transfer_one` and `transfer_one_message` are also NULL, this is an error, and the controller is not registered.

  - If not NULL, the SPI core assumes the controller does not support queuing and does not call `spi_controller_initialize_queue()`.

- `transfer_one` and `transfer_one_message` are mutually exclusive. If both are set, the former won't be invoked by the SPI core. `transfer_one` transfers a single SPI transfer and has no notion of `spi_message`. `transfer_one_message`, if provided by the driver, must work on the basis of `spi_message` and will be responsible for all the transfers in the messages. Controller drivers that need not bother with message-handling algorithms just have to set the `transfer_one` callback, in which case the SPI core will set `transfer_one_message` to `spi_transfer_one_message`. `spi_transfer_one_message` will take care of all the message logic, timings, CS, and other hardware-related properties prior to calling the driver provided `transfer_one` callback for each transfer in the message. CS remains active throughout the message transfers unless it is modified by a transfer that has `spi_transfer.cs_change = 1`. The message transfers will be performed using the clock and SPI mode parameters previously applied by `setup()` for this device.

- `kworker`: This is the kernel thread dedicated to the message pump processing.

- pump_messages: This is an abstraction of a work struct data structure for scheduling the function that processes the SPI message queue. It is scheduled in kworker. This work struct is backed by the spi_pump_messages() method, which checks if there are any SPI messages in the queue that need to be processed, and if so, it calls the driver to initialize the hardware and transfer each message.

- queue_lock: The spinlock to synchronize access to the message queue.

- queue: The message queue for this controller.

- idling: This indicates whether the controller device is entering an idle state.

- cur_msg: The currently in-flight SPI message.

- busy: This indicates the busyness of the message pump.

- running: This indicates that the message pump is running.

- Rt: This indicates whether kworker will run the message pump with real-time priority.

- dma_tx: The DMA transmit channel (when supported by the controller).

- dma_rx: The DMA receiving channel (when supported by the controller).

SPI transfers always read and write the same number of bytes, which means even when the client driver issues a half-duplex transfer, full duplex is emulated by the SPI core with dummy_rx and dummy_tx used to achieve this purpose:

- dummy_rx: This is a dummy receive buffer used for full-duplex devices, such that if a transfer's receive buffer is NULL, received data will be shifted to this dummy receive buffer before being discarded.

- dummy_tx: This is a dummy transmit buffer used for full-duplex devices, such that if a transfer's transmit buffer is NULL, this dummy transmit buffer will be zero-filled and used as a transmit buffer for the transfer.

Do note that the SPI core names the SPI message pump worker task with the controller device name (dev->name), set in spi_register_controller() as follows:

```
dev_set_name(&ctlr->dev, "spi%u", ctlr->bus_num);
```

Later, when the worker is created during the queue initialization (remember, spi_controller_initialize_queue()), it is given this name, as follows:

```
ctlr->kworker = kthread_create_worker(0, dev_name(&ctlr->dev));
```

To recognize the SPI message pump worker on your system, you can run the following command:

```
root@yocto-imx6:~# ps | grep spi
65 root          0 SW   [spi1]
```

In the preceding snippet, we can see the worker's name made up of the bus name along with the bus number.

In this section, we analyzed the concepts on the controller side to help get an understanding of the whole SPI slave implementation in the Linux kernel. The importance of this data structure is so great that I recommend you read this section whenever you feel you don't understand any mechanism in the coming sections. Now we can switch to SPI device data structures for real.

The struct spi_device structure

The first and most obvious data structure, `struct spi_device` represents an SPI device and is defined in `include/linux/spi/spi.h`:

```
struct spi_device {
    struct device dev;
    struct spi_controller  *controller;
    struct spi_master *master;
    u32        max_speed_hz;
    u8         chip_select;
    u8         bits_per_word;
    bool       rt;
    u16        mode;
    int         irq;
    [...]
    int cs_gpio; /* LEGACY: chip select gpio */
    struct gpio_desc *cs_gpiod; /* chip select gpio desc */
    struct spi_delay word_delay; /* inter-word delay */
    /* the statistics */
    struct spi_statistics statistics;
};
```

For the sake of readability, the number of fields listed is reduced to the strict minimum needed for the purpose of the book. The following list details the meaning of each element in this structure:

- `controller` represents the SPI controller this slave device belongs to. In other words, it represents the SPI controller (bus) on which the device is connected.

- The `master` element is still there for compatibility reasons and will be deprecated soon. It was the old name of the controller.

- `max_speed_hz` is the maximum clock rate to be used with this slave; this parameter can be changed from within the driver. We can override that parameter using `spi_transfer.speed_hz` for each transfer. We will discuss SPI transfer later.

- `chip_select` is the CS line assigned to this device. It is active low by default. This behavior can be changed in `mode` by adding the `SPI_CS_HIGH` flag.

- `rt`, if `true`, will make the message pump worker of the `controller` run as a real-time task

- `mode` defines how data should be clocked. The device driver may change this. The data clocking is MSB by default for each word in a transfer. This behavior can be overridden by specifying `SPI_LSB_FIRST`.

- `irq` represents the interrupt number (registered as a device resource in your board initialization file or through the device tree) you should pass to `request_irq()` to receive interrupts from this device.

- `cs_gpio` and `cs_gpiod` are both optional. The former is the legacy integer-based GPIO number of the CS line, while the latter is the new and recommended interface, based on the GPIO descriptor.

A word about SPI modes – they are built using two characteristics:

- CPOL, which is the initial clock polarity:

  - 0: The initial clock state is low, and the first edge is rising.

  - 1: The initial clock state is high, and the first state is falling.

- CPHA is the clock phase, determining at which edge the data will be sampled:

  - 0: Data is latched at the falling edge (high to low transition), whereas the output changes at the rising edge.

  - 1: Data is latched at rising edge (low to high transition), and the output changes at the falling edge.

This allows us to distinguish four SPI modes, which are derived macros made up of a mix of two main macros, defined in `include/linux/spi/spi.h` as follows:

```
#define    SPI_CPHA    0x01
#define    SPI_CPOL    0x02
```

The combinations of these macros give the following SPI modes:

Mode	CPOL	CPHA	Kernel macro	
0	0	0	`#define SPI_MODE_0`	`(0|0)`
1	0	1	`#define SPI_MODE_1`	`(0|SPI_CPHA)`
2	1	0	`#define SPI_MODE_2`	`(SPI_CPOL|0)`
3	1	1	`#define SPI_MODE_3`	`(SPI_CPOL|SPI_CPHA)`

Table 9.1 – SPI modes kernel definition

The following diagram is the representation of each SPI mode, in the same order as defined in the preceding array. That being said, only the MOSI line is represented, but the principle is the same for MISO.

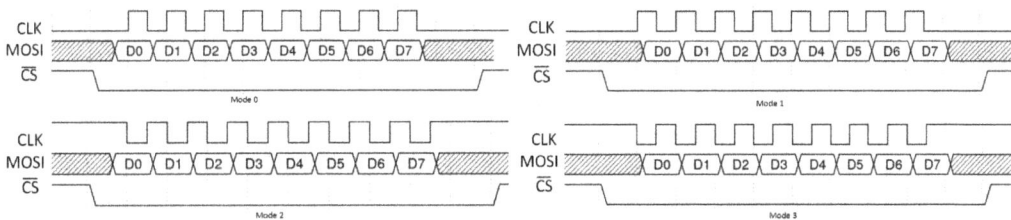

Figure 9.2 – SPI operating modes

Now that we are familiar with the SPI device data structure and the modes such a device can operate in, we can switch to the second-most important structure, the one representing the SPI device driver.

The spi_driver structure

Also called the protocol driver, an SPI device driver is responsible for driving devices sitting on the SPI bus. It is abstracted in the kernel by `struct spi_driver`, declared as follows:

```
struct spi_driver {
    const struct spi_device_id *id_table;
    int            (*probe)(struct spi_device *spi);
    int            (*remove)(struct spi_device *spi);
```

```
    void            (*shutdown)(struct spi_device *spi);
    struct device_driver    driver;
};
```

The following list outlines the meanings of the elements in this data structure:

- `id_table`: This is the list of SPI devices supported by this driver.

- `probe`: This method binds this driver to the SPI device. This function will be invoked on any device claiming this driver and will decide whether this driver is in charge of that device or not. If yes, the binding process occurs.

- `remove`: Unbinds this driver from the SPI device.

- `shutdown`: This is invoked during system state changes such as powering down and halting.

- `driver`: This is the low-level driver structure for the device and driver model.

This is all we can say for now on this data structure, except that each SPI device driver must fill and expose one instance of this type.

The message transfer data structures

The SPI I/O model consists of a set of queued messages, each of which can be made up of one or more SPI transfers. While a single message consists of one or more `struct spi_transfer` objects, each transfer represents a full duplex SPI transaction. Messages are submitted and processed either synchronously or asynchronously. The following is a diagram explaining the concept of message and transfer:

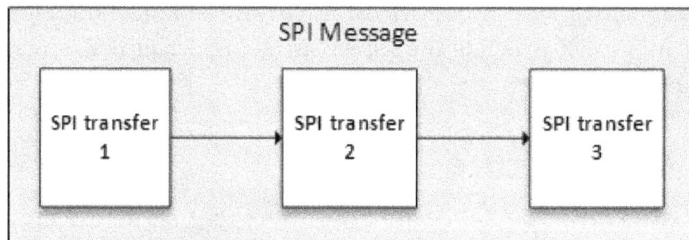

Figure 9.3 – Example SPI message structure

Now that we are familiar with the theoretical aspects, we can introduce the SPI transfer data structure, declared as follows:

```
struct spi_transfer {
    const void    *tx_buf;
```

```
    void           *rx_buf;
    unsigned       len;

    dma_addr_t     tx_dma;
    dma_addr_t     rx_dma;
    struct sg_table tx_sg;
    struct sg_table rx_sg;

    unsigned       cs_change:1;
    unsigned       tx_nbits:3;
    unsigned       rx_nbits:3;
#define     SPI_NBITS_SINGLE 0x01 /* 1bit transfer */
#define     SPI_NBITS_DUAL        0x02 /* 2bits transfer */
#define     SPI_NBITS_QUAD        0x04 /* 4bits transfer */
    u8         bits_per_word;
    u16        delay_usecs;
    struct     spi_delay     delay;
    struct spi_delay  cs_change_delay;
    struct spi_delay  word_delay;
    u32        speed_hz;
    u32        effective_speed_hz;
[...]
    struct list_head transfer_list;
#define SPI_TRANS_FAIL_NO_START  BIT(0)
    u16        error;
};
```

The following are the meanings of each element in the data structure:

- tx_buf is a pointer to the buffer that contains the data to be written. If set to NULL, this transfer will be considered as half duplex as a read-only transaction. It should be DMA-safe when you need to perform an SPI transaction through DMA.

- rx_buf is a buffer for data to be read (with the same properties as tx_buf), or NULL in a write-only transaction.

- tx_dma is the **Direct Memory Access (DMA)** address of tx_buf, in case spi_message.is_dma_mapped is set to 1.

- rx_dma is the same as tx_dma, but for rx_buf.

- `len` represents the size of the `rx` and `tx` buffers in bytes. Only `len` bytes shift out (or in) and attempting to shift out a partial word would result in an error.

- `speed_hz` supersedes the default speed specified in `spi_device.max_speed_hz`, but only for the current transfer. If 0, the default (from `spi_device`) is used.

- `bits_per_word`: A data transfer involves one or more words. A word is a unit of data whose size in bits varies according to the needs. Here, `bits_per_word` represents the size in bits of a word for this SPI transfer. This overrides the default value provided in `spi_device.bits_per_word`. If 0, the default (from `spi_device`) is used.

- `cs_change` determines whether the CS becomes inactive after this transfer completes. All SPI transfers begin with the appropriate CS signal active. Normally, it remains selected until the last transfer in the message is completed. Using `cs_change`, drivers can change the CS signal.

  This flag is used to make the CS temporarily inactive in the middle of the message (that is, before processing the `spi_transfer` on which it is specified) if the transfer isn't the last one in the message. Toggling CS in this way may be required to complete a chip command, allowing a single SPI message to handle the entire set of chip transactions.

- `delay_usecs` represents the delay (in microseconds) following this transfer before (optionally) changing the `chip_select` status, then starting the next transfer or completing this `spi_message`.

> **Note**
>
> SPI transfers always write the same number of bytes as they read, even in half-duplex transactions. The SPI core achieves this thanks to the controller's `dummy_rx` and `dummy_tx` elements. When the transmit buffer is null, `spi_transfer->tx_buf` will be set with the controller's `dummy_tx`. Then, zeroes will be shifted out while filling `rx_buf` with the data coming from the slave. If the receive buffer is null, `spi_transfer->rx_buf` will be set with the controller's `dummy_rx` and the data shifted in will be discarded.

struct spi_message

spi_message is used to atomically issue a sequence of transfers, each represented by a struct spi_transfer instance. We say *atomically* because no other spi_message may use that SPI bus until the ongoing sequence completes. Do however note that there are platforms that can handle many such sequences with a single programmed DMA transfer. An SPI message structure has the following declaration:

```
struct spi_message {
        struct list_head        transfers;
        struct spi_device       *spi;
        unsigned        is_dma_mapped:1;
        /* completion is reported through a callback */
        void                    (*complete)(void *context);
        void                    *context;
        unsigned        frame_length;
        unsigned        actual_length;
        int                     status;
    };
```

The following list outlines the meanings of elements in this data structure:

- transfers is the list of transfers that constitute the message. We will see later how to add a transfer to this list. Using the spi_transfer.cs_change flag on the last transfer in that atomic group may potentially save costs for chip deselect and select operations.

- is_dma_mapped informs the controller whether to use DMA (or not) to perform the transaction. Your code is then responsible for providing DMA and CPU virtual addresses for each transfer buffer.

- complete is a callback called when the transaction is done, and context is the parameter to be given to the callback.

- frame_length will be set automatically with the total number of bytes in the message.

- actual_length is the number of bytes transferred in all successful segments.

- status reports the transfer's status. This is 0 on success; otherwise, it's -errno.

`spi_transfer` elements in a message are processed in FIFO order. Until the message is completed (that is, until the completion callback is executed), you must make sure not to use transfer buffers in order to avoid data corruption. The code that submits a `spi_message` (and its `spi_transfers`) to the lower layers is responsible for managing its memory. Drivers must ignore the message (and its transfers) once submitted at least until its completion callback is invoked.

Accessing the SPI device

An SPI controller is able to communicate with one or more slaves, that is, one or more `struct spi_device`. They form a tiny bus that shares MOSI, MISO, and SCK signals but not CS. Because those shared signals are ignored unless the chip is selected, each device can be programmed to utilize a different clock rate. The SPI controller driver manages communication with those devices through a queue of `spi_message` transactions, moving data between CPU memory and an SPI slave device. For each message instance it queues, it calls the message's completion callback when the transaction completes.

Before a message can be submitted to the bus, it has to be initialized with `spi_message_init()`, which has the following prototype:

```
void spi_message_init(struct spi_message *message)
```

This function will zero each element in the structure and initialize the transfers list. For each transfer to be added to the message, you should call `spi_message_add_tail()` on that transfer, which will result in enqueuing the transfer into the message's transfers list. It has the following declaration:

```
spi_message_add_tail(struct spi_transfer *t, struct spi_message
*m)
```

Once this is done, you have two choices to start the transaction:

- **Synchronously**, using `int spi_sync(struct spi_device *spi, struct spi_message *message)`, which returns 0 on success, else a negative error code. This function may sleep and is not to be used in interrupt contexts. Do however note that this function may sleep in a non-interruptible manner, and does not allow specifying a timeout. A DMA-capable controller's driver may leverage this DMA feature to push/pull data directly into/from the message buffers.

The SPI device's CS is activated by the core during an entire message (from the first transfer to the last), and is then normally disabled between messages. There are drivers which, in order to minimize the impacts of selecting a chip (to save power for example), leave it selected, anticipating that the next message will go to the same chip.

- **Asynchronously**, using the `spi_async()` function, which can be used in an any context (atomic or not), and whose prototype is `int spi_async(struct spi_device *spi, struct spi_message *message)`. This function is context agnostic since only submission is done and the processing is asynchronous. However, the completion callback is invoked in a context that can't sleep. Before this callback is invoked, the value of `message->status` is undefined. At the time it is invoked, `message->status` holds the completion status, which is either `0` (to indicate complete success) or a negative error code.

After that callback returns, the driver that initiated the transfer request may deallocate the associated memory since it's no longer in use by any SPI core or controller driver code. Until the completion callback of the currently processed message returns, no subsequent `spi_message` queued to that device will be processed. This rule applies to synchronous transfer calls as well, since they are wrappers around this core asynchronous primitive. This function returns `0` on success, else a negative error code.

The following is an excerpt from a driver demonstrating SPI message and transfer initialization and submission:

```
Static int regmap_spi_gather_write(
                    void *context, const void *reg,
                    size_t reg_len, const void *val,
                    size_t val_len)
{
    struct device *dev = context;
    struct spi_device *spi = to_spi_device(dev);
    struct spi_message m;
    u32 addr;
    struct spi_transfer t[2] = {
       { .tx_buf = &addr, .len = reg_len, .cs_change = 0,},
       { .tx_buf = val, .len = val_len, },
    };
```

```
    addr = TCAN4X5X_WRITE_CMD  |
              (*((u16 *)reg) << 8) | val_len >> 2;

    spi_message_init(&m);
    spi_message_add_tail(&t[0], &m);
    spi_message_add_tail(&t[1], &m);
    return spi_sync(spi, &m);
}
```

The preceding excerpt however shows static initialization, on the fly, where both messages and transfers are discarded on the return path of the function. There may be cases where the driver would like to pre-allocate messages along with their transfers for the lifetime of the driver in order to avoid a frequent initialization overhead. In such cases, dynamic allocation can be used thanks to spi_message_alloc(), and freed using spi_message_free(). They have the following prototypes:

```
struct spi_message *spi_message_alloc(unsigned ntrans,
                                        gfp_t flags)
void spi_message_free(struct spi_message *m)
```

In the preceding snippet, ntrans is the number of transfers to allocate for this new spi_message, and flags represents the flags for the freshly allocated memory, where using GFP_KERNEL is enough. On success, this function returns the new allocated message structure along with its transfers. You can access transfer elements using kernel list-related macros such as list_first_entry, list_next_entry, or even list_for_each_entry. The following is an example showing the usage of these macros:

```
/* Completion handler for async SPI transfers */
static void my_complete(void *context)
{
    struct spi_message *msg = context;
    /* doing some other stuffs */
    [...]
    spi_message_free(m);
}

static int example_spi_async(struct spi_device *spi,
            struct my_fake_spi_reg *cmds, unsigned len)
```

```
{
    struct spi_transfer *xfer;
    struct spi_message *msg;

    msg = spi_message_alloc(len, GFP_KERNEL);
    if (!msg)
        return -ENOMEM;

    msg->complete = my_complete;
    msg->context = msg;

    list_for_each_entry(xfer, &msg->transfers,
                transfer_list) {
        xfer->tx_buf = (u8 *)cmds;
        /* feel free to handle .rx_buf, and so on */
        [...]
        xfer->len = 2;
        xfer->cs_change = true;
        cmds++;
    }

    return spi_async(spi, msg);
}
```

In the preceding excerpt, we have not only shown how to use dynamic message and transfer allocation. We have also seen how `spi_async()` is used. This example is quite useless since the allocated message and transfers are immediately freed upon completion. A best practice with dynamic allocation is to allocate Tx and Rx buffers dynamically as well, and keep them within arm's reach for the lifetime of the driver.

Note however that the device driver is responsible for organizing the messages and transfer in the most appropriate way for the device, as follows:

- When bidirectional reads and writes start and how its sequence of `spi_transfer` requests are arranged

- I/O buffer preparation, knowing that each `spi_transfer` wraps a buffer for each transfer direction, supporting full duplex transfers (even if one pointer is NULL, in which case the controller will use one of its dummy buffers)

- Optionally using `spi_transfer.delay_usecs` to define short delays after transfers
- Whether CS should change (becoming inactive) after a transfer and any delay by using the `spi_transfer.cs_change` flag

With `spi_async`, the device driver queues the messages, registers a completion callback, wakes the message pump, and immediately returns. The completion callback will be invoked when the transfers are complete. Because neither message queuing nor message pump scheduling can block, the `spi_async` function is considered context agnostic. However, it requires that you wait for the completion callback before you can access the buffers in the `spi_transfer` pointers you submitted. On the other hand, `spi_sync` queues the messages and blocks until they are complete. It does not require completion callback. When `spi_sync` returns, it is safe to access your data buffers. If you look at its implementation in `drivers/spi/spi.c`, you'll see it uses `spi_async` to put the calling thread to sleep until the completion callback is called. Since the 4.0 kernel there has been an improvement for `spi_sync` where, when there is nothing in the queue, the message pump will get executed in the context of the caller instead of the message pump thread, which avoids the cost of a context switch.

After the most important data structures and APIs of the SPI framework have been introduced, we can discuss the real driver implementation.

Dealing with the SPI driver abstraction and architecture

This is where the driver logic takes place. It consists of filling `struct spi_driver` with a set of driving functions that allow probing and controlling the underlying device.

Probing the device

The SPI device is probed by the `spi_driver.probe` callback. The probe callback is responsible for making sure the driver recognizes the given device before they can be bound together. This callback has the following prototype:

```
int probe(struct spi_device *spi)
```

This method must return 0 on success, or a negative error number otherwise. The only argument is the SPI device to be probed, whose structure has been pre-initialized by the core according to its description in the device tree.

However, most (if not all) of the properties of the SPI device can be overridden, as we have seen while describing its data structure. SPI protocol drivers may need to update the transfer mode if the device doesn't work with its default. They may likewise need to update clock rates or word sizes from their initial values. This is possible thanks to the spi_setup() helper, which has the following prototype:

```
int spi_setup(struct spi_device * spi)
```

This function must be called from a context that can sleep exclusively. It expects as a parameter an SPI device structure whose properties to override must have been set in their respective fields. Changes will be effective at the next device access (either for a read or write operation after it has been selected) except for SPI_CS_HIGH, which takes effect immediately. The SPI device is deselected on the return path of this function. This function returns 0 on success or a negative error on failure. It is worth paying attention to its return value because this call won't succeed if the driver provides an option that is not supported by the underlying controller or its driver. For instance, some hardware handles wire transfers using nine-bit words, **least significant bit** (**LSB**)-first wire encoding, or active-high CS, and others do not.

You likely want to call spi_setup() from probe() before submitting any I/O request to the device. However, it can be called anywhere in the code provided no message is pending for that device.

The following is a probing example that sets up the SPI device, checks its family ID, and returns 0 (device recognized) on success:

```
#define FAMILY_ID 0x57
static int fake_probe(struct spi_device *spi)
{
    int err;
    u8 id;

    spi->max_speed_hz =
              min(spi->max_speed_hz, DEFAULT_FREQ);
    spi->bits_per_word = 8;
    spi->mode = SPI_MODE_0;
    spi->rt = true;
    err = spi_setup(spi);
    if (err)
        return err;
```

```
    /* read family id */
    err = get_chip_version(spi, &id);
    if (err)
        return -EIO;

    /* verify family id */
    if (id != FAMILY_ID) {
        dev_err(&spi->dev"
    "chip family: expected 0x%02x but 0x%02x rea"\n",
            FAMILY_ID, id);
        return -ENODEV;
    }

    /* register with other frameworks */
    [...]

    return 0;
}
```

A real probing method would also probably deal with some driver state data structures or other per-device data structures. Regarding the get_chip_version() function, it may have the following body:

```
#define REG_FAMILY_ID 0x2445
#define DEFAULT_FREQ 10000000

static int get_chip_version(spi_device *spi, u8 *id)
{
    struct spi_transfer t[2];
    struct spi_message m;
    u16 cmd;
    int err;

    cmd = REG_FAMILY_ID;
    spi_message_init(&m);
    memset(&t, 0, sizeof(t));
```

```
    t[0].tx_buf = &cmd;
    t[0].len = sizeof(cmd);
    spi_message_add_tail(&t[0], &m);

    t[1].rx_buf = id;
    t[1].len = 1;
    spi_message_add_tail(&t[1], &m);

    return spi_sync(spi, &m);
}
```

Now that we have seen how to probe an SPI device, it will be useful to discuss how to tell the SPI core which devices the driver can support.

> **Note**
>
> The SPI core allows setting/getting driver state data using spi_get_drvdata() and spi_set_drvdata() in the same way as we did while discussing I2C device drivers in *Chapter 8, Writing I2C Device Drivers*.

Provisioning devices in the driver

As we need a list of i2c_device_id to tell I2C core what devices an I2C driver can support, we must provide an array of spi_device_id to inform the SPI core what devices our SPI driver supports. After that array has been filled, it must be assigned to the spi_driver.id_table field. Additionally, for device matching and module loading purposes, this same array needs to be given to the MODULE_DEVICE_TABLE macro. struct spi_device_id has the following declaration in include/linux/mod_devicetable.h:

```
struct spi_device_id {
    char name[SPI_NAME_SIZE];
    kernel_ulong_t driver_data;
};
```

In the preceding data structure, `name` is a descriptive name for the device, and `driver_data` is the driver state value. It can be set with a pointer to a per-device data structure. The following is an example:

```
#define ID_FOR_FOO_DEVICE    0
#define ID_FOR_BAR_DEVICE    1


static struct spi_device_id foo_idtable[] = {
  "{ ""oo", ID_FOR_FOO_DEVICE },
  "{ ""ar", ID_FOR_BAR_DEVICE },
   { },
};
MODULE_DEVICE_TABLE(spi, foo_idtable);
```

To be able to match the devices declared in the device tree, we need to define an array of `struct of_device_id` elements and both assign it to `spi_driver.of_match_table` and call the `MODULE_DEVICE_TABLE` macro on it. The following is an example, which also shows what the final `spi_driver` structure would look like when it is set up:

```
static const struct of_device_id foobar_of_match[] = {
        { .compatible"= "packtpub,foobar-dev"ce" },
        { .compatible"= "packtpub,barfoo-dev"ce" },
        {},
};
MODULE_DEVICE_TABLE(of, foobar_of_match);
```

The following excerpt shows the final `spi_driver` content:

```
static struct spi_driver foo_driver = {
    .driver         = {
        .name   "= ""oo",
        /* The below line adds Device Tree support */
        .of_match_table = of_match_ptr(foobar_of_match),
    },
    .probe          = my_spi_probe,
    .id_table       = foo_idtable,
};
```

In the preceding, we can see what an SPI driver structure looks like after it has been set up. There is however a missing element, the `spi_driver.remove` callback, which is used to undo what was done in the probing function.

Implementing the spi_driver.remove method

The `remove` callback must be used to release every resource grabbed and undo what was done at probing. This callback has the following prototype:

```
static int remove(struct spi_device *spi)
```

In the preceding snippet, `spi` is the SPI device data structure, the same given to the `probe` callback, which simplifies device state data structure tracking between the probing and the removal of the device. This method returns `0` on success or a negative error code on failure. You must make sure that the device is left in a coherent and stable state as well. The following is an example implementation:

```
static int mc33880_remove(struct spi_device *spi)
{
    struct mc33880 *mc;
    mc = spi_get_drvdata(spi); /* Get our data back */
    if (!mc)
        return -ENODEV;

    /*
     * unregister from frameworks with which we
     * registered in the probe function
     */
    gpiochip_remove(&mc->chip);
    [...]
    /* releasing any resource */
    mutex_destroy(&mc->lock);
    return 0;
}
```

In the preceding example, the code dealt with unregistering from the frameworks and releasing the resources. This is the classic case that you will face in 90% of cases.

Driver initialization and registration

At this step of implementation, your code is almost ready, and you would like to inform the SPI core of your SPI driver. This is driver registration. For SPI device drivers, the SPI core provides `spi_register_driver()` and `spi_unregister_driver()` both to register and unregister and SPI device driver with the SPI core. Those methods have the following prototypes:

```
int spi_register_driver(struct spi_driver *sdrv);
void spi_unregister_driver(struct spi_driver *sdrv);
```

In both functions, `sdrv` is the SPI driver structure that has been previously set up. The registration API returns zero on success or a negative error code on failure.

Driver registration and unregistering usually take place in the module initialization and module exit method. The following is a typical demonstration of SPI driver registration:

```
static int __init foo_init(void)
{
    [...] /*My init code */
    return spi_register_driver(&foo_driver);
}
module_init(foo_init);

static void __exit foo_cleanup(void)
{
    [...] /* My clean up code */
    spi_unregister_driver(&foo_driver);
}
module_exit(foo_cleanup);
```

If you do nothing at module initialization other than registering/unregistering the driver, you can use `module_spi_driver()` to factor your code as follows:

```
module_spi_driver(foo_driver);
```

This macro will populate module initialization and cleanup functions and will call `spi_register_driver` and `spi_unregister_driver` inside.

Instantiating SPI devices

SPI slave nodes must be children of the SPI controller node. In master mode, one or more slave nodes (up to the number of CSs) can be present.

The required properties are the following:

- `compatible`: The compatible string as defined in the driver for matching
- `reg`: The CS index of the device relative to the controller
- `spi-max-frequency`: The maximum SPI clocking speed of the device in Hz

All slave nodes can contain the following optional properties:

- `spi-cpol`: Boolean property which, if present, indicates that the device requires inverse **clock polarity** (**CPOL**) mode.
- `spi-cpha`: Boolean property indicating that this device requires shifted **clock phase** (**CPHA**) mode.
- `spi-cs-hi-h`: Empty property indicating that the device requires CS active high.
- `spi-3wire`: This is a Boolean property that indicates that this device requires 3-wire mode to work properly.
- `spi-lsb-first`: This is a Boolean property that indicates that this device requires LSB first mode.
- `spi-tx-bus-width`: This property indicates the bus width used for MOSI. If not present, it defaults to 1.
- `spi-rx-bus-width`: This property is used to indicate the bus width used for MISO. If not present, it defaults to 1.
- `spi-rx-delay--s`: This is used to specify a delay in microseconds after a read transfer.
- `spi-tx-delay-us`: This is used to specify a delay in microseconds after a write transfer.

The following is a real device tree listing for SPI devices:

```
ecspi1 {
    fsl,spi-num-CSs = <3>;
    cs-gpios = <&gpio5 17 0>, <&gpio5 17 0>, <&gpio5 17 0>;
    pinctrl-0 = <&pinctrl_ecspi1 &pinctrl_ecspi1_cs>;
    #address-cells = <1>;
```

```
    #size-cells = <0>;
    compatible"= "fsl,imx6q-ec"pi", "fsl,imx51-ec"pi";
    reg = <0x02008000 0x4000>;
    status"= "o"ay";


    ad7606r8_0: ad7606r8@0 {
        compatible"= "ad760"-8";
        reg = <0>;
        spi-max-frequency = <1000000>;
        interrupt-parent = <&gpio4>;
        interrupts = <30 0x0>;
    };
    label: fake_spi_device@1 {
        compatible"= "packtpub,foobar-dev"ce";
        reg = <1>;
        a-string-param"= "stringva"ue";
        spi-cs-high;
    };
    mcp2515can: can@2 {
        compatible"= "microchip,mcp2"15";
        reg = <2>;
        spi-max-frequency = <1000000>;
        clocks = <&clk8m>;
        interrupt-parent = <&gpio4>;
        interrupts = <29 IRQ_TYPE_LEVEL_LOW>;
    };
};
```

In the preceding device tree excerpt, ecspi1 represents the master SPI controller.
fake_spi_device and mcp2515can represent SPI slave devices, and their reg
properties represents their respective CS indices relative to the master.

Now that we are familiar with all the kernel aspects of the SPI slave-oriented framework,
let's see how we might avoid dealing with the kernel and try to implement everything in
the user space.

Learning how not to write SPI device drivers

The usual way to deal with SPI devices is to write kernel code to drive this device. Nowadays the `spidev` interface makes it possible to deal with such devices without even writing a line of kernel code. The use of this interface should be limited, however, to simple use cases such as talking to a slave microcontroller or for prototyping. Using this interface, you will not be able to deal with various **interrupts** (**IRQs**) the device may support nor leverage other kernel frameworks.

The `spidev` interface exposes a character device node in the form `/dev/spidevX.Y` where `X` represents the bus our device sits on, and `Y` represents the CS index (relative to the controller) assigned to the device node in the device tree. For example, `/dev/spidev1.0` means device `0` on SPI bus `1`. The same applies to the sysfs directory entry, which would be in the form `/sys/class/spidev/spidevX.Y`.

Prior to the character device appearing in the user space, the device node must be declared in the device tree as a child of the SPI controller node. The following is an example:

```
&ecspi2 {
    pinctrl-names"= "defa"lt";
    pinctrl-0 = <&pinctrl_teoulora_ecspi2>;
    cs-gpios = <&gpio2 26 1
                &gpio2 27 1>;
    num-cs = <2>;
    status"= "o"ay";

    spidev@0 {
        reg = <0>;
        compatib"e="semtech,sx1"01";
        spi-max-frequency = <20000000>;
    };
};
```

In the preceding snippet, `spidev@0` corresponds to our SPI device node. `reg = <0>` tells the controller that this device is using the first CS line (index starting from 0). The `compatible="semtech,sx1301"` property is used to match an entry in the `spidev` driver. It is no longer recommended to use `"spidev"` as a compatible string – you'll get a warning if you try. Finally, `spi-max-frequency = <20000000>` sets the default clock speed (20 MHz in this case) that our device will operate at, unless it is changed using the appropriate API.

From the user space, the required header files to deal with the `spidev` interface are as follows:

```
#include <fcntl.h>
#include <unistd.h>
#include <sys/ioctl.h>
#include <linux/types.h>
#include <linux/spi/spidev.h>
```

Because it is a character device, it is allowed (this is the only option, in fact) to use basic system calls such as `open()`, `read()`, `write()`, `ioctl()`, and `close()`. The following example shows some basic usage, with `read()` and `write()` operations only:

```c
#include <stdio.h>
#include <stdlib.h>

int main(int argc, char **argv)
{
    int i,fd;
    char *device = "/dev/spidev0.0";
    char wr_buf[]={0xff,0x00,0x1f,0x0f};
    char rd_buf[10];


    fd = open(device, O_RDWR);
    if (fd <= 0) {
        printf("Failed to open SPI device %s\n", device);
        exit(1);
    }

    if (write(fd, wr_buf, sizeof(wr_buf)) != sizeof(wr_buf))
        perror("Write Error");
    if (read(fd, rd_buf, sizeof(rd_buf)) != sizeof(rd_buf))
        perror("Read Error");
    else
        for (i = 0; i < sizeof(rd_buf); i++)
            printf("0x%02X ", rd_buf[i]);
    close(fd);
```

```
    return 0;
}
```

In the preceding code, you should note that the standard `read()` and `write()` operations are half-duplex only, and that the CS is deactivated between each operation. To be able to work in full duplex, you have no choice but to use the `ioctl()` interface, where you can pass both input and output buffers at your convenience. Moreover, with the `ioctl()` interface, you can use a set of `SPI_IOC_RD_*` and `SPI_IOC_WR_*` commands to get RD and set WR to override the device's current setting. The complete list and documentation for this can be found in `Documentation/spi/spidev` in the kernel sources.

The `ioctl()` interface allows composite operations without CS deactivation and is available using the `SPI_IOC_MESSAGE(N)` request. A new data structure takes place, the `struct spi_ioc_transfer`, which is the user space equivalent of `struct spi_transfer`. The following is an example of the ioctl commands:

```
#include <stdint.h>
#include <stdio.h>
#include <stdlib.h>
#include <string.h>
/* include required headers, listed early in the section */
[...]
static int pabort(const char *s)
{
    perror(s);
    return -1;
}

static int spi_device_setup(int fd)
{
    int mode, speed, a, b, i;
    int bits = 8;

    /* spi mode: mode 0 */
    mode = SPI_MODE_0;
    a = ioctl(fd, SPI_IOC_WR_MODE, &mode); /* set mode */
    b = ioctl(fd, SPI_IOC_RD_MODE, &mode); /* get mode */
    if ((a < 0) || (b < 0)) {
```

```
        return pabort("can't set spi mode");
    }


    /* Clock max speed in Hz */
    speed = 8000000; /* 8 MHz */
    a = ioctl(fd, SPI_IOC_WR_MAX_SPEED_HZ, &speed); /* set */
    b = ioctl(fd, SPI_IOC_RD_MAX_SPEED_HZ, &speed); /* get */
    if ((a < 0) || (b < 0))
        return pabort("fail to set max speed hz");
    /*
     * Set SPI to MSB first.
     * Here, 0 means "not to use LSB first".
     * To use LSB first, argument should be > 0
     */
    i = 0;
    a = ioctl(dev, SPI_IOC_WR_LSB_FIRST, &i);
    b = ioctl(dev, SPI_IOC_RD_LSB_FIRST, &i);
    if ((a < 0) || (b < 0))
        pabort("Fail to set MSB first\n");


    /* setting SPI to 8 bits per word */
    bits = 8;
    a = ioctl(dev, SPI_IOC_WR_BITS_PER_WORD, &bits); /* set */
    b = ioctl(dev, SPI_IOC_RD_BITS_PER_WORD, &bits); /* get */
    if ((a < 0) || (b < 0))
        pabort("Fail to set bits per word\n");


    return 0;
}
```

In the preceding example, getters are used for demonstration purposes only. It is not mandatory to issue the SPI_IOC_RD_* command after you have executed its SPI_IOC_WR_* equivalent. Now that we have seen most of those ioctl commands, let's see how to start transfers:

```
static void do_transfer(int fd)
{
```

```c
    int ret;
    char txbuf[] = {0x0B, 0x02, 0xB5};
    char rxbuf[3] = {0, };
    char cmd_buff = 0x9f;

    struct spi_ioc_transfer tr[2] = {
        0 = {
            .tx_buf = (unsigned long)&cmd_buff,
            .len = 1,
            .cs_change = 1;    /* We need CS to change */
            .delay_usecs = 50, /* wait after this transfer */
            .bits_per_word = 8,
        },
        [1] = {
            .tx_buf = (unsigned long)tx,
            .rx_buf = (unsigned long)rx,
            .len = txbuf(tx),
            .bits_per_word = 8,
        },
    };

    ret = ioctl(fd, SPI_IOC_MESSAGE(2), &tr);
    if (ret == 1){
        perror("can't send spi message");
        exit(1);
    }

    for (ret = 0; ret < sizeof(tx); ret++)
        printf("%.2X ", rx[ret]);
    printf("\n");
}
```

The preceding shows the concept of message and transfer transactions in the user space. Now that our helpers have been defined, we can write the main code to use them as follows:

```
int main(int argc, char **argv)
{
    char *device = "/dev/spidev0.0";
    int fd;
    int error;

    fd = open(device, O_RDWR);
    if (fd < 0)
        return pabort("Can't open device ");

    error = spi_device_setup(fd);
    if (error)
        exit (1);

    do_transfer(fd);

    close(fd);
    return 0;
}
```

We are now done with the main function. This section taught us to use the user space SPI APIs and commands to interact with the device. We are limited, however, in that we can't take advantage of device interrupt lines or other kernel frameworks.

Summary

In this chapter, we tackled SPI drivers and can now take the advantage of this serial (and full duplex) bus, which is way faster than I2C. We walked through all the data structures in this framework and discussed transferring over SPI, which is the most important section we covered. That said, the memory we accessed over those buses was off-chip – we may need more abstraction in order to avoid the SPI and I2C APIs.

This is where the next chapter comes in, dealing with the regmap API, which offers a higher and more unified level of abstraction so that SPI (and I2C) commands will become transparent to you.

Section 3 - Making the Most out of Your Hardware

This section will deal with Linux kernel memory management, advanced concepts in kernel interrupt management, and direct memory access. Then, in order to simplify memory access operations, we will use the memory access abstraction implemented in the Linux kernel, the regmap. At the end of this section, we will introduce the Linux device model to have a better understanding and overview of the device hierarchy on the system.

The following chapters will be covered in this section:

- *Chapter 10, Understanding the Linux Kernel Memory Allocation*
- *Chapter 11, Implementing Direct Memory Access (DMA) Support*
- *Chapter 12, Abstracting Memory Access – Introduction to the Regmap API: a Register Map Abstraction*
- *Chapter 13, Demystifying the Kernel IRQ Framework*
- *Chapter 14, Introduction to the Linux Device Model*

10
Understanding the Linux Kernel Memory Allocation

Linux systems use an illusion referred to as "virtual memory." This mechanism makes every memory address virtual, which means they do not point to any address in the RAM directly. This way, whenever we access a memory location, a translation mechanism is performed in order to match the corresponding physical memory.

In this chapter, we will deal with the whole Linux memory allocation and management system, covering the following topics:

- An introduction to Linux kernel memory-related terms
- Demystifying address translation and MMU
- Dealing with memory allocation mechanisms
- Working with I/O memory to talk with hardware
- Memory remapping

An introduction to Linux kernel memory-related terms

Though system memory (also known as RAM) can be extended in some computers that allow it, physical memory is a limited resource in computer systems.

Virtual memory is a concept, an illusion given to each process so that it thinks it has large and almost infinite memory, and sometimes more than the system really has. To set up everything, we will introduce the address space, virtual or logical address, physical address, and bus address terms:

- A physical address identifies a physical (RAM) location. Because of the virtual memory mechanism, the user or the kernel never directly deals with the physical address but can access it by its corresponding logical address.

- A virtual address does not necessarily exist physically. This address is used as a reference to access the physical memory location by CPU on behalf of the **Memory Management Unit** (**MMU**). The MMU sits between the CPU core and memory and is most often part of the physical CPU itself. That said, on ARM architectures, it's part of the licensed core. It is then responsible for converting virtual addresses into physical addresses every time memory locations are accessed. This mechanism is called **address translation**.

- A logical address is an address resulting from a linear mapping. It results from a mapping above PAGE_OFFSET. Such addresses are virtual addresses with a fixed offset from their physical addresses. Thus, a logical address is always a virtual address, and the reverse is not true.

- In computer systems, an address space is the amount of memory allocated for all possible addresses for a computational entity (in our case, the CPU). This address space may be virtual or physical. While physical address space can go up to the amount of RAM installed on the system (theoretically limited by the width of the CPU address bus and registers), the range of virtual addresses can extend to the highest address permitted by the RAM or by the operating system architecture (such as addressing up to 4 GB of virtual memory on a 1 GB RAM system).

As the MMU is the centerpiece of memory management, it organizes memory into logical units of fixed size called **pages**. The size of a page is a power of 2 in bytes and varies among systems. A page is backed by a page frame, and the size of the page matches a page frame. Before going further in our learning of memory management, let's introduce the other terms:

- A memory page, virtual page, or simply a page are terms used to refer to a fixed-length (PAGE_SIZE) block of virtual memory. The same term page is used as a kernel data structure to represent a memory page.

- On the other hand, a frame (or page frame) refers to a fixed-length block of physical memory (RAM) on top of which the operating system maps a page. The size of the page matches a page frame. Each page frame is given a number, called the **Page Frame Number (PFN)**.

- Then comes the term **page table**, which is a kernel and architecture data structure used to store the mappings between virtual addresses and physical addresses. The key pair page/frame describes a single entry in the page table and represents a mapping.

  Finally, the term "page-aligned" is used to qualify an address that starts exactly at the beginning of a page. It goes without saying that any memory whose address is a multiple of the system page size is said to be page-aligned. For example, on a 4 KB page size system, 4.096, 20.480, and 409.600 are instances of page-aligned memory addresses.

> **Note**
> The page size is fixed by the MMU, and the operating system can't modify it. Some processors allow for multiple page sizes (for example, ARMv8-A supports three different granule sizes: 4 KB, 16 KB, and 64 KB), and the OS can decide which one to use. 4 KB is a widely popular page granularity though.

Now that the terms frequently used while dealing with memory have been introduced, let's focus on memory management and organization by the kernel.

Linux is a virtual memory operating system. On a Linux running system, each process and even the kernel itself (and some devices) is allocated address space, which is some portion of the processor's virtual address space (note that neither the kernel nor processes deal with physical addresses – only MMU does). While this virtual address space is split between the kernel and user space, the upper part is used for the kernel and the lower part is used for user space.

The split varies between architectures and is held by the CONFIG_PAGE_OFFSET kernel configuration option. For 32-bit systems, the split is at 0xC0000000 by default. This is called the 3 GB/1 GB split, where the user space is given the lower 3 GB of virtual address space. The kernel can, however, be given a different amount of address space as desired by playing with the CONFIG_VMSPLIT_1G, CONFIG_VMSPLIT_2G, and CONFIG_VMSPLIT_3G_OPT kernel configuration options (see arch/x86/Kconfig and arch/arm/Kconfig). For 64-bit, the split varies by architecture, but it's high enough: 0x8000000000000000 for 64-bit ARM, and 0xffff880000000000 for x86_64.

A typical process's virtual address space layout looks like the following on a 32-bit system with the default splitting scheme:

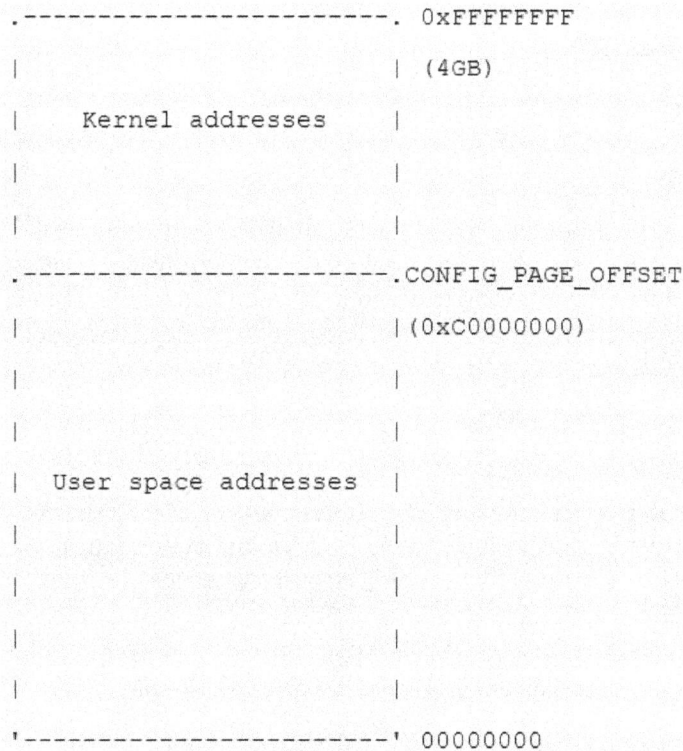

```
.-------------------------. 0xFFFFFFFF
|                         |
|                         | (4GB)
|    Kernel addresses     |
|                         |
|                         |
|                         |
.-------------------------.CONFIG_PAGE_OFFSET
|                         |
|                         |(0xC0000000)
|                         |
|                         |
|                         |
| User space addresses    |
|                         |
|                         |
|                         |
|                         |
|                         |
'-------------------------' 00000000
```

Figure 10.1 – 32-bit system memory splitting

While this layout is transparent on 64-bit systems, there are particularities on 32-bit machines that need to be introduced. In the next sections, we will study in detail the reason for this memory splitting, its usage, and where it applies.

Kernel address space layout on 32-bit systems – the concept of low and high memory

In an ideal world, all memory is permanently mappable. There are, however, some restrictions on 32-bit systems. This results in only a portion of RAM being permanently mapped. This part of memory can be accessed directly (by simple dereference) by the kernel and is called **low memory**, while the part of (physical) memory not covered by a permanent mapping is referred to as **high memory**. There are various architecture-dependent constraints on where exactly that border lies. For example, Intel cores can permanently map only up to the first 1 GB of RAM. This is a little bit less, 896 MiB of RAM, because part of this low memory is used to dynamically map high memory:

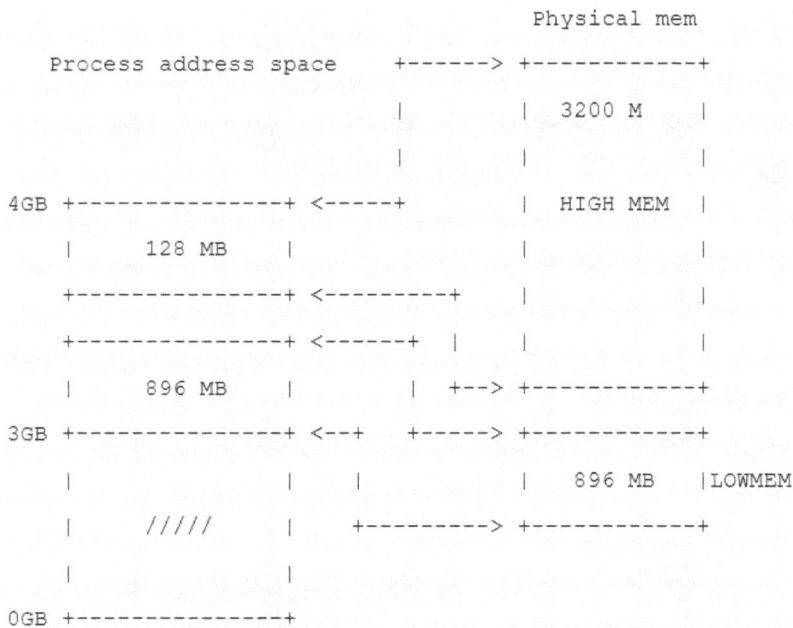

```
                                          Physical mem

      Process address space      +------> +------------+
                                 |        |  3200 M    |
                                 |        |            |
 4GB +---------------+ <-----+   |        | HIGH MEM   |
     |     128 MB    |       |   |        |            |
     +---------------+ <---------+   |    |            |
     +---------------+ <------+   |  |    |            |
     |     896 MB    |        |  | +--> +------------+
 3GB +---------------+ <--+   +-----> +------------+
     |               |    |   |        | 896 MB     |LOWMEM
     |   /////       |    |   +---------> +------------+
     |               |    |
 0GB +---------------+
```

Figure 10.2 – High and low memory splitting

In the preceding diagram, we can see that 128 MB of the kernel address space is used to map a high memory of RAM on the fly when needed. On the other hand, 896 MB of kernel address space is permanently and linearly mapped to a low 896 MB of RAM.

The high-memory mechanism can also be used on a 1 GB RAM system to dynamically map user memory whenever the kernel needs access. The fact that the kernel can map the whole RAM into its address space does not mean the user space can't access it. More than one mapping to a RAM page frame can exist; it can be both permanently mapped to the kernel memory space and mapped to some address in the user space when the process is chosen for execution.

> **Note**
>
> Given a virtual address, you can distinguish whether it is a kernel space or a user space address by using the process layout shown previously. Every address below `PAGE_OFFSET` comes from the user space; otherwise, it is from the kernel.

Low memory in detail

The first 896 MB of kernel address space constitutes the low memory region. Early in the boot process, the kernel permanently maps that 896 MB onto physical RAM. Addresses that result from that mapping are called **logical addresses**. These are virtual addresses but can be translated into physical addresses by subtracting a fixed offset, since the mapping is permanent and known in advance. Low memory matches with the lower bound of physical addresses. You can define low memory as being the memory for which logical addresses exist in the kernel space. The core of the kernel stays in low memory. Therefore, most of the kernel memory functions return low memory. In fact, to serve different purposes, kernel memory is divided into zones. Actually, the first 16 MB of LOWMEM is reserved for **Direct Memory Access (DMA)** usage. Hardware does not always allow you to treat all pages as identical because of limitations. We can then identify three different memory zones in the kernel space:

- ZONE_DMA: This contains page frames of memory below 16 MB, reserved for DMA.

- ZONE_NORMAL: This contains page frames of memory above 16 MB and below 896 MB, for normal use.

- ZONE_HIGHMEM: This contains page frames of memory at and above 896 MB.

However, on a 512 MB system, there will be no ZONE_HIGHMEM, 16 MB for ZONE_DMA, and 496 MB for ZONE_NORMAL.

From all the preceding, we can complete the definition of logical addresses, adding that these are addresses in kernel space mapped linearly on physical addresses, and that the corresponding physical address can be obtained by using an offset. Kernel virtual addresses are similar to logical addresses in that they are mappings from a kernel-space address to a physical address. However, the difference is that kernel virtual addresses do not always have the same linear, one-to-one mapping to physical locations as logical addresses do.

> **Note**
>
> You can convert a physical address into a logical address using the `__pa(address)` macro and the revert with the `__va(address)` macro.

Understanding high memory

The top 128 MB of the kernel address space is called a **high-memory region**. It is used by the kernel to temporarily map physical memory above 1 GB (above 896 MB in reality). When physical memory above 1 GB (or, more precisely, 896MB) needs to be accessed, the kernel uses that 128 MB to create a temporary mapping to its virtual address space, thus achieving the goal of being able to access all physical pages. You can define high memory as being memory for which logical addresses do not exist and which is not mapped permanently into a kernel address space. The physical memory above 896 MB is mapped on demand to the 128 MB of the `HIGHMEM` region.

Mapping to access high memory is created on the fly by the kernel and destroyed when done. This makes high memory access slower. However, the concept of high memory does not exist on 64-bit systems, due to the huge address range (264 TB), where the 3 GB/1 GB (or any similar split scheme) split does not make sense anymore.

An overview of a process address space from the kernel

On a Linux system, each process is represented in the kernel as an instance of `struct task_struct` (see `include/linux/sched.h`), which characterizes and describes this process. Before the process starts running, it is allocated a table of memory mapping, stored in a variable of the `struct mm_struct` type (see `include/linux/mm_types.h`). This can be verified by looking at the following excerpt of the `struct task_struct` definition, which embeds pointers to elements of the `struct mm_struct` type:

```
struct task_struct{
    [...]
    struct mm_struct *mm, *active_mm;
    [...]
}
```

In the kernel, there is a global variable that always points to the current process, `current`, and the `current->mm` field points to the current process memory-mapping table. Before going further in our explanation, let's have a look at the following excerpt of a `struct mm_struct` data structure:

```
struct mm_struct {
    struct vm_area_struct *mmap;
    unsigned long mmap_base;
```

```
    unsigned long task_size;
    unsigned long highest_vm_end;
    pgd_t * pgd;
    atomic_t mm_users;
    atomic_t mm_count;
    atomic_long_t nr_ptes;
#if CONFIG_PGTABLE_LEVELS > 2
    atomic_long_t nr_pmds;
#endif
    int map_count;
    spinlock_t page_table_lock;
    unsigned long total_vm;
    unsigned long locked_vm;
    unsigned long pinned_vm;
    unsigned long data_vm;
    unsigned long exec_vm;
    unsigned long stack_vm;
    unsigned long start_code, end_code, start_data, end_data;
    unsigned long start_brk, brk, start_stack;
    unsigned long arg_start, arg_end, env_start, env_end;

    /* ref to file /proc/<pid>/exe symlink points to */
    struct file __rcu *exe_file;
};
```

I intentionally removed some fields we are not interested in. There are some fields we will talk about later: pgd for example, which is a pointer to the process's base (first entry) level one table (**Page Global Directory**, abbreviated **PGD**), written in the translation table base address of the CPU at context switching. For a better understanding of this data structure, we can use the following diagram:

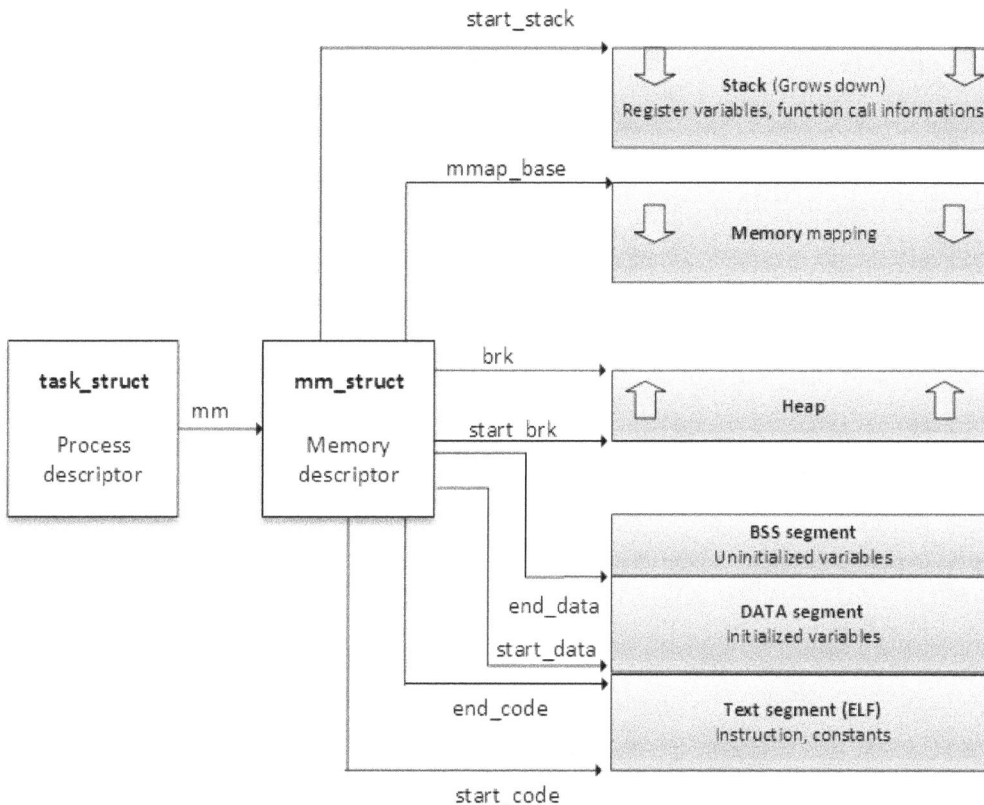

Figure 10.3 – A process address space

From a process point of view, a memory mapping can be seen as nothing but a set of page table entries dedicated to a consecutive virtual address range. That *"consecutive virtual address range"* is referred to as a memory area, or a **Virtual Memory Area** (**VMA**). Each memory mapping is described by a start address and length, permissions (such as whether the program can read, write, or execute from that memory), and associated resources (such as physical pages, swap pages, and file contents).

mm_struct has two ways to store process regions (VMAs):

- In a red-black tree (a self-balancing binary search tree), whose root element is pointed by the mm_struct->mm_rb field

- In a linked list, where the first element is pointed by the mm_struct->mmap field

Now that we have had an overview of a process address space and have seen that it is made of a set of virtual memory regions, let's dive into the details and study the mechanisms behind these memory regions.

Understanding the concept of VMA

In the kernel, process memory mappings are organized into areas, each referred to as a VMA. For your information, in each running process on a Linux system, the code section, each mapped file region (a library, for example), or each distinct memory mapping (if any) is materialized by a VMA. A VMA is an architecture-independent structure, with permissions and access control flags, defined by a start address and a length. Their sizes are always a multiple of the page size (PAGE_SIZE). A VMA consists of a few pages, each of which has an entry in the page table (the **Page Table Entry** (**PTE**)).

A VMA is represented in the kernel as an instance of struct vma_area, defined as the following:

```
struct vm_area_struct {
    unsigned long vm_start;
    unsigned long vm_end;
    struct vm_area_struct *vm_next, *vm_prev;
    struct mm_struct *vm_mm;
    pgprot_t vm_page_prot;
    unsigned long vm_flags;
    unsigned long vm_pgoff;
    struct file * vm_file;
    [...]
}
```

For the sake of readability and understandability of this section, only elements that are relevant for us have been listed. However, the following are the meanings of the remaining elements:

- vm_start is the VMA start address within the address space (vm_mm). It is the first address within this VMA.

- vm_end is the first byte after our end address within vm_mm. It is the first address outside this VMA.

- vm_next and vm_prev are used to implement a linked list of VM areas per task, sorted by address.

- vm_mm is the process address space that this VMA belongs to.

- `vm_page_prot` and `vm_flags` represent the access permission of the VMA. The former is an architecture-level data type, whose update is applied directly to the PTEs of the underlying architecture. It is a form of cached conversion from `vm_flags`, which stores the proper protection bits and the type of mapping in an architecture-independent manner.

- `vm_file` is the file backing this mapping. This can be NULL (for example, for anonymous mapping, such as a process's heap or stack).

- `vm_pgoff` is the offset (within `vm_file`) in page size unit. This offset is measured in number of pages.

The following diagram is an overview of a process memory mapping, highlighting each VMA and describing some of its structure elements:

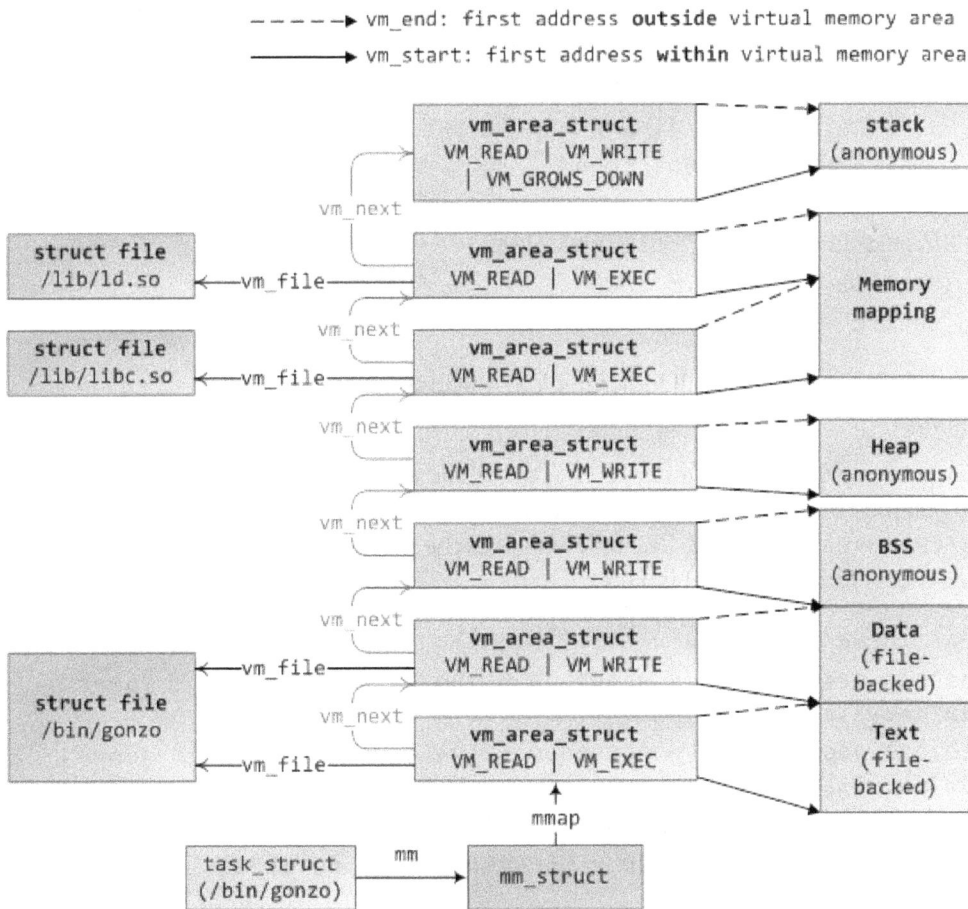

Figure 10.4 – Process memory mappings

The preceding image (from `http://duartes.org/gustavo/blog/post/how-the-kernel-manages-your-memory/`) describes a process's (started from `/bin/gonzo`) memory mappings (VMAs). We can see interactions between `struct task_struct` and its address space element (`mm`), which then lists and describes each VMA (the start, the end, and the backing file).

You can use the `find_vma()` function to find the VMA that corresponds to a given virtual address. `find_vma()` is declared in `linux/mm.h` as the following:

```
extern struct vm_area_struct * find_vma(
            struct mm_struct * mm, unsigned long addr);
```

This function searches and returns the first VMA that satisfies `vm_start <= addr < vm_end` or returns `NULL` if none is found. `mm` is the process address space to search in. For the current process, it can be `current->mm`. The following is an example:

```
struct vm_area_struct *vma =
                    find_vma(task->mm, 0x603000);
if (vma == NULL) /* Not found ? */
    return -EFAULT;
/* Beyond the end of returned VMA ? */
if (0x13000 >= vma->vm_end)
    return -EFAULT;
```

The preceding code excerpt will look for a VMA whose memory bounds contain `0x603000`.

Given a process whose identifier is `<PID>`, the whole memory mappings of this process can be obtained by reading the `/proc/<PID>/maps`, `/proc/<PID>/smaps`, and `/proc/<PID>/pagemap` files. The following lists the mappings of a running process, whose Process Identifier (PID) is `1073`:

```
# cat /proc/1073/maps
00400000-00403000 r-xp 00000000 b3:04 6438          /usr/
sbin/net-listener
00602000-00603000 rw-p 00002000 b3:04 6438          /usr/
sbin/net-listener
00603000-00624000 rw-p 00000000 00:00 0             [heap]
7f0eebe4d000-7f0eebe54000 r-xp 00000000 b3:04 11717 /usr/
lib/libffi.so.6.0.4
7f0eebe54000-7f0eec054000 ---p 00007000 b3:04 11717 /usr/
lib/libffi.so.6.0.4
```

```
7f0eec054000-7f0eec055000 rw-p 00007000 b3:04 11717    /usr/
lib/libffi.so.6.0.4
7f0eec055000-7f0eec069000 r-xp 00000000 b3:04 21629    /lib/
libresolv-2.22.so
7f0eec069000-7f0eec268000 ---p 00014000 b3:04 21629    /lib/
libresolv-2.22.so
[...]
7f0eee1e7000-7f0eee1e8000 rw-s 00000000 00:12 12532    /dev/
shm/sem.thk-mcp-231016-sema
[...]
```

Each line in the preceding excerpt represents a VMA, and the fields correspond to the {address (start-end)} {permissions} {offset} {device (major:minor)} {inode} {pathname (image)} pattern:

- address: Represents the starting and ending address of the VMA.

- permissions: Describes access rights of the region: r (read), w (write), and x (execute). p is if the mapping is private and s is for shared mapping.

- offset: If file mapping (the mmap system call), it is the offset in the file where the mapping takes place. It is 0 otherwise.

- major:minor: If file mapping, these represent the major and minor numbers of the devices in which the file is stored (the device holding the file).

- inode: If mapping from a file, this is the inode number of the mapped file.

- pathname: This is the name of the mapped file or left blank otherwise. There are other region names, such as [heap], [stack], or [vdso] (which stands for **virtual dynamic shared object**, a shared library mapped by the kernel into every process's address space, in order to reduce performance penalties when system calls switch to kernel mode).

Each page allocated to a process belongs to an area, thus any page that does not live in the VMA does not exist and cannot be referenced by the process.

High memory is perfect for user space because its address space must be explicitly mapped. Thus, most high memory is consumed by user applications. __GFP_HIGHMEM and GFP_HIGHUSER are the flags for requesting the allocation of (potentially) high memory. Without these flags, all kernel allocations return only low memory. There is no way to allocate contiguous physical memory from user space in Linux.

Now that VMAs have no more secrets for us, let's describe the hardware concepts involved in the translation to their corresponding physical addresses, if any, or their creation and allocation otherwise.

Demystifying address translation and MMU

MMU does not only convert virtual addresses into physical ones but also protects memory from unauthorized access. Given a process, any page that needs to be accessed from this process must exist in one of its VMAs and, thus, must live in the process's page table (every process has its own).

As a recall, memory is organized by chunks of fixed-size named pages for virtual memory and frames for physical memory. The size in our case is 4 KB. However, it is defined and accessible with the PAGE_SIZE macro in the kernel. Remember, however, that page size is imposed by the hardware. Considering a 4 KB page-sized system, bytes 0 to 4095 fall on page 0, bytes 4096 to 8191 fall on page 1, and so on.

The concept of a page table is introduced to manage mapping between pages and frames. Pages are spread over tables so that each PTE corresponds to a mapping between a page and a frame. Each process is then given a set of page tables to describe all of its memory regions.

To walk through pages, each page is assigned an index, called a **page number**. When it comes to a frame, it is a **Page Frame Number (PFN)**. This way, VMAs (logical addresses, more precisely) are composed of two parts: a page number and an offset. On 32-bit systems, the offset represents the 12 less significant bits of the address, whereas 13 less significant bits represent it on 8 KB page-size systems. The following diagram highlights this concept of addresses split into a page number and an offset:

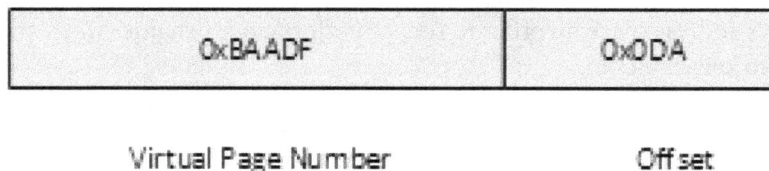

0xBAADF	0x0DA
Virtual Page Number	Offset

Figure 10.5 – Logical address representation

How does the OS or CPU know which physical address corresponds to a given logical address? They use a page table as a translation table and know that each entry's index is a virtual page number, and the value at this index is the PFN. To access physical memory given a virtual memory, the OS first extracts the offset, the virtual page number, and then walks through the process's page tables to match the virtual page number to the physical page. Once a match occurs, it is then possible to access data in that page frame:

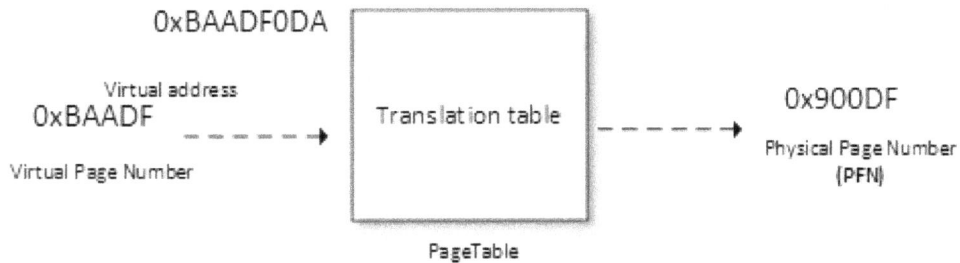

0xBAADF0DA

Virtual address
0xBAADF

Virtual Page Number

Translation table

0x900DF

Physical Page Number
(PFN)

PageTable

Figure 10.6 – Address translation

The offset is used to point to the right location in the frame. A page table not only holds mapping between physical and virtual page numbers but also accesses control information (read/write access, privileges, and so on).

The following diagram describes address decoding and page table lookup to point to the appropriate location in the appropriate frame:

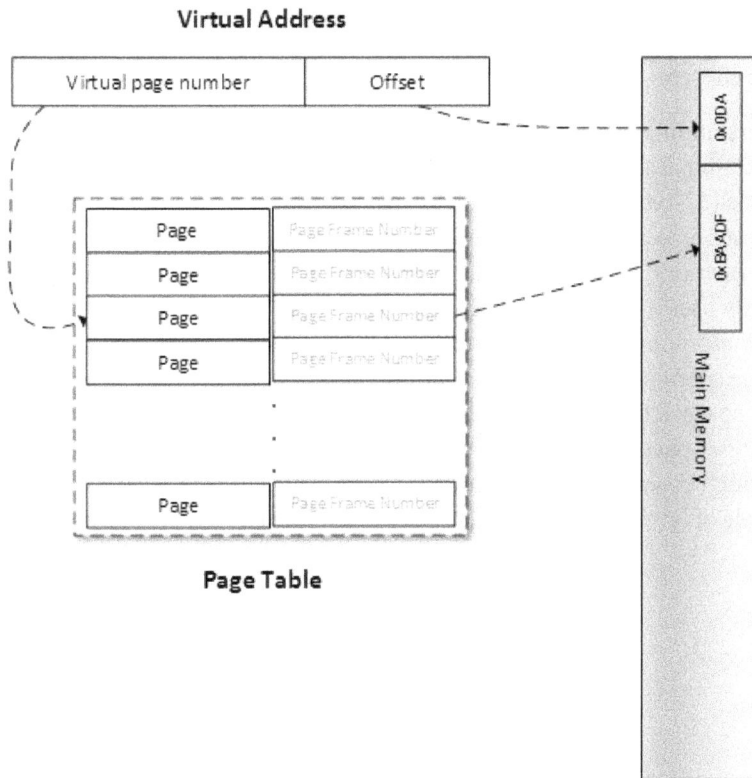

Virtual Address

Virtual page number	Offset

Page	Page Frame Number
Page	Page Frame Number
Page	Page Frame Number
Page	Page Frame Number
Page	Page Frame Number

Page Table

Main Memory

0x0DA

0xBAADF

Figure 10.7 – Virtual to physical address translation

The number of bits used to represent the offset is defined by the PAGE_SHIFT kernel macro. PAGE_SHIFT is the number of times needed to left-shift 1 bit to obtain the PAGE_SIZE value. It is also the number of times needed to right-shift a page's logical address to obtain its page number, which is the same for a physical address to obtain its page frame number. This macro is architecture-dependent and also depends on the page granularity. Its value could be considered as the following:

```
#ifdef CONFIG_ARM64_64K_PAGES
#define PAGE_SHIFT        16
#elif defined(CONFIG_ARM64_16K_PAGES)
#define PAGE_SHIFT        14
#else
#define PAGE_SHIFT        12
#endif
#define PAGE_SIZE         (_AC(1, UL) << PAGE_SHIFT)
```

The preceding states that by default (whether on ARM or ARM64), PAGE_SHIFT is 12, which means a 4 KB page size. On ARM64, it 14 or 16 when respectively a 16 KB or 64 KB page size is chosen.

With our understanding of address translation, the page table is a partial solution. Let's see why. Most 32-bit architectures require 32 bits (4 bytes) to represent a page table entry. On such systems (32-bit) where each process has its private 3 GB user address space, we need 786,432 entries to characterize and cover a process's address space. It represents too much physical memory spent per process just to store the memory mappings. In fact, a process generally uses a small but scattered portion of its virtual address space. To resolve that issue, the concept of a "level" was introduced. Page tables are hierarchized by level (page level). The space necessary to store a multi-level page table only depends on the virtual address space actually in use, instead of being proportional to the maximum size of the virtual address space. This way, unused memory is no longer represented, and the page table walk-through time is reduced. Moreover, each table entry in level N will point to an entry in the table of level N+1, level 1 being the higher level.

Linux can support up to four levels of paging. However, the number of levels to use is architecture-dependent. The following are descriptions of each lever:

- **Page Global Directory (PGD)**: This is the first level (level 1) page table. Each entry is of the `pgd_t` type in the kernel (generally, `unsigned long`) and points to an entry in the table at the second level. In the Linux kernel, `struct tastk_struct` represents a process's description, which in turn has a member (`mm`) whose type is `struct mm_struct`, which characterizes and represents the process's memory space. In `struct mm_struct`, there is a processor-specific field, `pgd`, which is a pointer to the first entry (entry 0) of the process's level-1 (PGD) page table. Each process has one and only one PGD, which may contain up to 1,024 entries.

- **Page Upper Directory (PUD)**: This represents the second level of indirection.

- **Page Middle Directory (PMD)**: This is the third indirection level.

- **Page Table Entry (PTE)**: This is the leaves of the tree. It is an array of `pte_t`, where each entry points to a physical page.

> **Note**
>
> All levels are not always used. The i.MX6's MMU only supports a two-level page table (PGD and PTE), which is the case for almost all 32-bit CPUs. In this case, PUD and PMD are simply ignored.

It is important to know that the MMU does not store any mapping. It is a data structure located in RAM. Instead, there is a special register in the CPU, called the **Page Table Base Register** (**PTBR**) or the **Translation Table Base Register 0** (**TTBR0**), which points to the base (entry 0) of the level-1 (top-level) page table (PGD) of the process. It is exactly where the `pdg` field of `struct mm_struct` points: `current->mm.pgd == TTBR0`.

At context switch (when a new process is scheduled and given the CPU), the kernel immediately configures the MMU and updates the PTBR with the new process's `pgd`. Now, when a virtual address is given to MMU, it uses the PTBR's content to locate the process's level-1 page table (PGD), and then it uses the level-1 index, extracted from the **Most-Significant Bits** (**MSBs**) of the virtual address to find the appropriate table entry, which contains a pointer to the base address of the appropriate level-2 page table. Then, from that base address, it uses the level-2 index to find the appropriate entry and so on, until it reaches the PTE. ARM architecture (i.MX6, in our case) has a two-level page table. In this case, the level-2 entry is a PTE and points to the physical page (PFN). Only the physical page is found at this step. To access the exact memory location in the page, the MMU extracts the memory offset, also part of the virtual address, and points to the same offset in the physical page.

For the sake of understandability, the preceding description has been limited to a two-level paging scheme but can easily extended. The following diagram is a representation of this two-level paging scheme:

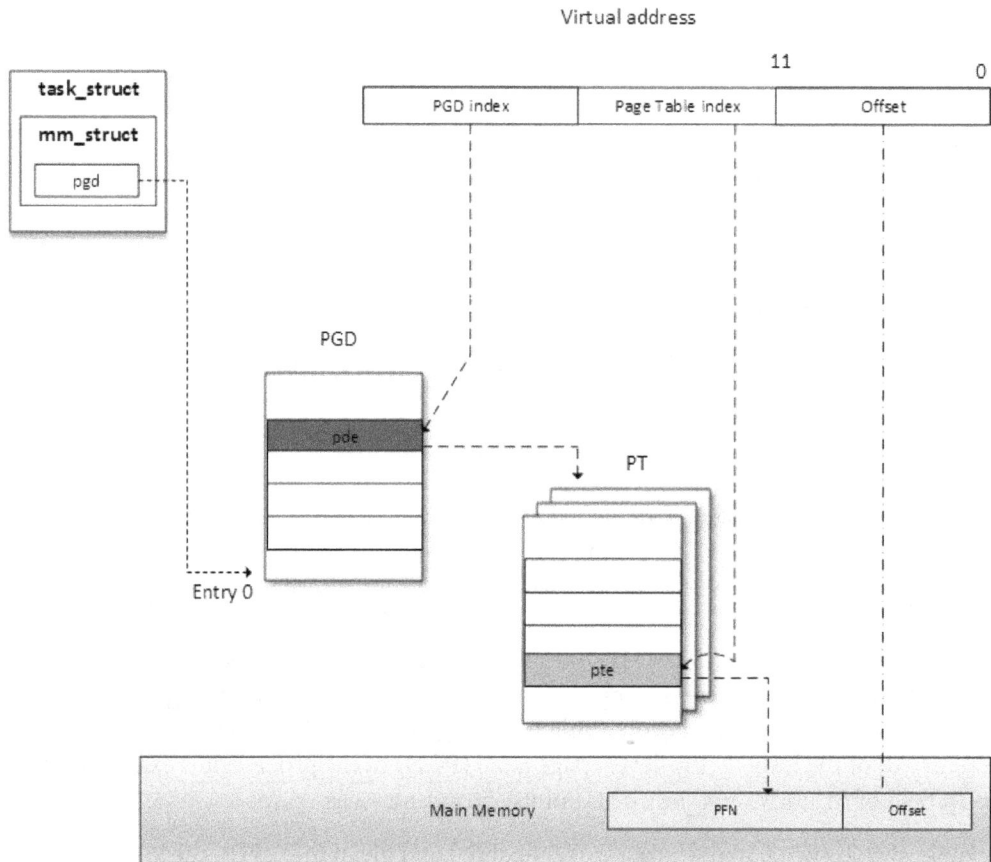

Figure 10.8 – A two-level address translation scheme

When a process needs to read from or write into a memory location (of course, we are talking about virtual memory), the MMU performs a translation into that process's page table to find the right entry (PTE). The virtual page number is extracted (from the virtual address) and used by the processor as an index into the process's page table to retrieve its page table entry. If there is a valid page table entry at that offset, the processor takes the page frame number from this entry. If not, it means the process accessed an unmapped area of its virtual memory. A page fault is then raised, and the OS should handle it.

In the real world, address translation requires a page-table walk, and it is not always a one-shot operation. There are at least as many memory accesses as there are table levels. A four-level page table would require four memory accesses. In other words, every virtual access would result in five physical memory accesses. The virtual memory concept would be useless if its access were four times slower than physical access. Fortunately, System-on-Chip (SoC) manufacturers worked hard to find a clever trick to address this performance issue: modern CPUs use a small associative and very fast memory called the **Translation Lookaside Buffer** (**TLB**), in order to cache the PTEs of recently accessed virtual pages.

Page lookup and the TLB

Before the MMU proceeds to address translation, there is another step involved. As there is a cache for recently accessed data, there is also a cache for recently translated addresses. As data cache speeds up the data accessing process, the TLB speeds up virtual address translation (yes, address translation is a time-consuming task). It is a **Content-Addressable Memory** (**CAM**), where the key is the virtual address and the value is the physical address. In other words, the TLB is a cache for the MMU. At each memory access, the MMU first checks for recently used pages in the TLB, which contains a few of the virtual address ranges to which physical pages are currently assigned.

How does the TLB work?

On memory access, the CPU walks through the TLB trying to find the virtual page number of the page that is being accessed. This step is called a **TLB lookup**. When a TLB entry is found (a match occurs), it is called TLB hit, and the CPU just keeps running and uses the PFN found in the TLB entry to calculate the target physical address. There is no page fault when a TLB hit occurs. If a translation can be found in the TLB, virtual memory access will be as fast as physical access. If there is no TLB hit, it is called TLB miss.

On a TLB miss, there are two possibilities. Depending on the processor type, the TLB miss event can be handled by the software, the hardware, or through the MMU:

- **Software handling**: The CPU raises a TLB miss interrupt, caught by the OS. The OS then walks through the process's page table to find the right PTE. If there is a matching and valid entry, then the CPU installs the new translation in the TLB. Otherwise, the page fault handler is executed.

- **Hardware handling**: It is up to the CPU (the MMU, in fact) to walk through the process's page table on hardware. If there is a match, the CPU adds the new translation in the TLB. Otherwise, the CPU raises a page fault interrupt, handled by the OS.

In both cases, the page fault handler is the same, `do_page_fault()`. This function is architecture-dependent; for ARM, it is defined in `arch/arm/mm/fault.c`.

The following is a diagram describing a TLB lookup, a TLB hit, or a TLB miss event:

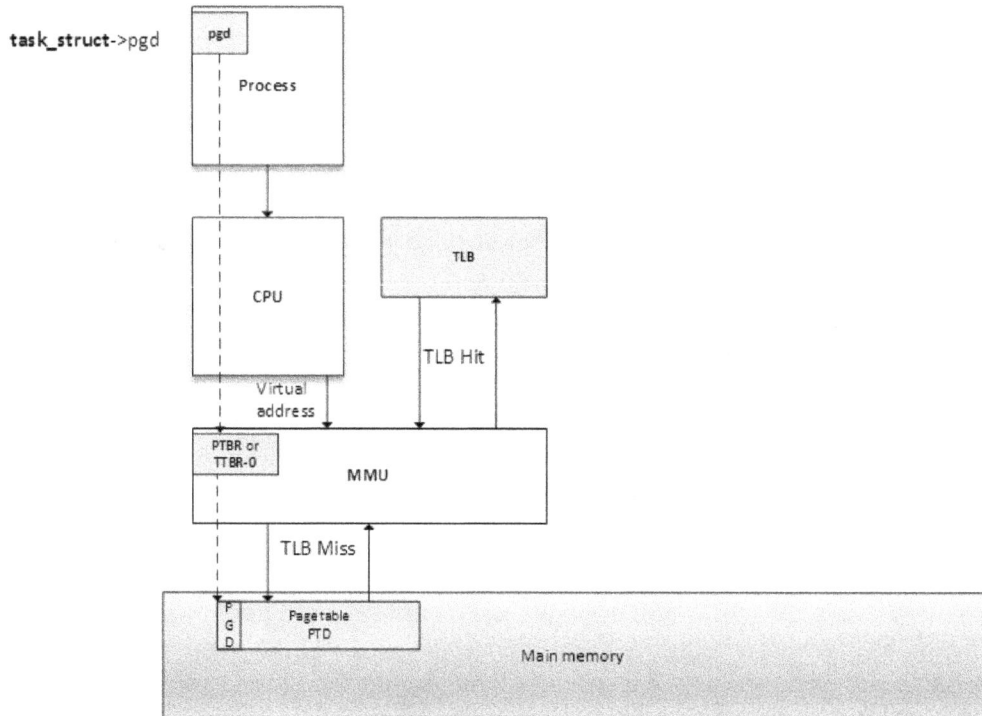

Figure 10.9 – The MMU and TLB walk-through process

Page table and page directory entries are architecture-dependent. An OS must make sure the structure of a table corresponds to a structure recognized by the MMU. On ARM processors, the location of the translation table must be written in the `control` coprocessor 15 (CP15) `c2` register, and then enable the caches and the MMU by writing to the CP15 `c1` register. Have a look at both `http://infocenter.arm.com/help/index.jsp?topic=/com.arm.doc.dui0056d/BABHJIBH.htm` and `http://infocenter.arm.com/help/index.jsp?topic=/com.arm.doc.ddi0433c/CIHFDBEJ.html` for detailed information.

Now that we are comfortable with the address translation schemes and their ease with the TLB, we can talk about memory allocation, which involves manipulating page entries under the hood.

Dealing with memory allocation mechanisms and their APIs

Before jumping to the list of APIs, let's start with the following figure, showing the different memory allocators that exist on a Linux-based system, which we will discuss later:

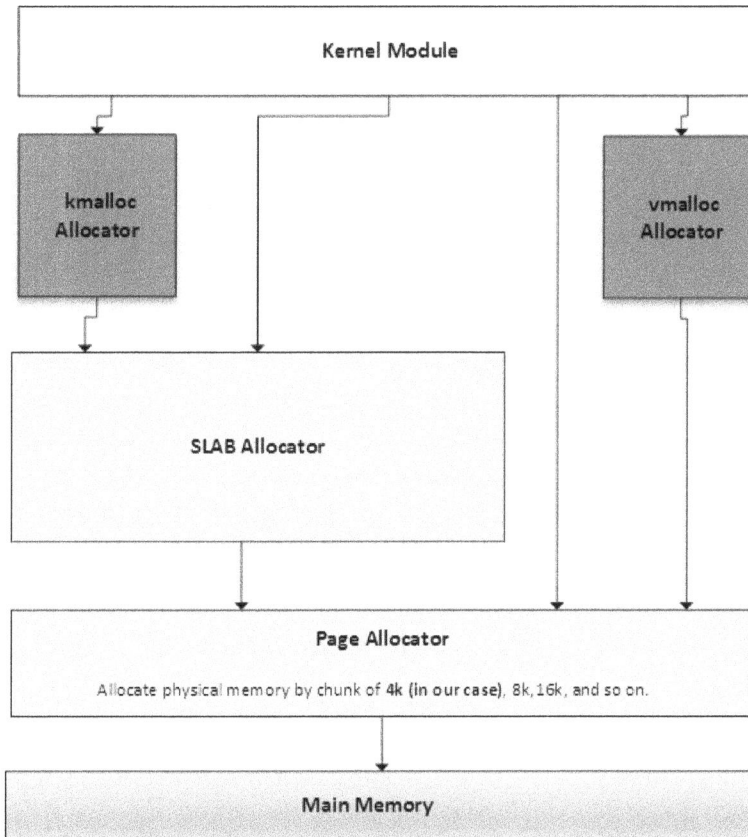

Figure 10.10 – Overview of kernel memory allocators

The preceding diagram is inspired by https://bootlin.com/doc/training/linux-kernel/linux-kernel-slides.pdf. What it shows is that there is an allocation mechanism to satisfy any kind of memory request. Depending on what you need memory for, you can choose the one closest to your goal. The main and lowest level allocator is the **page allocator**, which allocates memory in units of pages (a page being the smallest memory unit it can deliver). Then comes the **slab allocator**, which is built on top of the page allocator, getting pages from it and splitting them into smaller memory entities (by means of slabs and caches). This is the allocator on which the kmalloc API relies.

While `kmalloc` can be used to request memory from the slab allocator, we can directly talk to the slab to request memory from its caches, or even build our own caches.

Let's start this memory allocation journey with the main and lowest level allocator, the page allocator, from which the others derivate.

The page allocator

The page allocator is the low-level allocator on the Linux system, the one that serves as a basis for other allocators. This allocator brings with it the concept of the page (virtual) and page frame (physical). So, the system's physical memory is split into fixed-size blocks (called **page frames**), while virtual memory is organized into fixed-size blocks called pages. Page size is always the same as a page frame size. As this is the lowest-level allocator, a page is the smallest unit of memory that the OS will give to any memory request at a low level. In the Linux kernel, a page is represented as an instance of the `struct page` structure, which we will manipulate using dedicated APIs, introduced in the next section.

Page allocation APIs

This is the lowest-level allocator. It allocates and deallocates blocks of pages using the buddy algorithm. Pages are allocated in blocks that are to the power of 2 in size (to get the best from the buddy algorithm). That means it can allocate a block of 1 page, 2 pages, 4 pages, 8, 16, and so on. Pages returned from this allocation are physically contiguous. `alloc_pages()` is the main API and is defined as the following:

```
struct page *alloc_pages(gfp_t mask, unsigned int order)
```

The preceding function returns `NULL` when no page can be allocated. Otherwise, it allocates 2^{order} pages and returns a pointer to an instance of `struct page`, which points the first page of the reserved block. There is, however, a helper macro, `alloc_page()`, which can be used to allocate a single page. The following is its definition:

```
#define alloc_page(gfp_mask) alloc_pages(gfp_mask, 0)
```

This macro wraps `alloc_pages()` with an order parameter set with 0.

`__free_pages()` must be used to release memory pages allocated with the `alloc_pages()` function. It takes a pointer to the first page of the allocated block as a parameter, along with the order, the same that was used for allocation. It is defined as the following:

```
void __free_pages(struct page *page, unsigned int order);
```

There are other functions working in the same way, but instead of an instance of `struct page`, they return the (logical) address of the reserved block. These are `__get_free_pages()` and `__get_free_page()`, and the following are their definitions:

```
unsigned long __get_free_pages(gfp_t mask,
                                unsigned int order);
unsigned long get_zeroed_page(gfp_t mask);
```

`free_pages()` is used to free a page allocated with `__get_free_pages()`. It takes the kernel address representing the start region of allocated page(s), along with the order, which should be the same as that used for allocation:

```
free_pages(unsigned long addr, unsigned int order);
```

Whatever the allocation type is, `mask` specifies the memory zones from where the pages should be allocated and the behavior of the allocators. The following are possible values:

- `GFP_USER`: For user memory allocation.
- `GFP_KERNEL`: The commonly used flag for kernel allocation.
- `GFP_HIGHMEM`: This requests memory from the `HIGH_MEM` zone.
- `GFP_ATOMIC`: This allocates memory in an atomic manner that cannot sleep. It is used when we need to allocate memory from an interrupt context.

However, you should note that whether specifying the `GFP_HIGHMEM` flag with `__get_free_pages()` (or `__get_free_page()`) or not, it won't be considered. This flag is masked out in these functions to make sure that the returned address never represents high-memory pages (because of their non-linear/permanent mapping). If you need high memory, use `alloc_pages()` and then `kmap()` to access it.

`__free_pages()` and `free_pages()` can be mixed. The main difference between them is that `free_page()` takes a logical address as a parameter, whereas `__free_page()` takes a `struct page` structure.

Note

The maximum order that can be used varies between architectures. It depends on the `FORCE_MAX_ZONEORDER` kernel configuration option, which is `11` by default. In this case, the number of pages you can allocate is 1,024. It means that on a 4 KB-sized system, you can allocate up to *1,024 x 4 KB = 4 MB* at maximum. On ARM64, the maximum order varies with the selected page size. If it is a 16 KB page size, the maximum order is `12`, and if it is a 64 KB page size, the maximum order is `14`. These size limitations per allocation are valid for `kmalloc()` as well.

Page and addresses conversion functions

There are convenient functions exposed by the kernel to switch back and forth between the struct page instances and their corresponding logical addresses, which can be useful at different moments while dealing with memory. The page_to_virt() function is used to convert a struct page (as returned by alloc_pages(), for example) into a kernel logical address. Alternatively, virt_to_page() takes a kernel logical address and returns its associated struct page instance (as if it was allocated using the alloc_pages() function). Both virt_to_page() and page_to_virt() are declared in <asm/page.h> as the following:

```
struct page *virt_to_page(void *kaddr);
void *page_to_virt(struct page *pg)
```

There is another macro, page_address(), which simply wraps page_to_virt() and which is declared as the following:

```
void *page_address(const struct page *page)
```

It returns the logical address of the page passed in the parameter.

The slab allocator

The slab allocator is the one on which kmalloc() relies. Its main purposes are to eliminate fragmentation caused by memory (de)allocation, which is caused by the buddy system in case of small-size memory allocation, and to speed up memory allocation for commonly used objects.

Understanding the buddy algorithm

To allocate memory, the requested size is rounded up to the power of two, and the buddy allocator searches the appropriate list. If no entries exist on the requested list, an entry from the next upper list (which has blocks of twice the size of the previous list) is split into two halves (called **buddies**). The allocator uses the first half, while the other is added to the next list down. This is a recursive approach, which stops when either the buddy allocator successfully finds a block that can be split or reaches the largest block size and there are no free blocks available.

The following case study is heavily inspired by http://dysphoria.net/
OperatingSystems1/4_allocation_buddy_system.html. For example, if the
minimum allocation size is 1K bytes, and the memory size is 1 MB, the buddy allocator
will create an empty list for 1K byte holes, an empty list for 2K byte holes, one for 4K byte
holes, 8K, 16K, 32K, 64K, 128K, 256K, 512K, and one list for 1 MB holes. All of them are
initially empty, except for the 1 MB list, which has only one hole.

Let's now imagine a scenario where we want to allocate a 70K block. The buddy allocator
will round it up to 128K and will end up splitting the 1 MB into two 512K blocks, then
256K, and finally 128K, and then it will allocate one of the 128K blocks to the user. The
following are schemes that summarize this scenario:

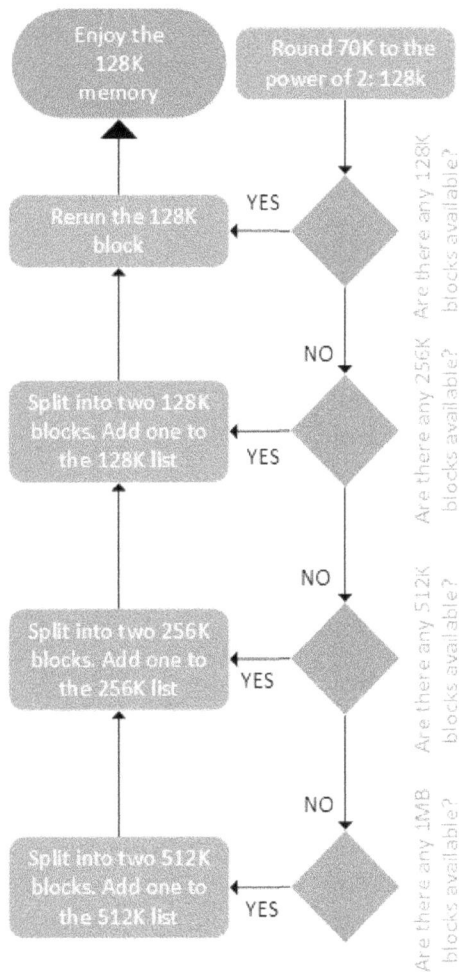

Figure 10.11 – Allocation using the buddy algorithm

The deallocation is as fast as allocation. The following is a figure that summarizes the deallocation algorithm:

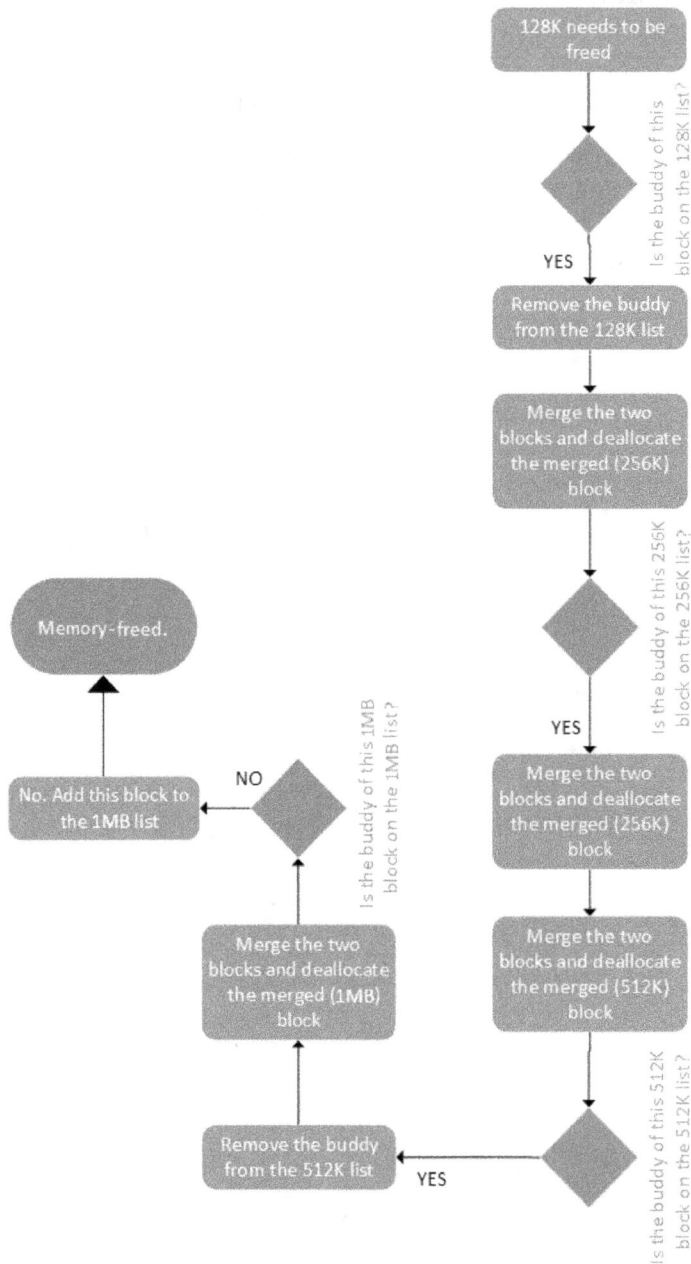

Figure 10.12 – Deallocation using the buddy algorithm

In the preceding diagram, we can see that memory deallocation using the buddy algorithm works. In the next section, we will study the slab allocator, built on top of this algorithm.

A journey into the slab allocator

Before we introduce the slab allocator, let's define some terms that it uses:

- **Slab**: A contiguous piece of physical memory made of several page frames. Each slab is divided into equal chunks of the same size, used to store specific types of kernel objects, such as `inode` and `mutexe` objects. A slab can be considered an array of identically sized blocks.

- **Cache**: This is made of one or more slabs in a linked list. The cache only stores objects of the same type (for example, `inode` objects only).

Slabs may be in one of the following states:

- **Empty**: Where all objects (chunks) on the slab are marked as free.

- **Partial**: Both used and free objects exist in the slab.

- **Full**: All objects on the slab are marked as used.

The memory allocator is responsible for building caches. Initially, each slab is empty and marked so. When one allocates memory for a kernel object, the allocator looks for a free location for that object on a partial/free slab in a cache for that type of object. If not found, the allocator allocates a new slab and adds it to the cache. The new object gets allocated from this slab, and the slab is marked as partial. When the code is done with the memory (memory-freed), the object is simply returned to the slab cache in its initialized state. This is the reason why the kernel also provides helper functions to obtain zeroed initialized memory, which allows us to get rid of previous content. The slab keeps a reference count of how many of its objects are being used so that when all slabs in a cache are full and another object is requested, the slab allocator is responsible for adding new slabs.

The following diagram illustrates the concept of slabs, caches, and their different states:

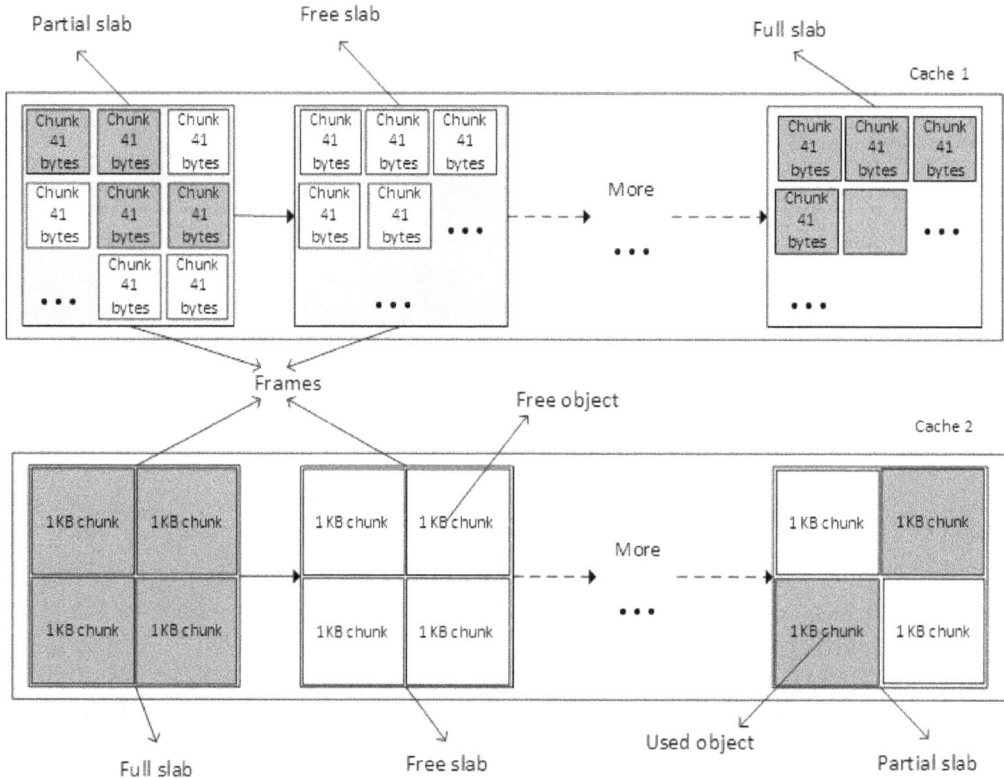

Figure 10.13 – Slabs and caches

It is a bit like creating a per-object allocator. The kernel allocates one cache per type of object, and only objects of the same type can be stored in a cache (for example, only `task_struct` structures).

There are different kinds of slab allocators in the kernel, depending on whether one needs compactness, cache-friendliness, or raw speed. These consist of the following:

- The **SLAB** (slab allocator), which is as cache-friendly as possible. This is the original memory allocator.

- The **SLOB** (simple list of blocks), which is as compact as possible, appropriate for systems with very low memory, mostly embedded systems with a few megabytes or tens of megabytes.

- The **SLUB** (unqueued allocator), which is quite simple and requires fewer instruction cost counts. This is the next-generation replacement memory allocator, enhancing and replacing the SLAB. It is based on the SLAB model but fixes several deficiencies in SLAB, especially on systems with a large number of processors. This has been (and still is) the default memory allocator in the Linux kernel since kernel 2.6.23 (`CONFIG_SLUB=y`). See this patch: `https://git.kernel.org/pub/scm/linux/kernel/git/torvalds/linux.git/commit/?id=a0acd820807680d2ccc4ef3448387fcdbf152c73`.

> **Note**
>
> The term **slab** has become a generic name referring to a memory allocation strategy employing an object cache, enabling efficient allocation and deallocation of kernel objects. It must not be confused with the allocator of the same name, SLAB, which nowadays has been replaced by SLUB.

kmalloc family allocation

`kmalloc()` is a kernel memory allocation function. It allocates physically contiguous (but not necessarily page-aligned) memory. The following image describes how memory is allocated and returned to the caller:

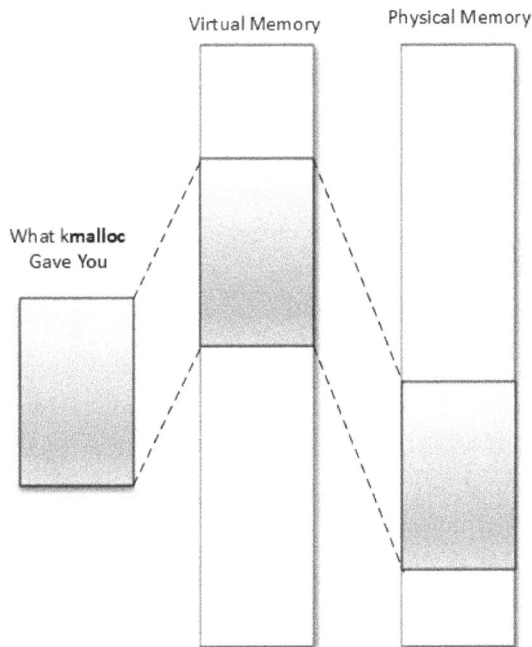

Figure 10.14 – kmalloc memory organization

This allocation API is the general and highest-level memory allocation API in the kernel, which relies on the SLAB allocator. Memory returned from `kmalloc()` has a kernel logical address because it is allocated from the LOW_MEM region unless HIGH_MEM is specified. It is declared in `<linux/slab.h>`, which is the header to include before using the API. It is defined as follows:

```
void *kmalloc(size_t size, int flags);
```

In the preceding code, `size` specifies the size of the memory to be allocated (in bytes). `flags` determines how and where memory should be allocated. Available flags are the same as the page allocator (GFP_KERNEL, GFP_ATOMIC, GFP_DMA, and so on) and the following are their definitions:

- GFP_KERNEL: This is the standard flag. We cannot use this flag in an interrupt handler because its code may sleep. It always returns memory from the LOM_MEM zone (hence, a logical address).

- GFP_ATOMIC: This guarantees the atomicity of the allocation. This flag is to be used when allocation is needed from an interrupt context. Because memory is allocated from an emergency pool or memory, you should not abuse its usage.

- GFP_USER: This allocates memory to a user space process. Memory is then distinct and separated from those allocated to the kernel.

- GFP_NOWAIT: This is to be used if the allocation is performed from within an atomic context, for example, interrupt handler use. This flag prevents direct reclaim and I/O and filesystem operations while doing allocation. Unlike GFP_ATOMIC, it does not use memory reserves. Consequently, under memory pressure, the GFP_NOWAIT allocation is likely to fail.

- GFP_NOIO: Like GFP_USER, this can block, but unlike GFP_USER, it will not start disk I/O. In other words, it prevents any I/O operation while doing allocation. This flag is mostly used in the block/disk layer.

- GFP_NOFS: This will use direct reclaim but will not use any filesystem interfaces.

- __GFP_NOFAIL: The virtual memory implementation must retry indefinitely because the caller is incapable of handling allocation failures. The allocation can stall indefinitely, but it will never fail. Consequently, it's useless to test for failure.

- GFP_HIGHUSER: This requests to allocate memory from the HIGH_MEMORY zone.

- GFP_DMA: This allocates memory from DMA_ZONE.

On successful allocation of memory, kmalloc() returns the virtual (logical, unless high memory is specified) address of the chunk allocated, guaranteed to be physically contiguous. On an error, it returns NULL.

For a device driver, however, it is recommended to use the managed version, devm_kmalloc(), which does not necessarily require freeing the memory, as it is handled internally by the memory core. The following is its prototype:

```
void *devm_kmalloc(struct device *dev, size_t size,
                   gfp_t gfp);
```

In the preceding prototype, dev is the device for which memory is allocated.

Note that kmalloc() relies on SLAB caches when allocating a small size of memory. For this reason, it can internally round the allocated area size up to the size of the smallest SLAB cache in which that memory can fit. This can result in returning more memory than requested. However, ksize() can be used to determine the actual amount (the size in bytes) of memory allocated. You can even use this additional memory, even though a smaller amount of memory was initially specified with the kmalloc() call.

The following is the ksize prototype:

```
size_t ksize(const void *objp);
```

In the preceding, objp is the object whose real size in bytes will be returned.

kmalloc() has the same size limitations as the page-related allocation API. For example, with the default FORCE_MAX_ZONEORDER set to 11, the maximum size per allocation with kmalloc() is 4 MB.

kfree function is used to free the memory allocated by kmalloc(). It is defined as the following:

```
void kfree(const void *ptr)
```

The following is an example of allocating and freeing memory using kmalloc() and kfree() respectively:

```
#include <linux/init.h>
#include <linux/module.h>
#include <linux/slab.h>
#include <linux/mm.h>

static void *ptr;
```

```c
static int alloc_init(void)
{
    size_t size = 1024; /* allocate 1024 bytes */
    ptr = kmalloc(size,GFP_KERNEL);
    if(!ptr) {
        /* handle error */
        pr_err("memory allocation failed\n");
        return -ENOMEM;
    } else {
        pr_info("Memory allocated successfully\n");
    }
    return 0;
}

static void alloc_exit(void)
{
    kfree(ptr);
    pr_info("Memory freed\n");
}

module_init(alloc_init);
module_exit(alloc_exit);

MODULE_LICENSE("GPL");
MODULE_AUTHOR("John Madieu");
```

The kernel provides other helpers based on `kmalloc()` as follows:

```c
void kzalloc(size_t size, gfp_t flags);
void kzfree(const void *p);
void *kcalloc(size_t n, size_t size, gfp_t flags);
void *krealloc(const void *p, size_t new_size,
               gfp_t flags);
```

krealloc() is the kernel equivalent of user space realloc() function. Because memory returned by kmalloc() retains the contents from its previous incarnation, you can request a zeroed kmalloc-allocated memory using kzalloc(). kzfree() is the freeing function for kzalloc(), whereas kcalloc() allocates memory for an array, and its n and size parameters respectively represent the number of elements in the array and the size of an element.

Since kmalloc() returns a memory area in the kernel permanent mapping, the logical address can be translated into a physical address using virt_to_phys(), or to a I/O bus address using virt_to_bus(). These macros internally call either __pa() or __va() if necessary. The physical address (virt_to_phys(kmalloc'ed address)), downshifted by PAGE_SHIFT, will produce a PFN (pfn) of the first page from which the chunk is allocated.

vmalloc family allocation

vmalloc() is the last kernel allocator we will discuss in the book. It returns memory that is exclusively contiguous in the virtual address space. The underlying frames are scattered, as we can see in the following diagram:

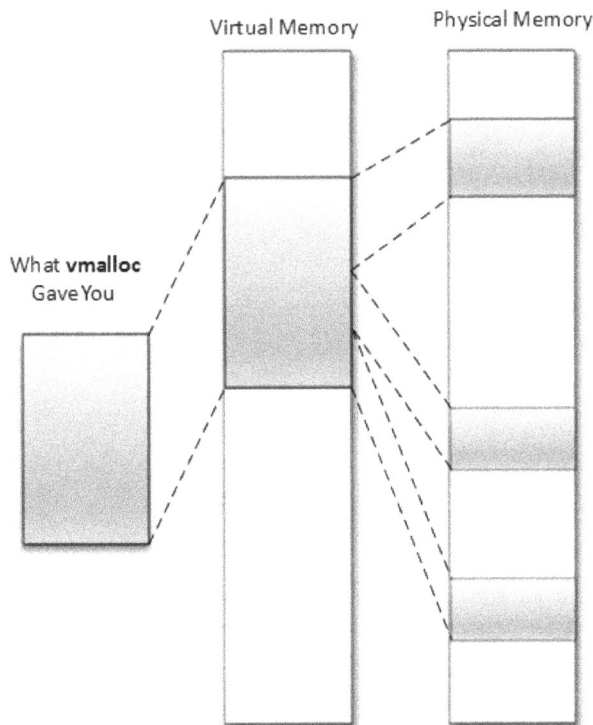

Figure 10.15 – vmalloc memory organization

In the preceding diagram, we can see that memory is not physically contiguous. Moreover, memory returned by vmalloc() always comes from the HIGH_MEM zone. Addresses returned are purely virtual (not logical) and cannot be translated into physical ones or bus addresses because there is no guarantee that the backing memory is physically contiguous. It means that memory returned by vmalloc() can't be used outside of the microprocessor (you cannot easily use it for a DMA purpose). It is correct to use vmalloc() to allocate memory for a large sequence of pages (it does not make sense to use it to allocate one page, for example) that exists only in software, such as a network buffer. It is important to note that vmalloc() is slower than kmalloc() and page allocator functions because it must both retrieve the memory and build the page tables, or even remap into a virtually contiguous range, whereas kmalloc() never does that.

Before using the vmalloc() API, you should include this header:

```
#include <linux/vmalloc.h>
```

The following are the vmalloc family prototypes:

```
void *vmalloc(unsigned long size);
void *vzalloc(unsigned long size);
void vfree(void *addr);
```

In the preceding prototypes, argument size is the size of memory you need to allocate. Upon successful allocation of memory, it returns the address of the first byte of the allocated memory block. On failure, it returns NULL. vfree() does the reverse and releases the memory allocated by vmalloc(). The vzalloc variant returns zeroed initialized memory.

The following is an example of using vmalloc:

```
#include<linux/init.h>
#include<linux/module.h>
#include <linux/vmalloc.h>

Static void *ptr;
static int alloc_init(void)
{
    unsigned long size = 8192; /* 2 x 4KB */
    ptr = vmalloc(size);
    if(!ptr)
    {
```

```
        /* handle error */
        pr_err("memory allocation failed\n");
        return -ENOMEM;
    } else {
        pr_info("Memory allocated successfully\n");
    }
    return 0;
}

static void my_vmalloc_exit(void)
{
    vfree(ptr);
    pr_info("Memory freed\n");
}
module_init(my_vmalloc_init);
module_exit(my_vmalloc_exit);

MODULE_LICENSE("GPL");
MODULE_AUTHOR("john Madieu, john.madieu@gmail.com");
```

vmalloc() will allocate non-contiguous physical pages and map them to a contiguous virtual address region. These vmalloc virtual addresses are limited in an area of kernel space, delimited by VMALLOC_START and VMALLOC_END, which are architecture-dependent. The kernel exposes /proc/vmallocinfo to display all vmalloc-allocated memory on the system.

A short story about process memory allocation under the hood

vmalloc() prefers the HIGH_MEM zone if it exists, which is suitable for processes, as they require implicit and dynamic mappings. However, because memory is a limited resource, the kernel will report the allocation of frame pages (physical pages) until necessary (when accessed, either by reading or writing). This on-demand allocation is called **lazy allocation**, eliminating the risk of allocating pages that will never be used.

Whenever a page is requested, only the page table is updated; in most cases, a new entry is created, which means only virtual memory is allocated. An interrupt called `page fault` is raised only when a user accesses the page. This interrupt has a dedicated handler, called `page fault handler`, and is called by the MMU in response to an attempt to access virtual memory, which did not immediately succeed.

In fact, a page fault interrupt is raised whatever the access type is (read, write, or execute) to a page whose entry in the page table has not got the appropriate permission bits set to allow that type of access. The response to that interrupt falls in one of the following three ways:

- **The hard fault**: When the page does not reside anywhere (neither in the physical memory nor a memory-mapped file), which means the handler cannot immediately resolve the fault. The handler will perform I/O operations in order to prepare the physical page needed to resolve the fault and may suspend the interrupted process and switch to another while the system works to resolve the issue.

- **The soft fault**: When the page resides elsewhere in memory (in the working set of another process). It means the fault handler may resolve the fault by immediately attaching a page of physical memory to the appropriate page table entry, adjusting the entry, and resuming the interrupted instruction.

- **The fault cannot be resolved**: This will result in a bus error or **segmentation violation (segv)**. A **Segmentation Violation Signal (SIGSEGV)** is sent to the faulty process, killing it (the default behavior), unless a signal handler has been installed for the SIGSEV to change the default behavior.

To summarize, memory mappings generally start out with no physical pages attached, only by defining the virtual address ranges without any associated physical memory. The actual physical memory is allocated later in response to a page fault exception when the memory is accessed, since the kernel provides some flags to determine whether the attempted access was legal and specifies the behavior of the page fault handler. Thus, the `brk()` user space, `mmap()`, and similar allocate (virtual) space, but physical memory is attached later.

Note

A page fault occurring in interrupt context causes a double fault interrupt, which usually panics the kernel (calling the `panic()` function). It is the reason why memory allocated in interrupt context is taken from a memory pool, which does not raise a page fault interrupt. If an interrupt occurs when a double fault is being handled, a triple fault exception is generated, causing the CPU to shut down and the OS to immediately reboot. This behavior is architecture-dependent.

The Copy on Write case

Let's consider a memory region or data that needs to be shared by two or more tasks. The **Copy on Write** (**CoW**) (heavily used with `fork()` system call) is a mechanism that allows the operating system not to immediately allocate memory and does not make a copy of it to each task that shares this data, until one of these tasks modifies (writes into) it – in this case, memory is allocated for its private copy (hence the name, CoW). Let's consider a shared memory page to describe in in the following how the `page fault handler` manages CoW:

1. When the page needs to be shared, a page table entry (whose target is marked as un-writable) pointing to this shared page is added to the process page table of each process accessing the shared page. This is an initial mapping.

2. The mapping will result the creation of a VMA per process, which is added to the VMA list of each process. The shared page is associated these VMAs (that is, the VMA previously created for each process), which are marked as writeable this time. Nothing else will happen as long as no process tries to modify the content of the shared page.

3. When one of the processes tries to write into the shared page (at its first write), the `fault handler` notices the difference between the PTE flag (previously marked as un-writable) and the VMA flag (marked as writable), which means, *"Hey, this is a CoW."*. It will then allocate a physical page, which is assigned to the PTE added previously (thus replacing the shared page previously assigned), updating the PTE flags (one of these flags will correspond to marking the PTE as writeable), flushing the TLB entry, and then will execute the `do_wp_page()` function, which will copy the content from the shared address to the new location, which is private to the process that issued the write. Subsequent writes from this process will be made to the private copy, not in the shared page.

We can now close our parenthesis on process memory allocation, with which we are now familiar. We have also learned the lazy allocation mechanism and what CoW is. We can also conclude our learning about in-kernel memory allocation. At this point, we can switch to I/O memory operations to talk with hardware devices.

Working with I/O memory to talk to hardware

So far, we have dealt with main memory, and we used to think of memory in terms of RAM. That said, RAM one a peripheral among many others, and its memory range corresponds to its size. RAM is unique in the way it is entirely managed by the kernel, transparently for users. The RAM controller is connected to the CPU data/control/address buses, which it shares with other devices. These devices are referred to as memory-mapped devices because of their locality regarding those buses, and communication (input/output operations) with those devices is called memory-mapped I/O. These devices include controllers for various buses provided by the CPU (USB, UART, SPI, I2C, PCI, and SATA), but also IPs such as VPU, GPU, Image Processing Unit (IPU), and Secure Non-Volatile Store (SNVS, a feature in i.MX chips from NXP).

On a 32-bit system, the CPU has up to 2^{32} choices of memory locations (from 0 to 0xFFFFFFFF). The thing is that not all those addresses address RAM. Some of these are reserved for peripheral access and are called I/O memory. This I/O memory is split into ranges of various sizes and assigned to those peripherals so that whenever the CPU receives a physical memory access request from the kernel, it can route it to the device whose address range contains the specified physical address. The address range assigned to each device (including the RAM controller) is described in the SoC data sheet, in a section called **memory map** most of the time.

Since the kernel exclusively works with virtual addresses (through page tables), accessing a particular address for any device would require this address to be mapped first (this is even more true if there is an IOMMU, the MMU equivalent for I/O devices). This mapping of memory addresses other than RAM modules causes a classic hole in the system address space (because address space is shared between memory and I/O).

The following diagram describes how I/O memory and main memory are seen by the CPU:

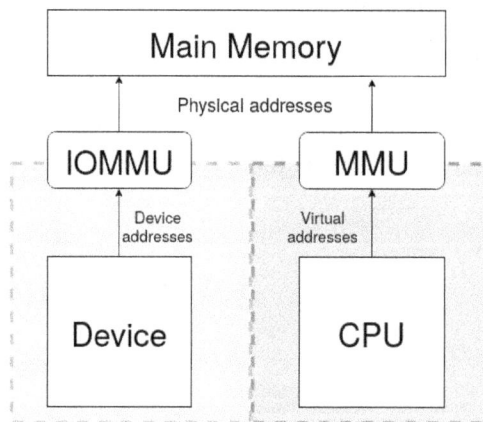

Figure 10.16 – (IO)MMU and main memory overview

> **Note**
>
> Always keep in mind that the CPU sees main memory (RAM) through the lenses of the MMU and devices through the lenses of IOMMU.

The main advantage with that is that the same instructions are used for transferring data to memory and I/O, which reduces software coding logic. There are some disadvantages, however. The first one is that the entire address bus must be fully decoded for every device, which increases the cost of adding hardware to the machine, leading to complex architecture.

The other inconvenience is that on a 32-bit system, even with 4 GB of RAM installed, the OS will never use the whole size because of the hole caused by memory-mapped devices, which stole part of the address space. x86 architectures adopted another approach called **Port Input Output** (**PIO**), with which registers are accessible via a dedicated bus by means of specific instructions (in and out, in an assembler, generally). In this case, device registers are not memory-mapped, and the system can address the whole address range for the RAM.

PIO device access

On a system where PIO is used, I/O devices are mapped into a separate address space. This is usually accomplished by having a different set of signal lines to indicate memory access versus device access. Such systems have two different address spaces, one for system memory, which we already discussed, and the other one for I/O ports, sometimes referred to as port address space and limited to 65,536 ports. This is an old method and very uncommon nowadays.

The kernel exports a few functions (symbols) to handle I/O ports. Prior to accessing any port regions, we must first inform the kernel that we are using a range of ports using the request_region() function, which will return NULL on error. Once done with the region, we must call release_region(). These are both declared in linux/ioport.h as the following:

```
struct ressource *request_region(unsigned long start,
                        unsigned long len, char *name);
void release_region(unsigned long start,
                        unsigned long len);
```

These are politeness functions that inform the kernel about your intention to make use/ release of a region of `len` ports, starting from `start`. The `name` parameter should be set with the name of the device or a meaningful one. Their use is not mandatory, however. It prevents two or more drivers from referencing the same range of ports. You can consult the ports currently in use on the system by reading the content of the `/proc/ioports` files.

After the region reservation has succeeded, the following APIs can be used to access the ports:

```
u8  inb(unsigned long addr)
u16 inw(unsigned long addr)
u32 inl(unsigned long addr)
```

The preceding functions respectively read 8, 16, or 32-bit-sized (wide) data from the `addr` ports. The write variants are defined as the following:

```
void outb(u8 b, unsigned long addr)
void outw(u16 b, unsigned long addr)
void outl(u32 b, unsigned long addr)
```

The preceding functions write b data, which can be 8, 16, or 32-bit-sized, into the `addr` port.

The fact that PIO uses a different set of instructions to access the I/O ports or MMIO is a disadvantage, as it requires more instructions than normal memory to accomplish the same task. For instance, 1-bit testing has only one instruction in MMIO, whereas PIO requires reading the data into a register before testing the bit, which is more than one instruction. One of the advantages of PIO is that it requires less logic to decode addresses, lowering the cost of adding hardware devices.

MMIO device access

Main memory addresses reside in the same address space as MMIO addresses. The kernel maps the device registers to a portion of the address space that would ordinarily be utilized by RAM so that instead of system memory (that is, RAM), I/O device registration takes place. As a result, talking with an I/O device is analogous to reading and writing to memory addresses dedicated to that device.

If we need to access, let's say, the 4 MB of I/O memory assigned to IPU-2 (from `0x02400000` to `0x027fffff`), the CPU (by means of the IOMMU) can assign to us the `0x10000000` to `0x103FFFFF` addresses, which are virtual of course. This is not consuming physical RAM (except for building and storing page table entry), just address space (do you see now why 32-bit systems run into issues with expansion cards such as high-end GPUs that have GB of RAM?), meaning that the kernel will no longer use this virtual memory range to map RAM. Now, a memory write/read to, say, `0x10000004` will be routed to the IPU-2 device. This is the basic premise of memory-mapped I/O.

Like PIO, there are MMIO functions to inform the kernel about our intention to use a memory region. Remember that this information is a pure reservation only. These are `request_mem_region()` and `release_mem_region()`, defined as the following:

```
struct ressource* request_mem_region(unsigned long start,
                           unsigned long len, char *name)
void release_mem_region(unsigned long start,
                     unsigned long len)
```

These are only politeness functions, though the former builds and returns an appropriate `resource` structure, corresponding to the start and length of the memory region, while the latter releases it.

For the device driver, however, it is recommended to use the managed variant, as it simplifies the code and takes care of releasing the resource. This managed version is defined as the following:

```
struct ressource* devm_request_region(
              struct device *dev, resource_size_t start,
              resource_size_t n, const char *name);
```

In the preceding, `dev` is the device owning the memory region, and the other parameters are the same as the non-managed version. Upon successful request, the memory region will be visible in `/proc/iomem`, which is a file that contains memory regions in use on the system.

Prior to accessing a memory region (and after you successfully request it), the region must be mapped into kernel address space by calling special architecture-dependent functions (which make use of IOMMU to build a page table and thus cannot be called from an interrupt handler). These are `ioremap()` and `iounmap()`, which handle cache coherency as well. The followings are their definitions:

```
void __iomem *ioremap(unsigned long phys_addr,
                      unsigned long size);
void iounmap(void __iomem *addr);
```

In the preceding functions, `phys_addr` corresponds to the device's physical address as specified in the device tree or in the board file. `size` corresponds to the size of the region to map. `ioremap()` returns a `__iomem void` pointer to the start of the mapped region. Once again, it is recommended to use the managed version, which has the following definition:

```
void __iomem *devm_ioremap(struct device *dev,
                           resource_size_t offset,
                           resource_size_t size);
```

> **Note**
>
> `ioremap()` builds new page tables, just as `vmalloc()` does. However, it does not actually allocate any memory but instead returns a special virtual address that can be used to access the specified I/O address. On 32-bit systems, the fact that MMIO steals physical memory address space to create a mapping for memory-mapped I/O devices is a disadvantage, since it prevents the system from using the stolen memory for general RAM purposes.

Because the mapping APIs are architecture-dependent, you should not deference (that is, getting/setting their value by reading/writing to the pointer) such pointers, even though on some architectures you can. The kernel provides portable functions to access memory-mapped regions. These are the following:

```
unsigned int ioread8(void __iomem *addr);
unsigned int ioread16(void __iomem *addr);
unsigned int ioread32(void __iomem *addr);
void iowrite8(u8 value, void __iomem *addr);
void iowrite16(u16 value, void __iomem *addr);
void iowrite32(u32 value, void __iomem *addr);
```

The preceding functions respectively read and write 8-, 16-, and 32-bit values.

> **Note**
>
> __iomem is a kernel cookie used by **Sparse**, a semantic checker used by the kernel to find possible coding faults. It prevents mixing normal pointer use (such as dereference) with I/O memory pointers.

In this section, we have learned how to map memory-mapped device memory into kernel address space to access its registers using dedicated APIs. This will serve in driving in-chip devices.

Memory (re)mapping

Kernel memory sometimes needs to be remapped, either from kernel to user space, or from high memory to a low memory region (from kernel to kernel space). The common case is remapping the kernel memory to user space, but there are other cases, such as when we need to access high memory.

Understanding the use of kmap

The Linux kernel permanently maps 896 MB of its address space to the lower 896 MB of the physical memory (low memory). On a 4 GB system, there is only 128 MB left to the kernel to map the remaining 3.2 GB of physical memory (high memory). However, low memory is directly addressable by the kernel because of the permanent and one-to-one mapping. When it comes to high memory (memory preceding 896 MB), the kernel has to map the requested region of high memory into its address space, and the 128 MB mentioned previously is especially reserved for this. The function used to perform this trick is kmap(). The kmap() function is used to map a given page into the kernel address space.

```
void *kmap(struct page *page);
```

page is a pointer to the struct page structure to map. When a high memory page is allocated, it is not directly addressable. kmap() is the function we call to temporarily map high memory into the kernel address space. The mapping will last until kunmap() is called:

```
void kunmap(struct page *page);
```

By *temporarily*, I mean the mapping should be undone as soon as it is no longer needed. A best programming practice is to unmap high memory mapping when it is no longer required.

This function works on both high and low memory. However, if a page structure resides in low memory, then just the virtual address of the page is returned (because low-memory pages already have permanent mappings). If the page belongs to high memory, a permanent mapping is created in the kernel's page tables, and the address is returned:

```
void *kmap(struct page *page)
{
    BUG_ON(in_interrupt());
    if (!PageHighMem(page))
        return page_address(page);

    return kmap_high(page);
}
```

`kmap_high()` and `kunmap_high()`, which are defined in `mm/highmem.c`, are at the heart of these implementations. However, `kmap()` maps pages into kernel space using a physically contiguous set of page tables allocated during the boot. Because the page tables are all connected, it's simple to move around without having to consult the page directory all the time. You should note that the `kmap` page tables correspond to kernel virtual addresses beginning with `PKMAP BASE`, which differs per architecture, and the reference count for its page table entries is kept in a separate array called `pkmap_count`.

The page frame of the page to map into kernel space is passed to `kmap()` as a `struct *page` argument, and this can be a regular or `HIGHMEM` page; in the first case, `kmap()` simply returns the direct-mapped address. For `HIGHMEM` pages, `kmap()` searches through the `kmap` page tables (which were allocated at boot time) for an unused entry – that is, an entry whose `pkmap_count` value is zero. If there are none, it goes to sleep and waits for another process to `kunmap` a page. When it finds an unused one, it inserts the physical page address of the page we want to map, incrementing at the same time the `pkmap_count` reference count corresponding to the page table entry, and returns the virtual address to the caller. The `page->virtual` for the page struct is also updated to reflect the mapped address.

`kunmap()` expects a `struct page*` representing the page to unmap. It finds the `pkmap_count` entry for the page's virtual address and decrements it.

Mapping kernel memory to user space

Mapping physical addresses is one of the most common operations, especially in embedded systems. Sometimes, you may want to share part of the kernel memory with user space. As mentioned earlier, the CPU runs in unprivileged mode when running in user space. To let a process access a kernel memory region, we need to remap that region into the process address space.

Using remap_pfn_range

remap_pfn_range() maps physical contiguous memory into a process address space by means of a VMA. It is particularly useful for implementing the mmap file operation, which is the backend of the mmap() system call.

After invoking the mmap() system call on a file descriptor (a device-backed file or not) given a region start and length, the CPU will switch to privileged mode. An initial kernel code will create an almost empty VMA as large as the requested mapping region and will run the corresponding file_operations.mmap callback, giving the VMA as a parameter. In turn, this callback should call remap_pfn_range(). This function will update the VMA and will derive the kernel's PTE of the mapped region before adding it to the process's page table, with different protection flags of course. The process's VMA list will be updated with the insertion of the VMA entry (with appropriate attributes), which will use the derived PTE to access the same memory. This way, the kernel and user space will both point to the same physical memory region, each through their own page tables but with different protection flags. Thus, instead of wasting memory and CPU cycles by copying, the kernel just duplicates the PTEs, each with their own attributes.

remap_pfn_range() is defined as the following:

```
int remap_pfn_range(struct vm_area_struct *vma,
                    unsigned long addr,
                    unsigned long pfn,
                    unsigned long size, pgprot_t flags);
```

A successful call will return 0, and a negative error code is returned on failure. Most of this function's arguments are provided when the mmap() system call is invoked. The following are their descriptions:

- vma: This is the virtual memory area provided by the kernel in case of the file_operations.mmap call. It corresponds to the user process's VMA, into which the mapping should be done.

- addr: This is the user (virtual) address where the VMA should start (vma->vm_start most of the time). It will result in a mapping from addr to addr + size.

- `pfn`: This represents the page frame number of the physical memory region to map. To obtain this page frame number, we must consider how the memory allocation was performed:

  - For memory allocated with `kmalloc()` or any other allocation API that returns a kernel logical address (`__get_free_pages()` with the `GFP_KERNEL` flag, for instance), `pfn` can be obtained as follows (obtaining the physical address and right-shifting this address's `PAGE_SHIFT` time):

    ```
    unsigned long pfn =
        virt_to_phys((void *)kmalloc_area)>>PAGE_SHIFT;
    ```

  - For memory allocated with `alloc_pages()`, we can use the following (where `page` is the pointer returned at allocation):

    ```
    unsigned long pfn = page_to_pfn(page)
    ```

  - Finally, for memory allocated with `vmalloc()`, the following can be used:

    ```
    unsigned long pfn = vmalloc_to_pfn(vmalloc_area);
    ```

- `size`: This is the dimension, in bytes, of the area being remapped. If it is not page-aligned, the kernel will take care of its alignment to the (next) page boundary.

- `flags`: This represents the protection requested for the new VMA. The driver can change the final values but should use the initial default values (found in `vma->vm_page_prot`) as a skeleton using the OR (`|` in the C language) operator. These default values are those which have been set by user space. Some of these flags are as follows:

  - `VM_IO`, which specifies a device's memory-mapped I/O.

  - `VM_PFNMAP`, to specify a page range managed without a baking `struct page`, just pure PFN. This is used most of the time for I/O memory mappings. In other words, it means that the base pages are just raw PFN mappings and do not have a struct page associated with them.

  - `VM_DONTCOPY`, which tells the kernel not to copy this VMA on a fork.

  - `VM_DONTEXPAND`, which prevents the VMA from expanding with `mremap()`.

  - `VM_DONTDUMP`, which prevents the VMA from being included in a core dump, even with `VM_IO` turned off.

Memory mapping works with memory regions that are multiples of PAGE_SIZE, so, for example, you should allocate an entire page instead of using a kmalloc-allocated buffer. kmalloc() can return (if requesting a non-multiple size of PAGE_SIZE) a pointer that isn't page-aligned, and in that case, it is a terribly bad idea to use such an unaligned address with remap_pfn_range(). Nothing will guarantee that the kmalloc()-returned address will be page-aligned, so you might corrupt slab internal data structures. Instead, you should be using kmalloc(PAGE_SIZE * npages or, even better, a page allocation API (or something similar because these functions always return a pointer that is page-aligned).

If your baking object (a file or device) supports an offset, then the VMA offset (the offset into the object where the mapping must start) should be considered to produce the PFN where mapping must start. vma->vm_pgoff will contain this offset (if specified by user space in the mmap()) value in units of the number of pages. The final PFN computation (or the position from where the mapping must start) will look like the following:

```
unsigned long pos
unsigned long off = vma->vm_pgoff;
/*compute the initial PFN according to the memory area */
[...]
/* Then compute the final position */
pos = pfn + off
[...]
return remap_pfn_range(vma, vma->vm_start,
        pos, vma->vm_end - vma->vm_start,
          vma->vm_page_prot);
```

In the preceding excerpt, the offset (specified in term of number of pages) has been included in the final position computation. This offset can, however, be ignored if the driver implementation does need its support.

> **Note**
>
> The offset can be used differently, by left-shifting PAGE_SIZE to obtain the offset by the number of bytes (offset = vma->vm_pgoff << PAGE_SHIFT), and then adding this offset to the memory start address before computing the final PFN (pfn = virt_to_phys(kmalloc_area + offset) >> PAGE_SHIFT).

Remapping vmalloc-allocated pages

Note that memory allocated with `vmalloc()` is not physically contiguous, so if you need to map a memory region allocated with `vmalloc()`, you must map each page individually and compute the physical address for each page. This can be achieved by looping over all pages in that `vmalloc`-allocated memory region and calling `remap_pfn_range()` as follows:

```
while (length > 0) {
    pfn = vmalloc_to_pfn(vmalloc_area_ptr);
    if ((ret = remap_pfn_range(vma, start, pfn,
      PAGE_SIZE, PAGE_SHARED)) < 0) {
        return ret;
    }
    start += PAGE_SIZE;
    vmalloc_area_ptr += PAGE_SIZE;
    length -= PAGE_SIZE;
}
```

In the preceding excerpt, `length` corresponds to the VMA size (`length = vma->vm_end - vma->vm_start`). `pfn` is computed for each page, and the starting address for the next mapping is incremented by `PAGE_SIZE` to map the next page in the region. The initial value of `start` is `start = vma->vm_start`.

That said, from within the kernel, the `vmalloc`-allocated memory can be used normally. The paginated use is necessary for remapping purposes only.

Remapping the I/O memory

Remapping the I/O memory requires a device's physical addresses, as specified in the device tree or the board file. In this case, for portability reasons, the appropriate function to use is `io_remap_pfn_range()`, whose parameters are the same as `remap_pfn_range()`. The only thing that changes is where the PFN comes from. Its prototype looks like the following:

```
int io_remap_page_range(struct vm_area_struct *vma,
                    unsigned long start,
                    unsigned long phys_pfn,
                    unsigned long size, pgprot_t flags);
```

In the preceding function, `vma` and `start` have the same meanings as `remap_pfn_range()`. `phys_pfn` is different, however, in the way it is obtained; it must correspond to the physical I/O memory address, as it will have been given to `ioremap()`, right-shifted `PAGE_SHIFT` times.

There is, however, a simplified `io_remap_pfn_range()` for common driver use: `vm_iomap_memory()`. This lite variant is defined as the following:

```
int vm_iomap_memory(struct vm_area_struct *vma,
                    phys_addr_t start, unsigned long len)
```

In the preceding function, `vma` is the user VMA to map to. `start` is the start of the I/O memory region to be mapped (as it would have been given to `ioremap()`), and `len` is the size of area. With `vm_iomap_memory()`, the driver just needs to give us the physical memory range to be mapped; the function will figure out the rest from the `vma` information. As with `io_remap_pfn_range()`, it returns 0 on success or a negative error code otherwise.

Memory remapping and caching issues

While caching is generally a good idea, it can introduce side effects, especially if, for a memory-mapped device (or even RAM), the values written to the mmap'ed registers must be instantaneously visible to the device.

You should note that, by default, the kernel remaps memory to user space with caching and buffering enabled. To change the default behavior, drivers must disable the cache on the VMA before invoking the remapping API. In order to do so, the kernel provides `pgprot_noncached()`. In addition to caching, this function also disables the bufferability of the specifier region. This helper takes an initial VMA access protection and returns an updated version with the cache disabled.

It is used as follows:

```
vma->vm_page_prot = pgprot_noncached(vma->vm_page_prot);
```

While testing a driver that I've developed for a memory-mapped device, I faced an issue where I had roughly 20 ms of latency (the time between when I updated the device register in user space through the mmap'ed area and the time when it was visible to the device) when caching was used.

After disabling the cache, this latency almost went away, as it fell below 200 µs. Amazing!

Implementing the mmap file operation

From user space, the `mmap()` system call is used to map physical memory into the address space of the calling process. In order to support this system call in a driver, this driver must implement the `file_operations.mmap` hook. After the mapping has been done, the user process will be able to write directly into the device memory via the returned address. The kernel will translate any accesses to that mapped region of memory through the usual pointer dereference into file operations.

The `mmap()` system call is declared as follows:

```
int mmap (void *addr, size_t len, int prot,
          int flags, int fd, ff_t offset);
```

From the kernel side, the `mmap` field in the driver's file operation structure (`struct file_operations` structure) has the following prototype:

```
int (*mmap)(struct file *filp,
            struct vm_area_struct *vma);
```

In the preceding file operation function, `filp` is a pointer to the open device file for the driver that results from the translation of the `fd` parameter (given in the system call). `vma` is allocated and given as parameter by the kernel. It points to the user process's VMA where the mapping should go. To understand how the kernel creates the new VMA, it uses the parameters given to the `mmap()` system call, which somehow affect some fields of the VMA as follows:

- `addr` is the user space's virtual address where the mapping should start. It has an impact on `vma>vm_start`. If `NULL` (the portable way), the kernel will automatically pick a free address.

- `len` specifies the length of the mapping and indirectly has an impact on `vma->vm_end`. Remember that the size of a VMA is always a multiple of `PAGE_SIZE`. It implies that `PAGE_SIZE` is the smallest size a VMA can have. If the `len` argument is not a page size multiple, it will be rounded up to the next highest page size multiple.

- `prot` affects the permission of the VMA, which the driver can find in `vma->vm_page_prot`.

- `flags` determines the type of mapping that the driver can find in `vma->vm_flags`. The mapping can be private or shared.

- `offset` specifies the offset within the mapped region. It is computed by the kernel so that it is stored in the `vma->vm_pgoff` in `PAGE_SIZE` unit.

With all these parameters defined, we can split the mmap file operation implementation into the following steps:

1. Get the mapping offset and check whether it is beyond our buffer size or not:

```
unsigned long offset = vma->vm_pgoff << PAGE_SHIFT;
if (offset >= buffer_size)
        return -EINVAL;
```

2. Check whether the mapping length is bigger than our buffer size:

```
unsigned long size = vma->vm_end - vma->vm_start;
if (buffer_size < (size + offset))
        return -EINVAL;
```

3. Compute the PFN that corresponds to the page where offset is located in the buffer. Note that the way the PFN is obtained depends on the way the buffer has been allocated:

```
unsigned long pfn;
pfn = virt_to_phys(buffer + offset) >> PAGE_SHIFT;
```

4. Set the appropriate flags, disabling caching if necessary:

   - Disable caching using vma->vm_page_prot = pgprot_noncached(vma->vm_page_prot);.

   - Set the VM_IO flag if necessary: vma->vm_flags |= VM_IO;. It also prevents the VMA from being included in the process's core dump.

   - Prevent the VMA from swapping out: vma->vm_flags |= VM_DONTEXPAND | VM_DONTDUMP. In kernel versions before 3.7, VM_RESERVED will be used.

5. Call remap_pfn_range() with the PFN calculated previously, size, and the protection flags. We will use vm_iomap_memory() in case of I/O memory mapping:

```
if (remap_pfn_range(vma, vma->vm_start, pfn,
                    size, vma->vm_page_prot)) {
    return -EAGAIN;
}
return 0;
```

6. Finally, pass the function to the `struct file_operations` structure:

```
static const struct file_operations my_fops = {
    .owner = THIS_MODULE,
    [...]
    .mmap = my_mmap,
    [...]
};
```

This file operation implementation closes our series on memory mappings. In this section, we have learned how mappings work under the hood and all the mechanisms involved, caching considerations included.

Summary

This chapter is one of the most important chapters. It demystifies memory management and allocation (how and where) in the Linux kernel. It teaches in detail how mapping and address translation work. Some other aspects, such as talking with hardware devices and remapping memory for user space (on behalf of the `mmap()` system call), were discussed in detail.

This provides a strong base to introduce and understand the next chapter, which deals with **Direct Memory Access** (**DMA**).

11
Implementing Direct Memory Access (DMA) Support

Direct Memory Access (DMA) is a feature of computer systems that allows devices to access the main system memory without CPU intervention, allowing the CPU to focus on other tasks. Examples of its usage include network traffic acceleration, audio data, or video frame grabbing, and its use is not limited to a particular domain. The peripheral responsible for managing the DMA transactions is the DMA controller, which is present in the majority of modern processors and microcontrollers.

The feature works in the following manner: When the driver needs to transfer a block of data, the driver sets up the DMA controller with the source address, the destination address, and the total number of bytes to copy. The DMA controller then transfers the data from the source to the destination automatically, without stealing CPU cycles. When the number of bytes remaining reaches zero, the block transfer ends, and the driver is notified.

> **Note**
>
> DMA does not always mean copy is going to be faster. It does not bring direct speed performance gains, but first, a true background operation, which leaves the CPU available to do other stuff, and then, performance gains due to sustaining the CPU cache/prefetcher state during DMA operation (which likely would be garbled when using plain old memcpy, executed on the CPU itself).

This chapter will deal with coherent and non-coherent DMA mappings, as well as coherency issues, the DMA engine's API, and DMA and DT bindings. More precisely, we will cover the following topics:

- Setting up DMA mappings
- Introduction to the concept of completion
- Working with the DMA engine's API
- Putting it all together – Single-buffer DMA mapping
- A word on cyclic DMA
- Understanding DMA and DT bindings

Setting up DMA mappings

For any type of DMA transfer, you need to provide source and destination addresses, as well as the number of words to transfer. In the case of peripheral DMA, this peripheral's FIFO acts as either the source or the destination, depending on the transfer direction. When the peripheral acts as the source, the destination address is a memory location (internal or external). When the peripheral acts as the destination, the source address is a memory location (internal or external).

In other words, a DMA transfer requires suitable memory mappings. This is what we will discuss in the following sections.

The concept of cache coherency and DMA

On a CPU equipped with a cache, copies of recently accessed memory areas are cached, even memory areas mapped for DMA. The reality is that memory shared between two independent devices is generally the source of cache coherency issues. Cache incoherency stems from the fact that other devices may not be aware of an update from another device writing. On the other hand, cache coherency ensures that every write operation appears to occur instantaneously, meaning that all devices sharing the same memory region see exactly the same sequence of changes.

A well-explained situation of coherency issues is illustrated in the following excerpt from the third edition of *Linux Device Drivers (LDD3)*:

> *"Let us imagine a CPU equipped with a cache and an external memory that can be accessed directly by devices using DMA. When the CPU accesses the location X in the memory, the current value will be stored in the cache. Subsequent operations on X will update the cached copy of X, but not the external memory version of X, assuming a write-back cache. If the cache is not flushed to the memory before the next time a device tries to access X, the device will receive a stale value of X. Similarly, if the cached copy of X is not invalidated when a device writes a new value to the memory, then the CPU will operate on a stale value of X."*

There are two ways to address this issue:

- A hardware-based solution. Such systems are coherent systems.

- A software-based solution, where the OS is responsible for ensuring cache coherency. Such systems are non-coherent systems.

Now that we are aware of the caching aspects of DMA, let's move a step forward and learn how to perform memory mappings for DMA.

Memory mappings for DMA

Memory buffers allocated for DMA purposes must be mapped accordingly. A DMA mapping consists of allocating a memory buffer suitable for DMA and generating a bus address for this buffer.

We distinguish between two types of DMA mappings – **coherent DMA mappings** and **streaming DMA mappings**. The former automatically addresses cache coherency issues, making it a good candidate for reuse over several transfers without unmapping in between transfers. This may entail considerable overhead on some platforms and, anyways, keeping memory synced has a cost. The streaming mapping has a lot of constraints in terms of coding and does not automatically address coherency issues, although there is a solution for that, which consists of several function calls between each transfer. Coherent mapping usually exists for the life of the driver, whereas one streaming mapping is usually unmapped once the DMA transfer completes.

> **Note**
>
> It is recommended to use streaming mapping when you can, and coherent mapping when you must. You should consider using coherent mapping if the buffer is accessed unpredictably by the CPU or the DMA controller since memory will always be synced. Otherwise, you should use streaming mapping because you know exactly when you need to access the buffer, in which case you'll first flush the cache (thereby syncing the buffer) before accessing the buffer.

The main header to include for handling DMA mappings is the following:

```
#include <linux/dma-mapping.h>
```

However, depending on the mapping, different APIs can be used. Before going further in the API, we need to understand the operations that are performed during DMA mappings:

1. Assuming the device supports DMA, if the driver sets up a buffer using `kmalloc()`, it will get a virtual address (let's call this X), which points nowhere yet.

2. The virtual memory system (helped by the **MMU**, the **Memory Management Unit**) will map X to a physical address (let's call this Y) in the system's RAM, assuming there is still free memory available.

    Because DMA does not flow through the CPU virtual memory system, the driver can use virtual address X to access the buffer at this point, but the device itself cannot.

3. In some simple systems (those without I/O MMU), the device can do DMA directly to physical the address Y. But in many others, devices see the main memory through the lenses of the I/O MMU; thus, there is I/O MMU hardware that translates DMA addresses to physical addresses, for example, it translates Z to Y.

4. This is where the DMA API intervenes:

    -   The driver can pass a virtual address X to a function such as `dma_map_single()` (which we will look at later in this chapter, in The *Single-buffer mapping* section), which sets up any appropriate I/O MMU mapping and returns the DMA address Z.

    -   The driver then instructs the device to do DMA into Z.

    -   The I/O MMU finally maps it to the buffer at address Y in the system's RAM.

Now that the concept of memory mapping for DMA has been introduced, we can start creating mappings, starting with the easiest ones – the coherent DMA mappings.

Creating coherent DMA mappings

Such mappings are most often used for long-lasting, bi-directional I/O buffers. The following function sets up a coherent mapping:

```
void *dma_alloc_coherent(struct device *dev, size_t size,
                         dma_addr_t *dma_handle, gfp_t flag)
```

This function is responsible for both the allocation and the mapping of the buffer. It returns a kernel virtual address for that buffer, which is `size` bytes wide and accessible by the CPU. The `size` parameter may be misleading as it is first given to `get_order()` APIs to get the page order that corresponds to this size. Consequently, this mapping is at least page-sized, and the number of pages the power of 2. `dev` is your device structure. The third argument is an output parameter that points to the associated bus address. Memory allocated for the mapping is guaranteed to be physically contiguous, and flags determine how memory should be allocated, which is usually `GFP_KERNEL`, or `GFP_ATOMIC` in an atomic context.

Do note that this mapping is said to be the following:

- Consistent (coherent) because the buffer content is always the same across all subsystems (either the device or the CPU)

- Synchronous, because a write by either the device or the CPU can immediately be read without worrying about cache coherency

To release the mapping, you can use the following API:

```
void dma_free_coherent(struct device *dev, size_t size,
                       void *cpu_addr, dma_addr_t dma_handle);
```

In the preceding prototype, `cpu_addr` and `dma_handle` correspond to the kernel virtual address and bus address returned by `dma_alloc_coherent()`. Those two parameters are required by the MMU (which returned the virtual address) and the I/O MMU (which returned the bus address) to release their mappings.

Creating streaming DMA mappings

Streaming DMA mapping memory buffers are typically mapped right before the transmission and unmapped afterward. Such mappings have more constraints and differ from coherent mappings for the following reasons:

- Mappings need to function with a buffer that has previously been allocated dynamically.

- Mappings may accept several non-contiguous and scattered buffers.

- For read transactions (device to CPU), buffers belong to the device, not to the CPU. Before the CPU can use the buffers, they should be unmapped first (after dma_unmap_{single,sg}()), or dma_sync_{single,sg}_for_cpu() must be invoked on those buffers. The main reason for this is caching purposes.

- For write transactions (CPU to device), the driver should place data in the buffer before establishing the mapping.

- The transfer direction has to be specified, and the data should move and should be used only based on this direction.

There are two forms of streaming mapping:

- Single-buffer mapping, which allows one physically contiguous buffer mapping

- Scatter/gather mapping, which allows several buffers to be passed (scattered over memory)

For both mappings, the transfer direction should be specified by a symbol of the enum dma_data_direction type, defined in include/linux/dma-direction.h, as follows:

```
enum dma_data_direction {
        DMA_BIDIRECTIONAL = 0,
        DMA_TO_DEVICE = 1,
        DMA_FROM_DEVICE = 2,
        DMA_NONE = 3,
};
```

In the preceding excerpt, each element is quite self-explanatory.

> **Note**
> Coherent mappings implicitly have a direction attribute setting set with
> `DMA_BIDIRECTIONAL`.

Now that we are aware of the two streaming DMA mapping methods, we can get the details of their implementations, starting with the single-buffer mappings.

Single-buffer mapping

Single-buffer mapping is a streaming mapping for occasional transfer. You can set up such a mapping using the `dma_map_single()` function, which has the following definition:

```
dma_addr_t dma_map_single(struct device *dev, void *ptr,
        size_t size, enum dma_data_direction direction);
```

The direction should be either `DMA_TO_DEVICE`, `DMA_FROM_DEVICE`, or `DMA_BIDIRECTIONAL`, respectively, when the CPU is the source (it writes to the device), when the CPU is the destination (it reads from the device), or when access is bi-directional for this mapping (implicitly used in coherent mappings). `dev` is the underlying `device` structure for your hardware device, `ptr` is an output parameter, and is the kernel virtual address of the buffer. This function returns an element of the `dma_addr_t` type, which is the bus address returned by the I/O MMU (if present) for the device so that the device can DMA into. You should use `dma_mapping_error()` (which must return 0 if no error occurred) to check whether the mapping returned a valid address and not go further in case of an error.

Such mapping can be released by the following function:

```
void dma_unmap_single(struct device *dev,
                      dma_addr_t dma_addr, size_t size,
                      enum dma_data_direction direction);

int dma_mapping_error(struct device *dev,
                      dma_addr_t dma_addr);
```

The other mapping is scatter/gather mappings, since memory buffers are spread (scattered) over the system on allocation and gathered by the driver.

Scatter/gather mappings

Scatter/gather mappings are a special type of streaming DMA mapping that allow the transfer of several memory buffers in a single shot, instead of mapping each buffer individually and transferring them one by one. Suppose you have several buffers that might not be physically contiguous, all of which need to be transferred at the same time to or from the device. This situation may occur due to the following:

- A readv or writev system call

- A disk I/O request

- Or simply a list of pages or a vmalloced region

Before you can issue such a mapping, you must set up an array of scatter elements, each of which should describe the mapping of an individual buffer. A scatter element is abstracted in the kernel as an instance of struct scatterlist, defined as follows:

```
struct scatterlist {
        unsigned long page_link;
        unsigned int     offset;
        unsigned int     length;
        dma_addr_t       dma_address;
        unsigned int     dma_length;
};
```

To set up a scatter list mapping, you should do the following:

- Allocate your scattered buffers.

- Create an array of scatter elements, initialize this array using sg_init_table() on it, and fill this array with allocated memory using sg_set_buf(). Note that each scatter element entry must be of page size, except the last one, which may not respect this rule.

- Call dma_map_sg() on the scatter list.

- Once done with DMA, call dma_unmap_sg() to unmap the scatter list.

The following is a diagram that describes most of the concepts of the scatter list:

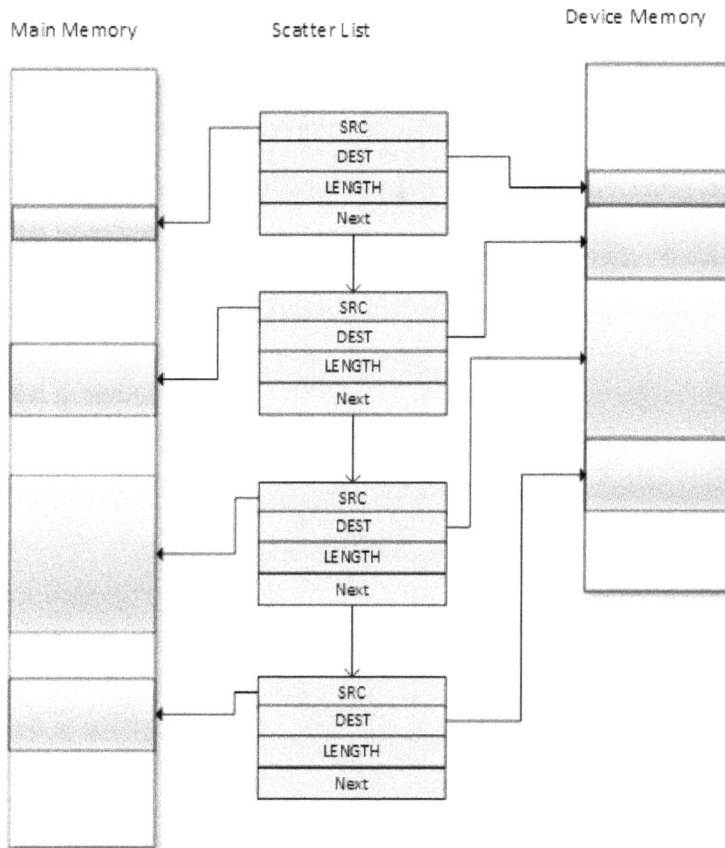

Figure 11.1 – Scatter/gather memory organization

While it is possible to DMA the content of several buffers individually, scatter/gather makes it possible to DMA the whole list at once by sending the pointer to the scatter list array to the device, along with its length, which is the number of entries in the array.

The prototypes of sg_init_table(), sg_set_buf(), and dma_map_sg() are as follows:

```
void sg_init_table(struct scatterlist *sgl,
                    unsigned int nents)
void sg_set_buf(struct scatterlist *sg, const void *buf,
                unsigned int buflen)
int dma_map_sg(struct device *dev,
               struct scatterlist *sglist, int nents,
               enum dma_data_direction dir);
```

In the preceding APIs, sgl is the scatterlist array to initialize and nents is the number of entries in this array. sg_set_buf() sets a scattlerlist entry to point at given data. In its parameters, sg is the scatterlist entry, data is the buffer corresponding to the entry, and buflen is the size of the buffer. dma_map_sg() returns the number of elements in the list that have been successfully mapped, which means it must never be less than zero. In the event of an error, this function returns zero.

The following is a code sample that demonstrates the principle of scatter/gather mapping:

```
u32 *wbuf, *wbuf2, *wbuf3;
wbuf = kzalloc(SDMA_BUF_SIZE, GFP_DMA);
wbuf2 = kzalloc(SDMA_BUF_SIZE, GFP_DMA);
wbuf3 = kzalloc(SDMA_BUF_SIZE/2, GFP_DMA);

struct scatterlist sg[3];
sg_init_table(sg, 3);
sg_set_buf(&sg[0], wbuf, SDMA_BUF_SIZE);
sg_set_buf(&sg[1], wbuf2, SDMA_BUF_SIZE);
sg_set_buf(&sg[2], wbuf3, SDMA_BUF_SIZE/2);
ret = dma_map_sg(dev, sg, 3, DMA_TO_DEVICE);
if (ret != 3) {
    /*handle this error*/
}
/* As of now you can use 'ret' or 'sg_dma_len(sgl)' to retrieve
the
 * length of the scatterlist array.
 */
```

The same rules described in the single-buffer mapping section apply to scatter/gather.

To unmap the list, you must use dma_unmap_sg(), which has the following definition:

```
void dma_unmap_sg_attrs(struct device *dev, struct scatterlist
*sg,
                         enum dma_data_direction dir, int
nents)
```

dev is a pointer to the same device that has been used for mapping, sg is the scatter list (actually a pointer to the first element in the list) to be unmapped, dir is the DMA direction, which should map the mapping direction, and nents is the number of elements in the list.

The following is an example that unmaps the previous implementation:

```
dma_unmap_sg(dev, sg, 3, DMA_TO_DEVICE);
```

In the preceding example, we used the same parameters that we used during the mapping.

Implicit and explicit cache coherency for streaming mapping

In either streaming mapping, dma_map_single()/dma_unmap_single() and dma_map_sg()/dma_unmap_sg() pairs take care of cache coherency when they are invoked. In the case of outgoing DMA transfer (CPU to device, DMA_TO_DEVICE direction flag set), since data must be in buffers before establishing the mapping, dma_map_sg()/dma_map_single() will handle cache coherency. In the case of device to CPU (DMA_FROM_DEVICE direction flag set), the mappings must be released first before the CPU can access the buffers. This is because dma_unmap_single()/dma_unmap_sg() implicitly take care of cache coherency as well.

However, if you need to use the same streaming DMA region numerous times and touch the data in between the DMA transfers, the buffer must be synced properly so that the device and CPU see the most up-to-date and correct copy of the DMA buffer. To avoid cache coherency issues, the driver must call dma_sync_{single,sg}_for_device() right before starting a DMA transfer from the RAM to the device (after you have put data in the buffer and before actually giving the buffer to the hardware). This function call will flush, if necessary, the cache lines corresponding to the DMA buffer. Similarly, the driver should not access the memory buffer immediately after completing the DMA transfer from the device to the RAM; instead, before reading the buffer, the driver should call dma_sync_{single,sg}_for_cpu(), which invalidates the associated hardware cache lines if necessary. In other words, when the source buffer is the device memory, the cache should be invalidated (cache data is not dirty as nothing has been written by the CPU to any buffer), whereas if the source is RAM (the destination is the device memory), this means the CPU may have written some data to the source buffer and the data may be in the cache line, hence the cache should be flushed.

The following are the prototypes of those syncing APIs:

```
void dma_sync_sg_for_cpu(struct device *dev,
                struct scatterlist *sg,
                int nents,
                enum dma_data_direction direction);
void dma_sync_sg_for_device(struct device *dev,
                struct scatterlist *sg, int nents,
                enum dma_data_direction direction);
```

```
void dma_sync_single_for_cpu(struct device *dev,
                    dma_addr_t addr, size_t size,
                    enum dma_data_direction dir)
void dma_sync_single_for_device(struct device *dev,
                    dma_addr_t addr, size_t size,
                    enum dma_data_direction dir)
```

In all of the preceding APIs, the direction parameter must remain the same as the direction specified during the mapping of the corresponding buffer.

In this section, we have learned to set up streaming DMA mappings. Now that we are done with mappings, let's introduce the concept of completion, which is used to notify a DMA transfer completion.

Introduction to the concept of completion

This section will briefly describe completion and the necessary part of its API that the DMA transfer uses. For a complete description, feel free to have a look at the kernel documentation at `Documentation/scheduler/completion.txt`. In kernel programming, a typical practice is to start some activity outside of the current thread and then wait for it to finish. Completions are good alternatives to waitqueues or sleeping APIs while waiting for a very commonly occurring process to complete. Completion variables are implemented using wait queues, with the only difference being that they make the developer's life easier as it does not require the wait queue to be maintained, which makes it very easy to see the intent of the code.

Working with completion requires this header:

```
#include <linux/completion.h>
```

A completion variable is represented in the kernel as an instance of struct completion structures that can be initialized statically as follows:

```
DECLARE_COMPLETION(my_comp);
```

Dynamic allocation of an initialization is done as follows:

```
struct completion my_comp;
init_completion(&my_comp);
```

When the driver initiates work whose completion must be awaited (a DMA transaction in our case), it just has to pass the completion event to the wait_for_completion() function, which has the following prototype:

```
void wait_for_completion(struct completion *comp);
```

When the completion occurs, the driver can wake the waiters using one of the following APIs:

```
void complete(struct completion *comp);
void complete_all(struct completion *comp);
```

complete() will wake up only one waiting task, while complete_all() will wake up every task waiting for that event. Completions are implemented in such a way that they will work properly even if complete() is called before wait_for_completion() is.

In this section, we have learned to implement a completion callback to notify the completeness status of a DMA transfer. Now that we are comfortable with all the common concepts of the DMA, we can start applying these concepts using the DMA engine APIs, which will also help us better understand how things work once everything is put together.

Working with the DMA engine's API

The DMA engine is a generic kernel framework used to develop DMA controller drivers and leverage this controller from the consumer side. Through this framework, the DMA controller driver exposes a set of channels that can be used by client devices. This framework then makes it possible for client drivers (also called slaves) to request and use DMA channels from the controller to issue DMA transfers.

The following diagram is the layering, showing how this framework is integrated with the Linux kernel:

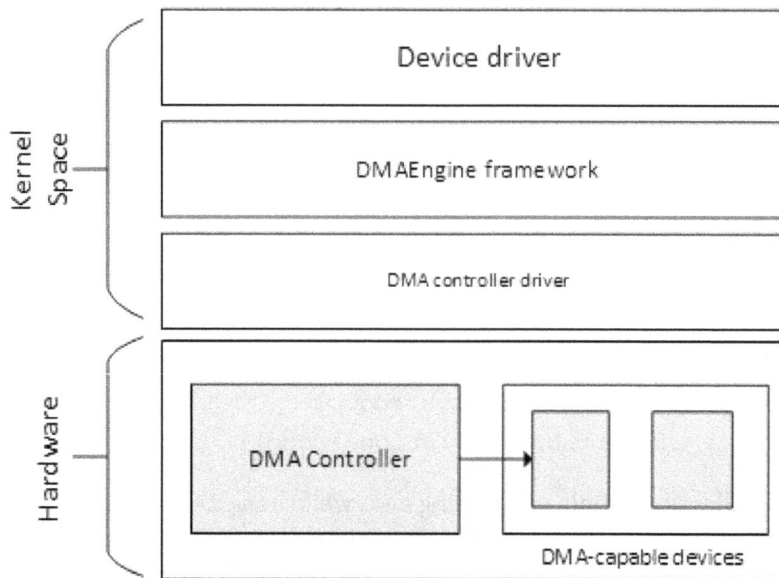

Figure 11.2 – DMA engine framework

Here we will simply walk through that (slave) API, which is applicable for slave DMA usage only. The mandatory header here is as follows:

```
#include <linux/dmaengine.h>
```

The slave DMA usage is straightforward, and consists of the following steps:

1. Informing the kernel about the device's DMA addressing capabilities.

2. Requesting a DMA channel.

3. If successful, configuring this DMA channel.

4. Preparing or configuring a DMA transfer. At this step, a transfer descriptor that represents the transfer is returned.

5. Submitting the DMA transfer using the descriptor. The transfer is then added to the controller's pending queue corresponding to the specified channel. This step returns a special cookie that you can use to check the progression of the DMA activity.

6. Starting the DMA transfers on the specified channel so that, if the channel is idle, the first transfer in the queue is started.

Now that we are aware of the steps needed to implement a DMA transfer, let's learn the data structures involved in the DMA engine framework before using the corresponding APIs.

A brief introduction to the DMA controller interface

The usage of DMA in Linux consists of two parts: the controllers, which perform memory transfer (without the CPU intervening), and the channels, which are the ways by which client drivers (that is, DMA-capable drivers) submit jobs to controllers. It goes without saying that both the controller and its channels are tightly coupled because the former exposes the latter to clients.

Although this chapter targets DMA client drivers, for the sake of understandability, we will be introducing some controller data structures and APIs.

The DMA controller data structure

The DMA controller is abstracted in the Linux kernel as an instance of `struct dma_device`. On its own, the controller is useless without clients, which would use the channels it exposes. Moreover, the controller driver must expose callbacks for channel configuration, as specified in its data structure, which has the following definition:

```
struct dma_device {
    unsigned int chancnt;
    unsigned int privatecnt;
    struct list_head channels;
    struct list_head global_node;
    struct dma_filter filter;
    dma_cap_mask_t  cap_mask;

    u32 src_addr_widths;
    u32 dst_addr_widths;
    u32 directions;
    int (*device_alloc_chan_resources)(
                                struct dma_chan *chan);
    void (*device_free_chan_resources)(
                                struct dma_chan *chan);
    struct dma_async_tx_descriptor
     *(*device_prep_dma_memcpy)(
        struct dma_chan *chan, dma_addr_t dst,
        dma_addr_t src, size_t len, unsigned long flags);
    struct dma_async_tx_descriptor
     *(*device_prep_dma_memset)(
        struct dma_chan *chan, dma_addr_t dest, int value,
```

```
        size_t len, unsigned long flags);
    struct dma_async_tx_descriptor
      *(*device_prep_dma_memset_sg)(
        struct dma_chan *chan, struct scatterlist *sg,
        unsigned int nents, int value,
        unsigned long flags);
    struct dma_async_tx_descriptor
      *(*device_prep_dma_interrupt)(
        struct dma_chan *chan, unsigned long flags);
    struct dma_async_tx_descriptor
      *(*device_prep_slave_sg)(
        struct dma_chan *chan, struct scatterlist *sgl,
          unsigned int sg_len,
          enum dma_transfer_direction direction,
          unsigned long flags, void *context);
    struct dma_async_tx_descriptor
      *(*device_prep_dma_cyclic)(
          struct dma_chan *chan, dma_addr_t buf_addr,
          size_t buf_len, size_t period_len,
          enum dma_transfer_direction direction,
          unsigned long flags);

    void (*device_caps)(struct dma_chan *chan,
                    struct dma_slave_caps *caps);
    int (*device_config)(struct dma_chan *chan,
                    struct dma_slave_config *config);
    void (*device_synchronize)(struct dma_chan *chan);

    enum dma_status (*device_tx_status)(
            struct dma_chan *chan, dma_cookie_t cookie,
            struct dma_tx_state *txstate);
    void (*device_issue_pending)(struct dma_chan *chan);
    void (*device_release)(struct dma_device *dev);
};
```

The complete definition of this data structure is available in `include/linux/ dmaengine.h`. For this chapter, only fields of our interest have been listed. Their meanings are as follows:

- `chancnt`: Specifies how many DMA channels are supported by this controller

- `channels`: The list of `struct dma_chan` structures, which corresponds to the DMA channels exposed by this controller

- `privatecnt`: How many DMA channels are requested by `dma_request_ channel()`, which is the DMA engine API to request a DMA channel

- `cap_mask`: One or more `dma_capability` flags, representing the capabilities of this controller

  The following are the possible values:

```
enum dma_transaction_type {
    DMA_MEMCPY,      /* Memory to memory copy */
    DMA_XOR,   /* Memory to memory XOR*/
    DMA_PQ,    /* Memory to memory P+Q computation */
    DMA_XOR_VAL, /* Memory buffer parity check using
                 * XOR */
    DMA_PQ_VAL,   /* Memory buffer parity check using
                 * P+Q */
    DMA_INTERRUPT,   /* The device can generate dummy
                     * transfer that will generate
                     * interrupts */
    DMA_MEMSET_SG,   /* Prepares a memset operation over a
                     * scatter list */
    DMA_SLAVE,       /* Slave DMA operation, either to or
                     * from a device */
    DMA_PRIVATE,     /* channels are not to be used
                     * for global memcpy. Usually
                     *used with DMA_SLAVE */
    DMA_SLAVE,       /* Memory to device transfers */
    DMA_CYCLIC,      /* can handle cyclic tranfers */
    DMA_INTERLEAVE, /* Memory to memory interleaved
                     * transfer */
}
```

As an example, this element is set in the i.MX DMA controller driver as follows:

```
dma_cap_set(DMA_SLAVE, sdma->dma_device.cap_mask);
dma_cap_set(DMA_CYCLIC, sdma->dma_device.cap_mask);
dma_cap_set(DMA_MEMCPY, sdma->dma_device.cap_mask);
```

- `src_addr_widths`: The bit mask of source address widths that the device supports. This width must be supplied in bytes; for example, if the device supports a width of 4, the mask should be set to `BIT(4)`.

- `dst_addr_widths`: The bit mask of destination address widths that the device supports.

- `directions`: The bit mask of slave directions supported by the device. Because enum `dma_transfer_direction` does not include a bit flag for each type, the DMA controller should set `BIT(<TYPE>)` and the same should be checked by the controller as well.

It is set in the i.MX SDMA controller driver as follows:

```
#define SDMA_DMA_DIRECTIONS (BIT(DMA_DEV_TO_MEM) | \
                             BIT(DMA_MEM_TO_DEV) | \
                             BIT(DMA_DEV_TO_DEV))
[...]
sdma->dma_device.directions = SDMA_DMA_DIRECTIONS;
```

- `device_alloc_chan_resources`: Allocates resources and returns the number of allocated descriptors. Invoked by the DMA engine core when requesting a channel on this controller.

- `device_free_chan_resources`: A callback allowing the release of the DMA channel's resources.

While the preceding was a generic callback, the following is a controller callback that depends on the controller capabilities and that must be provided if the associated capability bit masks are set in `cap_mask`.

- `device_prep_dma_memcpy` prepares a memcpy operation. If `DMA_MEMCPY` is set in `cap_mask`, then this element must be set. For each flag set, the corresponding callback must be provided, otherwise controller registration will fail. This is the case for all `device_prep_*` callbacks.

- `device_prep_dma_xor`: Prepares an XOR operation.

- `device_prep_dma_xor_val`: Prepares an xor validation operation.

- `device_prep_dma_memset`: Prepares a memset operation.

- `device_prep_dma_memset_sg`: Prepares a memset operation over a scatter list.

- `device_prep_dma_interrupt`: Prepares an end of chain interrupt operation.

- `device_prep_slave_sg`: Prepares a slave DMA operation.

- `device_prep_dma_cyclic`: Prepares a cyclic DMA operation. Such a DMA operation is frequently used in audio or UART drivers. A buffer of size `buf_len` is required by the function. The callback function will be called after `period_len` bytes have been transferred. We discuss such DMAs in the *A word on cyclic DMA* section.

- `device_prep_interleaved_dma`: Transfers expression in a generic way.

- `device_config`: Pushes a new configuration to a channel, with a return value of 0 in the event of success or an error code otherwise.

- `device_pause`: Pauses any current transfer on a channel and returns 0 or if the pausing is effective, or an error code otherwise.

- `device_resume`: Resumes any previously paused transfer on a channel. It returns 0 or an error code otherwise.

- `device_terminate_all`: A callback used to abort all the transfers on a channel, and which returns 0 in the event of success or an error code otherwise.

- `device_synchronize`: A callback allowing synchronization of the termination of a transfer to the current context.

- `device_tx_status`: Polls for transaction completion. The optional `txstate` parameter can be used to obtain a struct containing auxiliary transfer status information; otherwise, the call will just return a simple status code.

- `device_issue_pending`: A mandatory callback that pushes pending transactions to hardware. This is the backend of the `dma_async_issue_pending()` API.

While most drivers make a direct invocation of these callbacks (through `dma_chan->dma_dev->device_prep_dma_*`), you should be using the `dmaengine_prep_*` DMA engine APIs, which additionally do some sanity checks before invoking the appropriate callback. For example, for memory to memory, the driver should use the `device_prep_dma_memcpy()` wrapper.

The DMA channel data structure

A DMA channel is how a client driver submits DMA transactions (I/O data transfers) to the DMA controller. The way it works, a DMA-capable driver (client driver) requests one or more channels, reconfigures this channel, and asks the controller to use this channel to perform the submitted DMA transfer. A channel is defined as follows:

```
struct dma_chan {
    struct dma_device *device;
    struct device *slave;
    dma_cookie_t cookie;
    dma_cookie_t completed_cookie;
[...]
};
```

You can see a DMA channel as a highway for I/O data transfer. The following are the meanings of each element in this data structure:

- `device`: This is a pointer to the DMA device (the controller) that supplies this channel. This field can never be NULL if the channel has been requested successfully because a channel always belongs to a controller.

- `slave`: This is a pointer to the underlying `struct device` structure for the device using this channel (its driver is a client driver).

- `cookie`: This represents the last cookie value returned to the client by this channel.

- `Completed_cookie`: The last completed cookie for this channel.

The complete definition of this data structure can be found in `include/linux/dmaengine.h`.

> **Note**
> In the DMA engine framework, a cookie is nothing but a DMA transaction identifier that allows the status and progression of the transaction it identifies to be checked.

DMA transaction descriptor data structure

A transaction descriptor does nothing other than characterize and describe a DMA transaction (or DMA transfer by abuse of language). Such a descriptor is represented in the kernel using a `struct dma_async_tx_descriptor` data structure, which has the following definition:

```
struct dma_async_tx_descriptor {
    dma_cookie_t cookie;
    struct dma_chan *chan;
    dma_async_tx_callback callback;
    void *callback_param;
[...]
};
```

The meanings of each element we have retained in this data structure are set out here:

- `cookie`: A tracking cookie for this transaction. It allows the progression of this transaction to be checked.

- `chan`: The target channel for this operation.

- `callback`: A function that should be called once this operation is complete.

- `callback_param`: This is given as a parameter of the callback function.

You can find the complete data structure description in `include/linux/dmaengine.h`.

Handling device DMA addressing capabilities

The kernel considers that your device can handle 32-bit DMA addressing by default. However, the DMA memory address range your device can access may be limited, and this may be due to manufacturer or historical reasons. Some devices, for example, may only support the low order 24-bits of addressing. This limitation originated from the ISA bus, which was 24-bits wide and where DMA buffers could only live in the bottom 16 MB of the system's memory.

Nevertheless, you can use the concept of a DMA mask to inform the kernel of such limitations, which aims to inform the kernel of your device's DMA addressing capabilities.

This can be achieved using dma_set_mask_and_coherent(), which has the following prototype:

```
int dma_set_mask_and_coherent(struct device *dev,
                              u64 mask);
```

The preceding function will set the same mask for both streaming mappings and coherent mappings given that the DMA API guarantees that the coherent DMA mask can be set to the same or smaller than the streaming DMA mask.

However, for special requirements, you can use either dma_set_mask() or dma_set_coherent_mask() to set the mask accordingly. These APIs have the following prototypes:

```
int dma_set_mask(struct device *dev, u64 mask);
int dma_set_coherent_mask(struct device *dev, u64 mask);
```

In these functions, dev is the underlying device structure, while mask is a bit mask describing which bits of an address your device supports, which you can specify using the DMA_BIT_MASK macro along with the actual bit order.

Both dma_set_mask() and dma_set_coherent_mask() return zero to indicate that the device can perform DMA properly on the machine given the address mask specified. Any other return value would be an error, meaning that the given mask is too small to be supportable on the given system. In such a failure case, you can either fall back to non-DMA mode for data transfer in your driver or, if the DMA was mandatory, simply disable the feature in the device that required support for DMA or even not probe the device at all.

It is recommended that your driver prints a kernel warning (dev_warn() or pr_warn()) message when setting the DMA mask fails. The following is an example of pseudo-code for a sound card:

```
#define PLAYBACK_ADDRESS_BITS DMA_BIT_MASK(32)
#define RECORD_ADDRESS_BITS DMA_BIT_MASK(24)
struct my_sound_card *card;
struct device *dev;
...
if (!dma_set_mask(dev, PLAYBACK_ADDRESS_BITS)) {
    card->playback_enabled = 1;
} else {
    card->playback_enabled = 0;
    dev_warn(dev,
```

```
        "%s: Playback disabled due to DMA limitations\n",
        card->name);
}
if (!dma_set_mask(dev, RECORD_ADDRESS_BITS)) {
    card->record_enabled = 1;
} else {
    card->record_enabled = 0;
    dev_warn(dev,
            "%s: Record disabled due to DMA limitations\n",
            card->name);
}
```

In the preceding example, we have used the `DMA_BIT_MASK` macro to define the DMA mask. Then, we have disabled the features for which DMA support was mandatory when the required DMA mask was not supported. In either case, a warning is printed.

Requesting a DMA channel

A channel is requested using `dma_request_channel()`. Its prototype is as follows:

```
struct dma_chan *dma_request_channel(
                    const dma_cap_mask_t *mask,
                    dma_filter_fn fn, void *fn_param);
```

In the preceding, the mask must be a bit mask that represents the capabilities the channel must satisfy. It is essentially used to specify the type of transfer the driver needs to perform, which must be supported in `dma_device.cap_mask`.

The `dma_cap_zero()` and `dma_cap_set()` functions are used to clear the mask and set the capability we need; for example:

```
dma_cap_mask my_dma_cap_mask;
struct dma_chan *chan;
dma_cap_zero(my_dma_cap_mask);

/* Memory 2 memory copy */
dma_cap_set(DMA_MEMCPY, my_dma_cap_mask);
chan = dma_request_channel(my_dma_cap_mask, NULL, NULL);
```

fn is a callback pointer whose type has the following definition:

```
typedef bool (*dma_filter_fn)(struct dma_chan *chan,
                  void *filter_param);
```

Actually, dma_requaest_channel() walks through the available DMA controllers in the system (dma_device_list, defined in drivers/dma/dmaengine.c) and for each of them, it looks for a channel that corresponds to the request. If the filter_fn parameter (which is optional) is NULL, dma_request_channel() will simply return the first channel that satisfies the capability mask. Otherwise, when the mask parameter is insufficient for specifying the necessary channel, you can use the filter_fn routine as a filter so that each available channel in the system will be given to this callback for acceptance or not. The kernel calls the filter_fn routine once for each free channel in the system. Upon seeing a suitable channel, filter_fn should return DMA_ACK, which will tag the given channel to be the return value from dma_request_channel().

A channel allocated through this interface is exclusive to the caller until dma_release_channel() is called. It has the following definition:

```
void dma_release_channel(struct dma_chan *chan)
```

This API releases the DMA channel and makes it available for request by other clients.

By way of additional information, available DMA channels on a system can be listed in user space using the ls /sys/class/dma/ command as follows:

```
root@raspberrypi4-64:~# ls /sys/class/dma/
dma0chan0    dma0chan1    dma0chan2    dma0chan3    dma0chan4
dma0chan5    dma0chan6    dma0chan7    dma1chan0    dma1chan1
```

In the preceding snippet, the chan<chan-index> channel name is concatenated with the DMA controller, dma<dma-index>, to which it belongs. Whether a channel is in use or not can be seen by printing the in_use file value in the corresponding channel directory as follows:

```
root@raspberrypi4-64:~# cat /sys/class/dma/dma0chan0/in_use
1
root@raspberrypi4-64:~# cat /sys/class/dma/dma0chan1/in_use
1
root@raspberrypi4-64:~# cat /sys/class/dma/dma0chan2/in_use
1
root@raspberrypi4-64:~# cat /sys/class/dma/dma0chan3/in_use
```

```
0
root@raspberrypi4-64:~# cat /sys/class/dma/dma0chan4/in_use
0
root@raspberrypi4-64:~# cat /sys/class/dma/dma0chan5/in_use
0
root@raspberrypi4-64:~# cat /sys/class/dma/dma0chan6/in_use
0
root@raspberrypi4-64:~#
```

In the preceding, we can see, for example, that dma0chan1 is in use, while dma0chan6 is not.

Configuring the DMA channel

For the DMA transfer to operate normally on a channel, a client-specific configuration must be applied to this channel. Thereby, the DMA engine framework allows this configuration by using a struct dma_slave_config data structure, which represents the runtime configuration of a DMA channel. This allows clients to specify parameters such as the DMA direction, DMA addresses (source and destination), bus width, and DMA burst lengths, for the peripheral. This configuration is then applied to the underlying hardware using the dmaengine_slave_config() function, which is defined as follows:

```
int dmaengine_slave_config(struct dma_chan *chan,
                           struct dma_slave_config *config)
```

The chan parameter represents the DMA channel to configure, and config is the configuration to be applied.

To better fine-tune this configuration, we must look at the struct dma_slave_config structure, which is defined as follows:

```
struct dma_slave_config {
    enum dma_transfer_direction direction;
    phys_addr_t src_addr;
    phys_addr_t dst_addr;
    enum dma_slave_buswidth src_addr_width;
    enum dma_slave_buswidth dst_addr_width;
    u32 src_maxburst;
    u32 dst_maxburst;
```

```
    [...]
};
```

Here is the meaning of each element in the structure:

- `direction` indicates whether the data should go in or out on this slave channel, right now. The possible values are as follows:

```
/* dma transfer mode and direction indicator */
enum dma_transfer_direction {
    DMA_MEM_TO_MEM, /* Async/Memcpy mode */
    DMA_MEM_TO_DEV, /* From Memory to Device */
    DMA_DEV_TO_MEM, /* From Device to Memory */
    DMA_DEV_TO_DEV, /* From Device to Device */
    DMA_TRANS_NONE,
};
```

- `src_addr`: This is the physical address (the bus address actually) of the buffer where the DMA slave data should be read (RX). This element is ignored if the source is memory. `dst_addr` is the physical address (the bus address) of the buffer where the DMA slave data should be written (TX), which is ignored if the source is memory. `src_addr_width` is the width in bytes of the source (RX) register where the DMA data should be read. If the source is memory, this may be ignored depending on the architecture. In the same manner, `dst_addr_width` is the same as `src_addr_width`, but for the destination target (TX).

  Any bus width must be one of the following enumerations:

```
enum dma_slave_buswidth {
    DMA_SLAVE_BUSWIDTH_UNDEFINED = 0,
    DMA_SLAVE_BUSWIDTH_1_BYTE = 1,
    DMA_SLAVE_BUSWIDTH_2_BYTES = 2,
    DMA_SLAVE_BUSWIDTH_3_BYTES = 3,
    DMA_SLAVE_BUSWIDTH_4_BYTES = 4,
    DMA_SLAVE_BUSWIDTH_8_BYTES = 8,
    DMA_SLAVE_BUSWIDTH_16_BYTES = 16,
    DMA_SLAVE_BUSWIDTH_32_BYTES = 32,
    DMA_SLAVE_BUSWIDTH_64_BYTES = 64,
};
```

- `src_maxburs`: This is the maximum number of words that can be sent to the device in a single burst (consider words as units of the `src_addr_width` member, not bytes). On I/O peripherals, typically half the FIFO depth is used so that it does not overflow. On memory sources, this may or may not be applicable. `dst_maxburst` is similar to `src_maxburst`, but it is used for the destination target.

The following is an example of DMA channel configuration:

```
struct dma_chan *my_dma_chan;
dma_addr_t dma_src_addr, dma_dst_addr;
struct dma_slave_config channel_cfg = {0};

/* No filter callback, neither filter param */
my_dma_chan = dma_request_channel(my_dma_cap_mask,
                                    NULL, NULL);

/* scr_addr and dst_addr are ignored for mem to mem copy */
channel_cfg.direction = DMA_MEM_TO_MEM;
channel_cfg.dst_addr_width = DMA_SLAVE_BUSWIDTH_32_BYTES;

dmaengine_slave_config(my_dma_chan, &channel_cfg);
```

In the preceding excerpt, `dma_request_channel()` is used to request a DMA channel, which is then configured using `dmaengine_slave_config()`.

Configuring the DMA transfer

This step allows the type of transfer to be defined. A DMA transfer is configured (or should we say prepared) thanks to one of the `device_prep_dma_*` callbacks of the controller associated with the DMA channel to which the transfer will be submitted. Each of these APIs returns a transfer descriptor, represented by the `struct dma_async_tx_descriptor` data structure, which can be used later for customization before submitting the transfer.

For a memory-to-memory transfer, for example, you should be using the `device_prep_dma_memcpy` callback, as in the following code:

```
struct dma_device *dma_dev = my_dma_chan->device;
struct dma_async_tx_descriptor *tx_desc = NULL;
tx_desc = dma_dev->device_prep_dma_memcpy(
```

```
                            my_dma_chan, dma_dst_addr,
                            dma_src_addr, BUFFER_SIZE, 0);

if (!tx_desc) {
    /* dma_unmap_* the buffer */
    handle_error();
}
```

In the preceding code sample, we dereference the controller callback for invocation while
we could have checked for its existence first. However, for sanity and portability reasons, it is
recommended to use the dmaengine_prep_* DMA engine APIs instead of invoking the
controller callback directly. Our tx_desc assignation will then have the following form:

```
tx_desc = dmaengine_prep_dma_memcpy(my_dma_chan,
            dma_dst_addr, dma_src_addr, BUFFER_SIZE, 0);
```

This last approach is safer and portable regarding the controller data structure that may be
subject to changes.

Additionally, the client driver can use the callback element of the dma_async_tx_
descriptor structure (returned by the dmaengine_prep_* function) to supply a
completion callback.

Submitting the DMA transfer

To put the transaction in the driver pending queue, dmaengine_submit() is used,
which has the following prototype:

```
dma_cookie_t dmaengine_submit(
                    struct dma_async_tx_descriptor *desc)
```

This API is the frontend of the controller's device_issue_pending callback. This
function returns a cookie that you can use to check the progression of DMA activity
through other DMA engines. To check whether the returned cookie is valid, you can
use the dma_submit_error() helper, as we will see in the example. Assuming the
completion callback has not yet been provided, it can be set up before submitting the
transfer, as in the following excerpt:

```
struct completion transfer_ok;
init_completion(&transfer_ok);
/*
```

```
 * you can also set the parameter to be given to this
 * callback in tx->callback_param
 */
Tx_desc->callback = my_dma_callback;

/* Submitting our DMA transfer */
dma_cookie_t cookie = dmaengine_submit(tx);
if (dma_submit_error(cookie)) {
    /* handle error */
    [...]
}
```

The preceding excerpt is quite short and self-explanatory. For a parameter to be passed to the callback, it must be set in the descriptor's `callback_param` field. It can be a device state structure, for example.

> **Note**
>
> An interrupt (from the DMA controller) is raised after each DMA transfer has been completed, after which the next transfer in the queue is initiated and a tasklet is activated. If the client driver has provided a completion callback, the tasklet will call it when it is scheduled. Thus, the completion callback runs in an interrupt context.

Issuing pending DMA requests and waiting for callback notification

Starting the transaction is the last step of the DMA transfer setup. Transactions in the pending queue of a channel are activated by calling `dma_async_issue_pending()` on that channel. If the channel is idle, then the first transaction in the queue is started and subsequent ones are queued up. Upon completion of a DMA operation, the next one in the queue is started and a tasklet triggered. This tasklet is in charge of calling the client driver completion callback routine for notification, if set:

```
void dma_async_issue_pending(struct dma_chan *chan);
```

This function is a wrapper around the controller's `device_issue_pending` callback. An example of its usage would look like the following:

```
dma_async_issue_pending(my_dma_chan);
wait_for_completion(&transfer_ok);

/* may be unmap buffer if necessary and if it is not
 * done in the completion callback yet
 */
[...]
/* Process buffer through rx_data and tx_data virtual
addresses. */
[...]
```

The `wait_for_completion()` function will block, putting the current task to sleep until our DMA callback gets called to update (complete) our completion variable in order to resume the blocked code. It is a good alternative to `while (!done) msleep(SOME_TIME);`. The following is an example:

```
static void my_dma_complete_callback (void *param)
{
    complete(transfer_ok);
[...]
}
```

This is all in our DMA transfer implementation. When the completion callback returns, the main code will resume and continue its normal workflow.

Now that we have gone through the DMA engine APIs, we can summarize the knowledge in a complete example, as we see in the next section.

Putting it all together – Single-buffer DMA mapping

Let's consider the following case where we would like to map a single buffer (streaming mapping) and DMA data from the source, `src`, to the destination, `dst`. We will use a character device so that any write operation in this device will trig the DMA and any read operation will compare both the source and destination to check whether they match.

First, let's enumerate the header files required to pull the necessary APIs:

```
#define pr_fmt(fmt) "DMA-TEST: " fmt

#include <linux/module.h>
#include <linux/slab.h>
#include <linux/init.h>
#include <linux/dma-mapping.h>
#include <linux/fs.h>
#include <linux/dmaengine.h>
#include <linux/device.h>
#include <linux/io.h>
#include <linux/delay.h>
```

Let's now define some global variables for the driver:

```
/* we need page aligned buffers */
#define DMA_BUF_SIZE  2 * PAGE_SIZE

static u32 *wbuf;
static u32 *rbuf;
static int dma_result;
static int gMajor; /* major number of device */
static struct class *dma_test_class;
static struct completion dma_m2m_ok;
static struct dma_chan *dma_m2m_chan;
```

In the preceding, wbuf represents the source buffer, and rbuf represents the destination buffer. Since our implementation is based on a character device, gMajor and dma_test_class are used to represent the major number and the class of the character device.

Because DMA mappings need to be given a device structure as the first parameter, let's create a dummy one:

```
static void dev_release(struct device *dev)
{
    pr_info( "releasing dma capable device\n");
}
```

```
static struct device dev = {
    .release = dev_release,
    .coherent_dma_mask = ~0, // allow any address
    .dma_mask = &dev.coherent_dma_mask, // use the same mask
};
```

Because we have used a static device, we set the device's DMA mask in the device structure. In a platform driver, we would have used dma_set_mask_and_coherent() to achieve that.

The time has come to implement our first file operation, the open method, which in our case, simply allocates buffers:

```
int dma_open(struct inode * inode, struct file * filp)
{
    init_completion(&dma_m2m_ok);
    wbuf = kzalloc(DMA_BUF_SIZE, GFP_KERNEL | GFP_DMA);
    if(!wbuf) {
        pr_err("Failed to allocate wbuf!\n");
        return -ENOMEM;
    }

    rbuf = kzalloc(DMA_BUF_SIZE, GFP_KERNEL | GFP_DMA);
    if(!rbuf) {
        kfree(wbuf);
        pr_err("Failed to allocate rbuf!\n");
        return -ENOMEM;
    }
    return 0;
}
```

The preceding character device's open operation does nothing other than allocate the buffer that will be used for our transfer. These buffers will be freed when the device file is closed, which will result in invoking our device's release function, implemented as follows:

```
int dma_release(struct inode * inode, struct file * filp)
{
    kfree(wbuf);
    kfree(rbuf);
```

```
        return 0;
}
```

We arrive at the implementation of the read method. This method will simply add
an entry to the kernel message buffer, reporting the result of the DMA operation.
It is implemented as follows:

```
ssize_t dma_read (struct file *filp, char __user * buf,
                      size_t count, loff_t * offset)
{
        pr_info("DMA result: %d!\n", dma_result);
        return 0;
}
```

Now comes the DMA-related part. We first implement the completion callback, which
does nothing other than invoke complete() on our completion structure and add a
trace in the kernel log buffer. It is implemented as follows:

```
static void dma_m2m_callback(void *data)
{
        pr_info("in %s\n",__func__);
        complete(&dma_m2m_ok);
}
```

The choice has been made to implement all the DMA logic in the write method. There is
no technical reason behind this choice. A user is free to adapt the code architecture, based
on the following implementation:

```
ssize_t dma_write(struct file * filp,
                      const char __user * buf,
                      size_t count, loff_t * offset)
{
        u32 *index, i;
        size_t err = count;
        dma_cookie_t cookie;
        dma_cap_mask_t dma_m2m_mask;
        dma_addr_t dma_src, dma_dst;
        struct dma_slave_config dma_m2m_config = {0};
        struct dma_async_tx_descriptor *dma_m2m_desc;
```

In the preceding, there are variables we will require in order to perform our memory-to-memory DMA transfer.

Now that our variables are defined, we initialize the source buffer with some content that will later be copied to the destination with the DMA operation:

```
pr_info("Initializing buffer\n");
index = wbuf;
for (i = 0; i < DMA_BUF_SIZE/4; i++) {
    *(index + i) = 0x56565656;
}
data_dump("WBUF initialized buffer", (u8*)wbuf,
        DMA_BUF_SIZE);
pr_info("Buffer initialized\n");
```

The source buffer is ready, and we can now start the DMA-related code. At this first step, we initialize capabilities and request a DMA channel:

```
dma_cap_zero(dma_m2m_mask);
dma_cap_set(DMA_MEMCPY, dma_m2m_mask);
dma_m2m_chan = dma_request_channel(dma_m2m_mask,
                                    NULL, NULL);
if (!dma_m2m_chan) {
        pr_err("Error requesting the DMA channel\n");
        return -EINVAL;
} else {
        pr_info("Got DMA channel %d\n",
                dma_m2m_chan->chan_id);
}
```

In the preceding, the channel could have also registered with dma_m2m_chan = dma_request_chan_by_mask(&dma_m2m_mask);. The advantage of using this method is that only the mask has to be specified in a parameter, and the driver need not bother with other arguments.

In the second step, we set slave- and controller-specific parameters, and then we create the mappings for both source and destination buffers:

```
dma_m2m_config.direction = DMA_MEM_TO_MEM;
dma_m2m_config.dst_addr_width =
```

```
                          DMA_SLAVE_BUSWIDTH_4_BYTES;
    dmaengine_slave_config(dma_m2m_chan,
                                &dma_m2m_config);
    pr_info("DMA channel configured\n");


    /* Grab bus addresses to prepare the DMA transfer */
    dma_src = dma_map_single(&dev, wbuf, DMA_BUF_SIZE,
                            DMA_TO_DEVICE);
    if (dma_mapping_error(&dev, dma_src)) {
            pr_err("Could not map src buffer\n");
            err = -ENOMEM;
            goto channel_release;
    }
    dma_dst = dma_map_single(&dev, rbuf, DMA_BUF_SIZE,
                            DMA_FROM_DEVICE);
    if (dma_mapping_error(&dev, dma_dst)) {
            dma_unmap_single(&dev, dma_src,
                            DMA_BUF_SIZE, DMA_TO_DEVICE);
            err = -ENOMEM;
            goto channel_release;
    }
    pr_info("DMA mappings created\n");
```

In the third step, we grab a descriptor for the transaction:

```
    dma_m2m_desc =
        dmaengine_prep_dma_memcpy(dma_m2m_chan,
                        dma_dst, dma_src, DMA_BUF_SIZE,0);
    if (!dma_m2m_desc) {
            pr_err("error in prep_dma_sg\n");
            err = -EINVAL;
            goto dma_unmap;
    }
    dma_m2m_desc->callback = dma_m2m_callback;
```

Calling dmaengine_prep_dma_memcpy() results in invoking dma_m2m_chan->device->device_prep_dma_memcpy(). It is, however, recommended to use the DMA engine method since it is more portable.

In the fourth step, we submit the DMA transaction:

```
cookie = dmaengine_submit(dma_m2m_desc);
if (dma_submit_error(cookie)) {
        pr_err("Unable to submit the DMA coockie\n");
        err = -EINVAL;
        goto dma_unmap;
}
pr_info("Got this cookie: %d\n", cookie);
```

Now that the transaction has been submitted, we can move to the fifth and final step, where we issue pending DMA requests and wait for callback notification:

```
dma_async_issue_pending(dma_m2m_chan);
pr_info("waiting for DMA transaction...\n");


/* you also can use wait_for_completion_timeout() */
wait_for_completion(&dma_m2m_ok);
```

At this point in the code, the DMA transaction has run until completion, and we can check whether source and destination buffers have the same content. However, before accessing the buffers, they must be synced; luckily, the unmapping methods perform an implicit buffer sync:

```
dma_unmap:
    /* we do not care about the source anymore */
    dma_unmap_single(&dev, dma_src, DMA_BUF_SIZE,
                        DMA_TO_DEVICE);
    /* unmap the DMA memory destination for CPU access.
     * This will sync the buffer */
    dma_unmap_single(&dev, dma_dst, DMA_BUF_SIZE,
                        DMA_FROM_DEVICE);
    /*
     * if no error occured, then we are safe to access
     * the buffer. The buffer must be synced first, and
     * thanks to dma_unmap_single(), it is.
     */
    if (err >= 0) {
        pr_info("Checking if DMA succeed ...\n");
```

```
        for (i = 0; i < DMA_BUF_SIZE/4; i++) {
            if (*(rbuf+i) != *(wbuf+i)) {
                pr_err("Single DMA buffer copy falled!,
                        r=%x,w=%x,%d\n",
                        *(rbuf+i), *(wbuf+i), i);
                return err;
            }
        }


        pr_info("buffer copy passed!\n");
        dma_result = 1;
        data_dump("RBUF DMA buffer", (u8*)rbuf,
                DMA_BUF_SIZE);
    }
channel_release:
    dma_release_channel(dma_m2m_chan);
    dma_m2m_chan = NULL;

    return err;
}
```

In the preceding write operation, we have gone through the five steps required to perform our DMA transfer: requesting a DMA channel; configuring this channel; preparing a DMA transfer; submitting this transfer; and then triggering the transfer providing a completion callback in the meantime.

After we are done with operation definitions, we can set up a file operation data structure as follows:

```
struct file_operations dma_fops = {
    .open = dma_open,
    .read = dma_read,
    .write = dma_write,
    .release = dma_release,
};
```

Now that the file operation has been set up, we can implement the module's `init` function, where we create and register the character device as follows:

```c
int __init dma_init_module(void)
{
    int error;
    struct device *dma_test_dev;
    /* register a character device */
    error = register_chrdev(0, "dma_test", &dma_fops);
    if (error < 0) {
      pr_err("DMA test driver can't get major number\n");
        return error;
    }
    gMajor = error;
    pr_info("DMA test major number = %d\n",gMajor);
    dma_test_class = class_create(THIS_MODULE,
                                    "dma_test");
    if (IS_ERR(dma_test_class)) {
        pr_err("Error creating dma test module class.\n");
        unregister_chrdev(gMajor, "dma_test");
        return PTR_ERR(dma_test_class);
    }
    dma_test_dev = device_create(dma_test_class, NULL,
                    MKDEV(gMajor, 0), NULL, "dma_test");
    if (IS_ERR(dma_test_dev)) {
        pr_err("Error creating dma test class device.\n");
        class_destroy(dma_test_class);
        unregister_chrdev(gMajor, "dma_test");
        return PTR_ERR(dma_test_dev);
    }

    dev_set_name(&dev, "dmda-test-dev");
    device_register(&dev);
    pr_info("DMA test Driver Module loaded\n");
    return 0;
}
```

The module initialization will create and register a character device. This operation must be reverted when the module is unloaded, that is, in the module's exit method, implemented as follows:

```
static void dma_cleanup_module(void)
{
    unregister_chrdev(gMajor, "dma_test");
    device_destroy(dma_test_class, MKDEV(gMajor, 0));
    class_destroy(dma_test_class);
    device_unregister(&dev);
    pr_info("DMA test Driver Module Unloaded\n");
}
```

At this point, we can register our module's init and exit methods with the driver core and provide metadata for our module. This is done as follows:

```
module_init(dma_init_module);
module_exit(dma_cleanup_module);


MODULE_AUTHOR("John Madieu, <john.madieu@laabcsmart.com>");
MODULE_DESCRIPTION("DMA test driver");
MODULE_LICENSE("GPL");
```

The full code is available in the repository of the book in the chapter-12/ directory.

Now that we are familiar with the DMA engine APIs and have summarized our skills in a concrete example, we can discuss a particular DMA transfer, the Cyclic DMA, mostly used in UART drivers.

A word on cyclic DMA

Cyclic mode is a particular DMA transfer mode where an I/O peripheral drives the data transaction, triggering transfers repeatedly on a periodic basis. While dealing with callbacks that the DMA controller can expose, we have seen dma_device.device_prep_dma_cyclic, which is the backend for dmaengine_prep_dma_cyclic(), which has the following prototype:

```
struct dma_async_tx_descriptor
    *dmaengine_prep_dma_cyclic(
        struct dma_chan *chan, dma_addr_t buf_addr,
```

```
            size_t buf_len, size_t period_len,
            enum dma_transfer_direction dir,
            unsigned long flags)
```

The preceding API takes in five parameters: chan, which is the allocated DMA channel structure; buf_addr, the handle to the mapped DMA buffer; buf_len, which is the size of the DMA buffer; period_len, the size of one cyclic period; dir, the direction of the DMA transfer; and flags, the control flags for this transfer. In the event of success, this function returns a DMA channel descriptor structure, which can be used to assign a completion function to the DMA transfer. Most of the time, flags correspond to DMA_PREP_INTERRUPT, which means that the DMA transfer callback should be invoked upon each cycle completion.

Cyclic mode is mostly used in TTY drivers, where the data is fed into a **First In First Out (FIFO)** ring buffer. In this mode, the allocated DMA buffer is divided into periods equal in size (often referenced as cyclic periods) so that every time one such transfer is finished, the callback function is invoked.

The callback function that has been implemented is used to keep track of the state of the ring buffer and buffer management is implemented using the kernel ring buffer API (so you need to include <linux/circ_buf.h>):

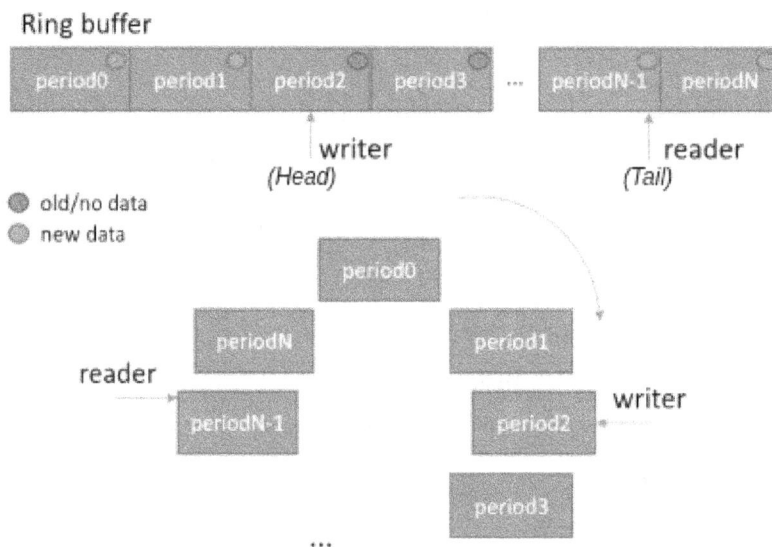

Figure 11.3 – Cyclic DMA ring buffer

The following is an example from the Atmel serial driver in drivers/tty/serial/atmel_serial.c, which demonstrates this principle of cyclic DMA quite well.

The driver first prepares the DMA resources as in the following:

```
static int atmel_prepare_rx_dma(struct uart_port *port)
{
    struct atmel_uart_port *atmel_port =
                        to_atmel_uart_port(port);
    struct device *mfd_dev = port->dev->parent;
    struct dma_async_tx_descriptor *desc;
    dma_cap_mask_t          mask;
    struct dma_slave_config config;
    struct circ_buf         *ring;
    int ret, nent;

    ring = &atmel_port->rx_ring;
    dma_cap_zero(mask);
    dma_cap_set(DMA_CYCLIC, mask);

    atmel_port->chan_rx =
                dma_request_slave_channel(mfd_dev, "rx");
    sg_init_one(&atmel_port->sg_rx, ring->buf,
                    sizeof(struct atmel_uart_char) *
                    ATMEL_SERIAL_RINGSIZE);
    nent = dma_map_sg(port->dev, &atmel_port->sg_rx, 1,
                        DMA_FROM_DEVICE);

    /* Configure the slave DMA */
    [...]
    ret = dmaengine_slave_config(atmel_port->chan_rx,
                            &config);
    /* Prepare a cyclic dma transfer, assign 2
     * descriptors, each one is half ring buffer size */
    desc =
        dmaengine_prep_dma_cyclic(atmel_port->chan_rx,
            sg_dma_address(&atmel_port->sg_rx),
            sg_dma_len(&atmel_port->sg_rx),
            sg_dma_len(&atmel_port->sg_rx)/2,
            DMA_DEV_TO_MEM, DMA_PREP_INTERRUPT);
```

```
    desc->callback = atmel_complete_rx_dma;
    desc->callback_param = port;
    atmel_port->desc_rx = desc;
    atmel_port->cookie_rx = dmaengine_submit(desc);

    dma_async_issue_pending(chan);
    return 0;

chan_err:
[...]
}
```

For the sake of readability, error checking has been omitted. The function starts by setting the appropriate DMA capability mask (using dma_set_cap()) before requesting the DMA channel. After the channel has been requested, the mapping (a streaming one) is created and the channel is configured using dmaengine_slave_config(). Thereafter, a cyclic DMA transfer descriptor is obtained thanks to dmaengine_prep_dma_cyclic() and DMA_PREP_INTERRUPT is there to instruct the DMA engine core to invoke the callback at the end of each cycle transfer. The descriptor obtained is then configured with the callback along with its parameter before being submitted to the DMA controller using dmaengine_submit() and fired with dma_async_issue_pending().

The atmel_complete_rx_dma() callback will schedule a tasklet whose handler is atmel_tasklet_rx_func() and which will invoke the real DMA completion callback, atmel_rx_from_dma(), implemented as follows:

```
static void atmel_rx_from_dma(struct uart_port *port)
{
    struct atmel_uart_port *atmel_port =
                                to_atmel_uart_port(port);
    struct tty_port *tport = &port->state->port;
    struct circ_buf *ring = &atmel_port->rx_ring;
    struct dma_chan *chan = atmel_port->chan_rx;
    struct dma_tx_state state;
    enum dma_status dmastat;
    size_t count;
    dmastat = dmaengine_tx_status(chan,
                atmel_port->cookie_rx, &state);
```

```c
/* CPU claims ownership of RX DMA buffer */
dma_sync_sg_for_cpu(port->dev, &atmel_port->sg_rx, 1,
                    DMA_FROM_DEVICE);
/* The current transfer size should not be larger
 * than the dma buffer length.
 */
ring->head =
    sg_dma_len(&atmel_port->sg_rx) - state.residue;

/* we first read from tail to the end of the buffer
 * then reset tail */
if (ring->head < ring->tail) {
    count =
        sg_dma_len(&atmel_port->sg_rx) - ring->tail;
    tty_insert_flip_string(tport,
                        ring->buf + ring->tail, count);
    ring->tail = 0;
    port->icount.rx += count;
}

/* Finally we read data from tail to head */
if (ring->tail < ring->head) {
    count = ring->head - ring->tail;
    tty_insert_flip_string(tport,
                        ring->buf + ring->tail, count);
    /* Wrap ring->head if needed */
    if (ring->head >= sg_dma_len(&atmel_port->sg_rx))
        ring->head = 0;
    ring->tail = ring->head;
    port->icount.rx += count;
}

/* USART retrieves ownership of RX DMA buffer */
dma_sync_sg_for_device(port->dev, &atmel_port->sg_rx,
                        1, DMA_FROM_DEVICE);
```

```
    [...]
    tty_flip_buffer_push(tport);
[...]
}
```

In the DMA completion callback, we can see that before the buffer is being accessed by the CPU, dma_sync_sg_for_cpu() is invoked to invalidate the corresponding hardware cache lines. Then, some ring buffers and TTY-related operations are performed (respectively, reading the received data and forwarding it to the TTY layer). And finally, the buffer is given back to the device after dma_sync_sg_for_device() is invoked.

To summarize, the preceding example did not only show how cyclic DMA works but also showed how to address coherency issues when the buffer is used and reused between transfers, either by the CPU or the device.

Now that we are familiar with the cyclic DMA, we have concluded our series on DMA transfer and DMA engine APIs. We have learned how to set up transfers, initiate them, and await their completion.

In the next section, we will learn how to specify and grab DMA channels from the device tree and the code.

Understanding DMA and DT bindings

DT binding for the DMA channel depends on the DMA controller node, which is SoC-dependent, and some parameters (such as DMA cells) may vary from one SoC to another. This example only focuses on the i.MX SDMA controller, which can be found in the kernel source, at Documentation/devicetree/bindings/dma/fsl-imx-sdma.txt.

Consumer binding

According to the SDMA event-mapping table, the following code shows the DMA request signals for peripherals in i.MX 6Dual/6Quad:

```
uart1: serial@02020000 {
    compatible = "fsl,imx6sx-uart", "fsl,imx21-uart";
    reg = <0x02020000 0x4000>;
    interrupts = <GIC_SPI 26 IRQ_TYPE_LEVEL_HIGH>;
    clocks = <&clks IMX6SX_CLK_UART_IPG>,
             <&clks IMX6SX_CLK_UART_SERIAL>;
```

```
      clock-names = "ipg", "per";

      dmas = <&sdma 25 4 0>, <&sdma 26 4 0>;

      dma-names = "rx", "tx";

      status = "disabled";

};
```

The second cells (25 and 26) in the dma property correspond to the DMA request/event ID. Those values come from the SoC manuals (i.MX53 in our case). You can have a look at https://community.nxp.com/servlet/JiveServlet/download/614186-1-373516/iMX6_Firmware_Guide.pdf and the Linux reference manual at https://community.nxp.com/servlet/JiveServlet/download/614186-1-373515/i.MX_Linux_Reference_Manual.pdf.

The third cell indicates the priority of use. The driver code to request a specified parameter is defined next. You can find the complete code in drivers/tty/serial/imx.c in the kernel source tree. The following is the excerpt of the code grabbing elements from the device tree:

```
static int imx_uart_dma_init(struct imx_port *sport)
{
    struct dma_slave_config slave_config = {};
    struct device *dev = sport->port.dev;
    int ret;

    /* Prepare for RX : */
    sport->dma_chan_rx =
                dma_request_slave_channel(dev, "rx");
    if (!sport->dma_chan_rx)
        /* cannot get the DMA channel. handle error */
        [...]

    [...] /* configure the slave channel */
    ret = dmaengine_slave_config(sport->dma_chan_rx,
                                 &slave_config);
[...]

    /* Prepare for TX */
    sport->dma_chan_tx =
                dma_request_slave_channel(dev, "tx");
```

```
    if (!sport->dma_chan_tx) {
        /* cannot get the DMA channel. handle error */
        [...]

    [...] /* configure the slave channel */
    ret = dmaengine_slave_config(sport->dma_chan_tx,
                                    &slave_config);
    if (ret) {
        [...] /* handle error */
    }
    [...]
}
```

The magic call here is dma_request_slave_channel(), which will parse the device node (in the DT) using of_dma_request_slave_channel() to gather channel settings, according to the DMA channel name (refer to the named resource in *Chapter 6, Understanding and Leveraging the Device Tree*).

Summary

DMA is a feature that is found in many modern CPUs. This chapter gives you the necessary steps to get the most out of this device, using the kernel DMA mapping and DMA engine APIs. After this chapter, I have no doubt you will be able to set up at least a memory-to-memory DMA transfer. Further information can be found at Documentation/dmaengine/, in the kernel source tree.

However, the next chapter deals with the regmap, which introduces memory-oriented abstractions, and which unify access to memory-oriented devices (I2C, SPI, or memory-mapped).

12

Abstracting Memory Access – Introduction to the Regmap API: a Register Map Abstraction

Before the Regmap API was developed, there was redundant code for the device drivers dealing with SPI, I2C, or memory-mapped devices. Many of these drivers contained some very similar code for accessing hardware device registers.

The following figure shows how SPI, I2C, and memory-mapped related APIs were used standalone before Regmap was introduced:

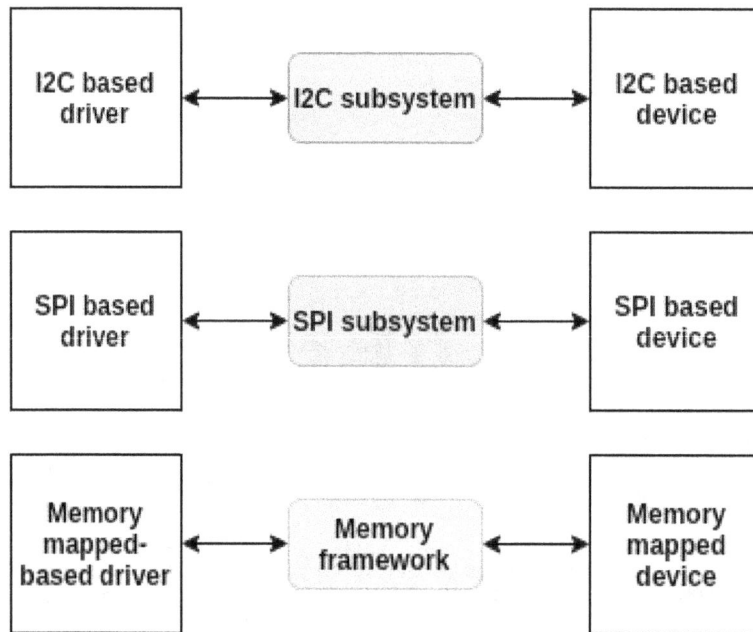

Figure 12.1 – I2C, SPI, and memory-mapped access before Regmap

The Regmap API was introduced in version v3.1 of the Linux kernel and proposes a solution that factors out and unifies these similar register access codes, saving code and making it much easier to share infrastructure. It is then just a matter of how to initialize and to configure a regmap structure, and process any read/write/modify operations fluently, whether it is SPI, I2C, or memory-mapped.

The following diagram depicts this API unification:

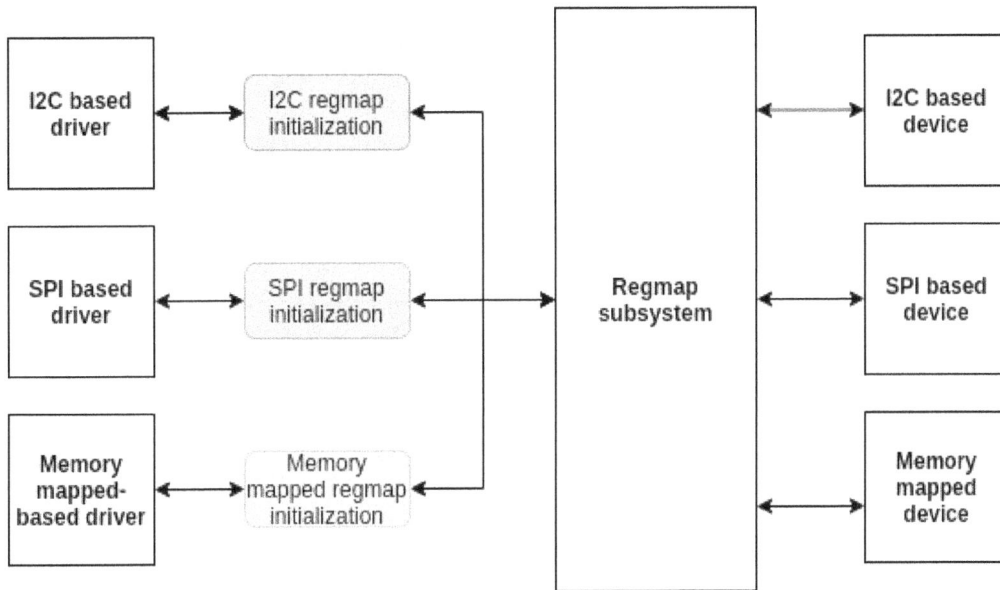

Figure 12.2 - I2C, SPI, and memory-mapped access after Regmap

The previous figure shows how Regmap unified transactions between devices and their respective bus frameworks. In this chapter, we will cover as much as possible of the whole aspect of the APIs this framework offers, from initialization to complex use cases.

This chapter will walk through the Regmap framework via the following topics:

- Introduction to the Regmap data structures
- Handling Regmap initialization
- Using Regmap register access functions
- Regmap-based SPI driver example – putting it all together
- Leveraging Regmap from the userspace

Introduction to the Regmap data structures

The Regmap framework, which is enabled via the `CONFIG_REGMAP` kernel configuration option, is made of a few data structures, among which the most important are `struct regmap_config`, which represents the Regmap configuration, and `struct regmap`, which is the Regmap instance itself. That said, all of the Regmap data structures are defined in `include/linux/regmap.h`. It then goes without saying that this header must be included in all Regmap-based drivers:

```
#include <linux/regmap.h>
```

Including the preceding header is sufficient to make the most out of the Regmap framework. With this header, a lot of data structures will be made available, among which, `struct regmap_config` is the most important, which we will describe in the next section.

Understanding the struct regmap_config structure

`struct regmap_config` stores the configuration of the register map during the driver's lifetime. What you set there affects the memory read/write operations. This is the most important structure, which is defined as follows:

```
struct regmap_config {
    const char *name;
    int reg_bits;
    int reg_stride;
    int pad_bits;
    int val_bits;
    bool (*writeable_reg)(struct device *dev,
                            unsigned int reg);
    bool (*readable_reg)(struct device *dev,
                            unsigned int reg);
    bool (*volatile_reg)(struct device *dev,
                            unsigned int reg);
    bool (*precious_reg)(struct device *dev,
                            unsigned int reg);
    bool disable_locking;
    regmap_lock lock;
    regmap_unlock unlock;
    void *lock_arg;
```

```
        int (*reg_read)(void *context, unsigned int reg,
                        unsigned int *val);
        int (*reg_write)(void *context, unsigned int reg,
                        unsigned int val);
        bool fast_io;
        unsigned int max_register;
        const struct regmap_access_table *wr_table;
        const struct regmap_access_table *rd_table;
        const struct regmap_access_table *volatile_table;
        const struct regmap_access_table *precious_table;
[...]
        const struct reg_default *reg_defaults;
        unsigned int num_reg_defaults;
        enum regcache_type cache_type;
        const void *reg_defaults_raw;
        unsigned int num_reg_defaults_raw;

        unsigned long read_flag_mask;
        unsigned long write_flag_mask;

        bool use_single_rw;
        bool can_multi_write;

        enum regmap_endian reg_format_endian;
        enum regmap_endian val_format_endian;
        const struct regmap_range_cfg *ranges;
        unsigned int num_ranges;
}
```

Don't be afraid of how big this structure is. All the elements are self-explanatory. However, for more clarity, let's expand on their meanings here:

- `reg_bits` is a mandatory field, which is the number of valid bits in a register's address. This is the size in bits of register addresses.

- `reg_stride` represents a value that valid register addresses must be a multiple of. If set to 0, a value of 1 will be used. If set to 4, for example, an address will be considered valid only if this address is a multiple of 4.

- `pad_bits` is the number of bits of padding between the register and value. This is the number of bits to (left) shift the register value when formatting.

- `val_bits` represents the number of bits used to store a register's value. It is a mandatory field.

- `writeable_reg` is an optional callback function. If provided, it is used by the Regmap subsystem when a register needs to be written. Before writing into a register, this function is automatically called to check whether the register can be written to or not. The following is an example of using such a function:

```
static bool foo_writeable_register(struct device *dev,
                                    unsigned int reg)
{
    switch (reg) {
    case 0x30 ... 0x38:
    case 0x40 ... 0x45:
    case 0x50 ... 0x57:
    case 0x60 ... 0x6e:
    case 0x70 ... 0x75:
    case 0x80 ... 0x85:
    case 0x90 ... 0x95:
    case 0xa0 ... 0xa5:
    case 0xb0 ... 0xb2:
        return true;
    default:
        return false;
    }
}
```

- `readable_reg` is the same as `writeable_reg` but for all register read operations.

- volatile_reg is an optional callback function called every time a register needs to be read or written through the Regmap cache. If the register is volatile, the function should return true. A direct read/write is then performed on the register. If false is returned, it means the register is cacheable. In this case, the cache will be used for a read operation, and the cache will be written in the case of a write operation:

```
static bool foo_volatile_register(struct device *dev,
                                  unsigned int reg)
{
    switch (reg) {
    case 0x24 ... 0x29:
    case 0xb6 ... 0xb8:
        return true;
    default:
        return false;
    }
}
```

- precious_reg: Some devices are sensitive to reads on some of their registers, especially for things such as clear on read interrupt status registers. With this set, this optional callback must return true if the specified register falls in this case, which will present the core (debugfs, for example) from internally generating any reads of it. This way, only explicit reads by the driver will be allowed.

- disable_locking tells whether the following lock/unlock callbacks should be used or not. If false, it means not to use any locking mechanisms. It means this regmap object is either protected by external means or is guaranteed not to be accessed from multiple threads.

- lock/unlock are optional lock/unlock callbacks, overriding default lock/unlock functions of regmap, based on spinlock or mutex, depending on whether accessing the underlying device may put the caller to sleep or not.

- lock_arg will be used as the only argument of lock/unlock functions (ignored if regular lock/unlock functions are not overridden).

- reg_read: Your device may not support simple I2C/SPI read operations. You'll then have no choice but to write your own customized read function. reg_read should then point to that function. That said, most devices do not need that.

- reg_write is the same as reg_read but for write operations.

- `fast_io` indicates that the register IO is fast. If set, the `regmap` will use a spinlock instead of a mutex to perform locking. This field is ignored if custom lock/unlock (not discussed here) functions are used (see the fields `lock`/`unlock` of `struct regmap_config` in the kernel sources). It should be used only for "nobus" cases (MMIO devices), since accessing I2C, SPI, or similar buses may put the caller to sleep.

- `max_register`: This optional element specifies the maximum valid register address above which no operation is permitted.

- `wr_table`: Instead of providing a `writeable_reg` callback, you could provide a `regmap_access_table` object, which is a structure holding a `yes_ranges` and a `no_range` field, both pointers to `struct regmap_range`. Any register that belongs to a `yes_range` entry is considered as writeable and is considered as not writeable if belonging to `no_range`.

- `rd_table` is the same as `wr_table`, but for any read operation.

- `volatile_table`: Instead of `volatile_reg`, you could provide `volatile_table`. The principle is then the same as `wr_table` or `rd_table`, but for caching mechanisms.

- `precious_table`: As above, for precious registers.

- `reg_defaults` is an array of elements of type `reg_default`, where each `reg_default` element is a `{reg, value}` structure that represents power-on reset values for a register. This is used with the cache so that a read of an address that exists in this array, and that has not been written since power-on reset, will return the default register value in this array without performing any read transactions on the device. An example of this is the IIO device driver, whose link is the following: `https://elixir.bootlin.com/linux/v5.10/source/drivers/iio/light/apds9960.c`.

- `num_reg_defaults` is the number of elements in `reg_defaults`.

- `cache_type`: The actual cache type, which can be either `REGCACHE_NONE`, `REGCACHE_RBTREE`, `REGCACHE_COMPRESSED`, or `REGCACHE_FLAT`.

- `read_flag_mask`: This is the mask to be applied in the top bytes of the register when doing a read. Normally, the highest bit in the top byte of a write or read operation in SPI or I2C is set to differentiate write and read operations.

- `write_flag_mask`: The mask to be set in the top bytes of the register when doing a write.

- `use_single_rw` is a Boolean that, if set, will instruct the register map to convert any bulk write or read operation on the device into a series of single write or read operations. This is useful for devices that do not support bulk read or write, or either.

- `can_multi_write` only targets write operations. If set, it indicates that this device supports the multi-write mode of bulk write operations. If clear, multi-write requests will be split into individual write operations.

You should look at `include/linux/regmap.h` for more details on each element. The following is an example of the initialization of `regmap_config`:

```
static const struct regmap_config regmap_config = {
    .reg_bits       = 8,
    .val_bits       = 8,
    .max_register   = LM3533_REG_MAX,
    .readable_reg   = lm3533_readable_register,
    .volatile_reg   = lm3533_volatile_register,
    .precious_reg   = lm3533_precious_register,
};
```

The preceding example shows how to build a basic register map configuration. Though only a few elements are set in the configuration data structure, enhanced configuration can be set up by learning about each element that we have described.

Now that we have learned about Regmap configuration, let's see how to use this configuration with the initialization API that corresponds to our needs.

Handling Regmap initialization

As we said earlier, the Regmap API supports SPI, I2C, and memory-mapped register access. Their respective support can be enabled in the kernel thanks to the CONFIG_REGMAP_SPI, CONFIG_REGMAP_I2C, and CONFIG_REGMAP_MMIO kernel configuration options. It can go far beyond that and managing IRQs as well, but this is out of the scope of this book. Depending on the memory access method you need to support in the driver, you will have to call either devm_regmap_init_i2c(), devm_regmap_init_spi(), or devm_ regmap_init_mmio() in the probe function. To write generic drivers, Regmap is the best choice you can make.

The Regmap API is generic and homogenous, and initialization only changes between bus types. Other functions are the same. It is a good practice to always initialize the register map in the probe function, and you must always fill the `regmap_config` elements prior to initializing the register map using one of the following APIs:

```
struct regmap *devm_regmap_init_spi(struct spi_device *spi,
                              const struct regmap_config);
struct regmap *devm_regmap_init_i2c(struct i2c_client *i2c,
                              const struct regmap_config);
struct regmap * devm_regmap_init_mmio(
                    struct device *dev,
                    void __iomem *regs,
                    const struct regmap_config *config)
```

These are resource-managed APIs whose allocated resources are automatically freed when the device leaves the system or when the driver is unloaded. In the preceding prototypes, the return value will be a pointer to a valid `struct regmap` object or an `ERR_PTR()` error on failure. `regs` is a pointer to a memory-mapped IO region (returned by `devm_ioremap_resource()` or any `ioremap*` family function). `dev` is the device (`struct device`) to interact with in the case of a memory-mapped `regmap`, and `spi` and `i2c` are respectively SPI or I2C devices to interact with in the case of SPI- or I2C-based `regmap`.

Calling one of these functions is sufficient to start interacting with the underlying device. Whether the Regmap is an I2C, SPI, or a memory-mapped register map, if it has not been initialized with a resource managed API variant, it must be freed with the `regmap_exit()` function:

```
void regmap_exit(struct regmap *map)
```

This function simply releases a previously allocated register map.

Now that the register access method has been defined, we can jump to the device access functions, which allow reading from or writing into device registers.

Using Regmap register access functions

Remap register access methods handle data parsing, formatting, and transmission. In most cases, device accesses are performed with `regmap_read()`, `regmap_write()`, and `regmap_update_bits()`, which are the three important APIs when it comes to writing/reading data into/from the device. Their respective prototypes are the following:

```
int regmap_read(struct regmap *map, unsigned int reg,
                unsigned int *val);
int regmap_write(struct regmap *map, unsigned int reg,
                unsigned int val);
int regmap_update_bits(struct regmap *map,
                unsigned int reg, unsigned int mask,
                unsigned int val);
```

`regmap_write()` writes data to the device. If set in `regmap_config`, `max_register` will be used to check whether the register address that needs to be accessed is greater or lower. If the register address passed is lower or equal to `max_register`, then the next operation will be performed; otherwise, the Regmap core will return an invalid I/O error (`-EIO`). Right after, the `writeable_reg` callback is called. The callback must return true before going to the next step. If it returns false, then `-EIO` is returned, and the write operation is stopped. If `wr_table` is set instead of `writeable_reg`, then the following happens:

- If the register address lies in `no_ranges`, then `-EIO` is returned.

- If the register address lies in `yes_ranges`, the next step is performed.

- If the register address is not present in `yes_range` or `no_range`, then `-EIO` is returned, and the operation is terminated.

If `cache_type != REGCACHE_NONE`, then caching is enabled. In this case, the cache entry is first updated with the new value, and then a write to the hardware is performed. Otherwise, no caching action is performed. If the `reg_write` callback is provided, it is used to perform the write operation. Otherwise, the generic Regmap's write function will be executed to write the data into the specified register address.

`regmap_read()` reads data from the device. It works exactly like `regmap_write()` with appropriate data structures (`readable_reg` and `rd_table`). Therefore, if provided, `reg_read` is used to perform the read operation; otherwise, the generic register map read function will be performed.

`regmap_update_bits()` is a three-in-one function. It performs a read/modify/write cycle on the specified register address. It is a wrapper on `_regmap_update_bits`, which looks like the following:

```
static int _regmap_update_bits(struct regmap *map,
             unsigned int reg, unsigned int mask,
             unsigned int val, bool *change,
             bool force_write)
{
    int ret;
    unsigned int tmp, orig;
    if (change)
        *change = false;
    if (regmap_volatile(map, reg) &&
                    map->reg_update_bits) {
        ret = map->reg_update_bits(map->bus_context,
                                    reg, mask, val);
        if (ret == 0 && change)
            *change = true;
    } else {
        ret = _regmap_read(map, reg, &orig);
        if (ret != 0)
            return ret;

        tmp = orig & ~mask;
        tmp |= val & mask;
        if (force_write || (tmp != orig)) {
            ret = _regmap_write(map, reg, tmp);
            if (ret == 0 && change)
                *change = true;
        }
    }
    return ret;
}
```

This way, the bits you need to update must be set to 1 in `mask`, and corresponding bits should be set to the value you need to give to them in `val`.

As an example, to set the first and third bits to 1, mask should be 0b00000101, and the value should be 0bxxxxx1x1. To clear the seventh bit, mask must be 0b01000000 and the value should be 0bx0xxxxxx, and so on.

Bulk and multiple registers reading/writing APIs

regmap_multi_reg_write() is one of the APIs that allows you to write multiple registers to the device. Its prototype looks like this:

```
int regmap_multi_reg_write(struct regmap *map,
                           const struct reg_sequence *regs,
                           int num_regs)
```

In this prototype, regs is an array of elements of type reg_sequence, which represents register/value pairs for sequences of writes with an optional delay in microseconds to be applied after each write. The following is the definition of this data structure:

```
struct reg_sequence {
    unsigned int reg;
    unsigned int def;
    unsigned int delay_us;
};
```

In the preceding data structure, reg is the register address, def is the register value, and delay_us is the delay to be applied after the register write in microseconds.

The following is a usage of such a sequence:

```
static const struct reg_sequence foo_default_regs[] = {
    { FOO_REG1,      0xB8 },
    { BAR_REG1,      0x00 },
    { FOO_BAR_REG1,  0x10 },
    { REG_INIT,      0x00 },
    { REG_POWER,     0x00 },
    { REG_BLABLA,    0x00 },
};
static int probe ( ...)
{
    [...]
    ret = regmap_multi_reg_write(my_regmap,
```

```
                                foo_default_regs,
                                ARRAY_SIZE(foo_default_regs));
    [...]
}
```

In the preceding, we have learned how to use our first multi-register write API, which takes a set of registers along with their values.

There are also `regmap_bulk_read()` and `regmap_bulk_write()`, which can be used to read/write multiple registers from/to the device. Their usage fits for large blocks of data and they are defined as follows:

```
int regmap_bulk_read(struct regmap *map,
                    unsigned int reg, void *val,
                    size_tval_count);
int regmap_bulk_write(struct regmap *map,
                    unsigned int reg,
                    const void *val, size_t val_count);
```

In the parameters in the preceding functions, `map` is the register map to operate on and `reg` is the register address from where the read/write operation must start. In the case of the read, `val` will contain the read value; it must be allocated to store at least the `count` value in the native register size for the device. In the case of the write operation, `val` must point to the data array to write to the device. Finally, `count` is the number of elements in `val`.

Understanding the Regmap caching system

Obviously, Regmap supports data caching. Whether the cache system is used or not depends on the value of the `cache_type` field in `regmap_config`. Looking at `include/linux/regmap.h`, accepted values are as follows:

```
/* An enum of all the supported cache types */
enum regcache_type {
    REGCACHE_NONE,
    REGCACHE_RBTREE,
    REGCACHE_COMPRESSED,
    REGCACHE_FLAT,
};
```

The cache type is set to REGCACHE_NONE by default, meaning that the cache is disabled. Other values simply define how the cache should be stored.

Your device may have a predefined power-on reset value in certain registers. Those values can be stored in an array so that any read operation returns the value contained in the array. However, any write operation affects the real register in the device and updates the content in the array. It is a kind of a cache that we can use to speed up access to the device. That array is reg_defaults. Looking at the source, its structure looks like this:

```
struct reg_default {
    unsigned int reg;
    unsigned int def;
};
```

In the preceding data structure, reg is the register address and def is the register default value. reg_defaults is ignored if cache_type is set to none. If the default_reg element is not set but you still enable the cache, the corresponding cache structure will be created for you.

It is quite simple to use. Just declare it and pass it as a parameter to the regmap_config structure. Let's have a look at the LTC3589 regulator driver in drivers/regulator/ltc3589.c:

```
static const struct reg_default ltc3589_reg_defaults[] = {
{ LTC3589_SCR1,   0x00 },
{ LTC3589_OVEN,   0x00 },
{ LTC3589_SCR2,   0x00 },
{ LTC3589_VCCR,   0x00 },
{ LTC3589_B1DTV1, 0x19 },
{ LTC3589_B1DTV2, 0x19 },
{ LTC3589_VRRCR,  0xff },
{ LTC3589_B2DTV1, 0x19 },
{ LTC3589_B2DTV2, 0x19 },
{ LTC3589_B3DTV1, 0x19 },
{ LTC3589_B3DTV2, 0x19 },
{ LTC3589_L2DTV1, 0x19 },
{ LTC3589_L2DTV2, 0x19 },
};
static const struct regmap_config ltc3589_regmap_config = {
        .reg_bits = 8,
```

```
        .val_bits = 8,
        .writeable_reg = ltc3589_writeable_reg,
        .readable_reg = ltc3589_readable_reg,
        .volatile_reg = ltc3589_volatile_reg,
        .max_register = LTC3589_L2DTV2,
        .reg_defaults = ltc3589_reg_defaults,
        .num_reg_defaults = ARRAY_SIZE(ltc3589_reg_defaults),
        .use_single_rw = true,
        .cache_type = REGCACHE_RBTREE,
};
```

Any read operation on any one of the registers present in the array will immediately return the value in the array. However, a write operation will be performed on the device itself and will update the affected register in the array. This way, reading the LTC3589_ VRRCR register will return 0xff and write any value in that register, and it will update its entry in the array so that any new read operation will return the last written value directly from the cache.

Now that we are able to use the Regmap APIs to access the device registers whatever the underlying bus is these devices sit on, the time has come for us to summarize the knowledge we have learned so far in a practical example.

Regmap-based SPI driver example – putting it all together

All the steps involved in setting up Regmap, from configuration to device register access, can be enumerated as follows:

- Setting up a struct regmap_config object according to the device characteristics. Defining the register range if needed, default values if any, cache_ type if needed, and so on. If custom read/write functions are needed, pass them to the reg_read/reg_write fields.

- In the probe function, allocating a register map using devm_regmap_init_ i2c(), devm_regmap_init_spi(), or devm_regmap_init_mmio() depending on the connection with the underlying device – I2C, SPI, or memory-mapped.

- Whenever you need to read/write from/into registers, calling remap_ [read|write] functions.

- When done with the register map, assuming you used resource-managed APIs, you have nothing else to do as devres core will take care of releasing the Regmap resources; otherwise, you'll have to call `regmap_exit()` to free the register map allocated in the probe.

Let's now materialize these steps in a real driver example that takes advantage of the Regmap framework.

A Regmap example

To achieve our goal, let's first describe a fake SPI device for which we can write a driver using the Regmap framework. For understandability, let's use the following characteristics:

- The device supports 8-bit register addressing and 8-bit register values.

- The maximum address that can be accessed in this device is `0x80` (it does not necessarily mean that this device has `0x80` registers).

- The write mask is `0x80`, and the valid address ranges are as follows:

  - `0x20` to `0x4F`

  - `0x60` to `0x7F`

- Since the device supports simple SPI read/write operations, there is no need to provide a custom read/write function.

Now that we are done with the device and the Regmap specifications, we can start writing the code.

The following includes the required header to deal with Regmap:

```
#include <linux/regmap.h>
```

Depending on the APIs needed in the driver, other headers might be included.

Then, we define our private data structure as follows:

```
struct private_struct
{
    /* Feel free to add whatever you want here */
    struct regmap *map;
    int foo;
};
```

Then, we define a read/write register range, that is, registers that are allowed to be accessed:

```
static const struct regmap_range wr_rd_range[] =
{
    {
            .range_min = 0x20,
            .range_max = 0x4F,
    },{
            .range_min = 0x60,
            .range_max = 0x7F
    },
};
struct regmap_access_table drv_wr_table =
{
    .yes_ranges =   wr_rd_range,
    .n_yes_ranges = ARRAY_SIZE(wr_rd_range),
};
struct regmap_access_table drv_rd_table =
{
    .yes_ranges =   wr_rd_range,
    .n_yes_ranges = ARRAY_SIZE(wr_rd_range),
};
```

However, it must be noted that if writeable_reg and/or readable_reg are set, there is no need to provide wr_table and/or rd_table.

After that, we define the callback that will be called any time a register is accessed for a write or a read operation. Each callback must return true if it is allowed to perform the specified operation on the register:

```
static bool writeable_reg(struct device *dev,
                            unsigned int reg)
{
    if (reg>= 0x20 &&reg<= 0x4F)
        return true;
    if (reg>= 0x60 &&reg<= 0x7F)
        return true;
    return false;
```

```
}

static bool readable_reg(struct device *dev,
                         unsigned int reg)
{
    if (reg>= 0x20 &&reg<= 0x4F)
        return true;
    if (reg>= 0x60 &&reg<= 0x7F)
        return true;
    return false;
}
```

Now that all Regmap-related operations have been defined, we can implement the driver's probe method as follows:

```
static int my_spi_drv_probe(struct spi_device *dev)
{
    struct regmap_config config;
    struct private_struct *priv;
    unsigned char data;

    /* setup the regmap configuration */
    memset(&config, 0, sizeof(config));
    config.reg_bits = 8;
    config.val_bits = 8;
    config.write_flag_mask = 0x80;
    config.max_register = 0x80;
    config.fast_io = true;
    config.writeable_reg = drv_writeable_reg;
    config.readable_reg = drv_readable_reg;

    /*
     * If writeable_reg and readable_reg are set,
     * there is no need to provide wr_table nor rd_table.
     * Uncomment below code only if you do not want to use
     * writeable_reg nor readable_reg.
     */
```

```
    //config.wr_table = drv_wr_table;
    //config.rd_table = drv_rd_table;


    /* allocate the private data structures */
    /* priv = kzalloc */


    /* Init the regmap spi configuration */
    priv->map = devm_regmap_init_spi(dev, &config);
    /* Use devm_regmap_init_i2c in case of i2c bus */


    /*
     * Let us write into some register
     * Keep in mind that, below operation will remain same
     * whether you use SPI, I2C, or memory mapped Regmap.
     * It is and advantage when you use regmap.
     */
    regmap_read(priv->map, 0x30, &data);
    [...] /* Process data */


    data = 0x24;
    regmap_write(priv->map, 0x23, data); /* write new value */


    /* set bit 2 (starting from 0) and bit 6
     * of register 0x44 */
    regmap_update_bits(priv->map, 0x44,
                        0b00100010, 0xFF);
    [...] /* Lot of stuff */
    return 0;
}
```

In the preceding probe method, there are even more commands than code. We simply needed to demonstrate how the device specification could be translated into a register map configuration and used as main access functions to the device registers.

Now that we are done with Regmap from within the kernel, let's see how user space can make the most out of this framework in the next section.

Leveraging Regmap from the user space

Register maps can be monitored from the user space via the debugfs file system. First, debugfs needs to be enabled via the CONFIG_DEBUG_FS kernel configuration option. Then, debugfs can be mounted using the following command:

```
mount -t debugfs none /sys/kernel/debug
```

After that, the debugfs register map implementation can be found under /sys/kernel/debug/regmap/. This debugfs view implemented by drivers/base/regmap/regmap-debugfs.c in kernel sources contains a register cache (mirror) for drivers/peripherals based on the Regmap API.

From the Regmap main debugfs directory, we can get the list of devices whose drivers are based on the Regmap API using the following command:

```
root@jetson-nano-devkit:~# ls -l /sys/kernel/debug/regmap/
drwxr-xr-x    2 root    root    0 Jan  1  1970 4-003c-power-slave
drwxr-xr-x    2 root    root    0 Jan  1  1970 4-0068
drwxr-xr-x    2 root    root    0 Jan  1  1970 700e3000.mipical
drwxr-xr-x    2 root    root    0 Jan  1  1970 702d3000.amx
drwxr-xr-x    2 root    root    0 Jan  1  1970 702d3100.amx
drwxr-xr-x    2 root    root    0 Jan  1  1970 hdaudioC0D3-hdaudio
drwxr-xr-x    2 root    root    0 Jan  1  1970 tegra210-admaif
drwxr-xr-x    2 root    root    0 Jan  1  1970 tegra210-adx.0
[...]
root@jetson-nano-devkit:~#
```

In each directory, there can be one or more of the following files:

- access: An encoding of the various access permissions to each register in respect of the pattern readable writable volatile precious:

```
root@jetson-nano-devkit:~# cat /sys/kernel/debug/
regmap/4-003c-power-slave/access
00: y y y n
01: y y y n
02: y y y n
03: y y y n
04: y y y n
05: y y y n
```

```
06: y y y n
07: y y y n
08: y y y n
[...]
5c: y y y n
5d: y y y n
5e: y y y n
```

For example, the line 5e: y y y n means that the register at address 5e is readable, writeable, volatile, but not precious.

- name: The driver name associated with the register map. Check for the corresponding driver. For example, the 702d3000.amx register map entry:

```
root@jetson-nano-devkit:~# cat /sys/kernel/debug/
regmap/702d3000.amx/name
tegra210-amx
```

There are, however, Regmap entries starting with dummy- as follows:

```
root@raspberrypi4-64:~# ls -l /sys/kernel/debug/regmap/
drwxr-xr-x  2 root root 0 Jan 1 1970 dummy-avs-monitor@
fd5d2000
root@raspberrypi4-64:~#
```

This kind of entry is set when there are no associated /dev entries (devtmpfs). You can check this by printing the underlying device name, which will be nodev as follows:

```
root@raspberrypi4-64:~# cat /sys/kernel/debug/regmap/
dummy-avs-monitor\@fd5d2000/name
nodev
root@raspberrypi4-64:~#
```

You can look for the suffix name after dummy- to find the relevant node in the device tree, for example, dummy-avs-monitor@fd5d2000:

```
avs_monitor: avs-monitor@7d5d2000 {
    compatible = "brcm,bcm2711-avs-monitor",
                 "syscon", "simple-mfd";
    reg = <0x7d5d2000 0xf00>;
[...]
};
```

- `cache_bypass`: Puts the register map into cache-only mode. If enabled, writes to the register map will only update the hardware and not the cache directly:

```
root@jetson-nano-devkit:~# cat /sys/kernel/debug/
regmap/702d3000.amx/cache_bypass
N
```

To enable cache bypassing, you should echo Y in this file as follows:

```
root@jetson-nano-devkit:~# echo Y > /sys/kernel/debug/
regmap/702d30[579449.571475] tegra210-amx tegra210-amx.0:
debugfs cache_bypass=Y forced
00.amx/cache_bypass
root@jetson-nano-devkit:~#
```

This will additionally print a message in the kernel log buffer.

- `cache_dirty`: Indicates that HW registers were reset to default values and that hardware registers do not match the cache state. The read value can be either Y or N.

- `cache_only`: Echoing N in this file will disable caching for this register map and in the meantime, will trigger cache syncing, while writing Y will force registers in this register map to be cached only. Reading this file value will return Y or N according to the current caching enabled state. Any write happening while this value is true will be cached (only the register cache will be updated, no hardware changes will occur).

- `range`: The valid register ranges of the register map:

```
root@jetson-nano-devkit:~# cat /sys/kernel/debug/
regmap/4-003c-power-slave/range
0-5e
```

- `rbtree`: Provides how much memory overhead the `rbtree` cache adds:

```
root@jetson-nano-devkit:~# cat /sys/kernel/debug/
regmap/4-003c-power-slave/rbtree
0-5e (95)
1 nodes, 95 registers, average 95 registers, used 175
bytes
```

- `registers`: The file used to read and write the actual registers associated with the register map:

```
root@jetson-nano-devkit:~# cat /sys/kernel/debug/
regmap/4-003c-power-slave/registers
00: d2
01: 1f
02: 00
03: dc
04: 0f
05: 00
06: 00
07: 00
08: 02
[...]
5c: 35
5d: 81
5e: 00
#
```

In the preceding output, the register address is shown first, then its content after.

This section is quite short but is a concise description of Regmap monitoring from the user space. It allows reading and writing register content, and in some circumstances, changing the behavior of the underlying Regmap.

Summary

This chapter is all about register access related Regmap APIs. Its simplicity should give you an idea of how useful and widely used it is. This chapter has shown everything you need to know about the Regmap API. Now you should be able to convert any standard SPI/I2C/memory-mapped driver into Regmap.

The next chapter will cover IRQ management under Linux, however, two chapters after, we will cover IIO devices, a framework for analog-to-digital converters. Those kinds of devices always sit on top of SPI/I2C buses. It could be a challenge for us, at the end of that chapter, to write an IIO driver using the Regmap API.

13
Demystifying the Kernel IRQ Framework

Linux is a system on which devices notify the kernel about events by means of **interrupt requests (IRQs)**, though some devices are polled. The CPU exposes IRQ lines, shared or not, used by connected devices so that when a device needs the CPU, it sends a request to the CPU. When the CPU gets this request, it stops its actual job and saves its context, in order to serve the request issued by the device. After serving the device, its state is restored back to exactly where it stopped when the interruption occurred.

In this chapter, we will deal with the APIs that the kernel offers to manage IRQs and the ways in which multiplexing can be done. Moreover, we will analyze and look closer at **interrupt controller** driver writing.

To summarize, in this chapter, the following topics will be covered:

- Brief presentation of interrupts
- Understanding interrupt controllers and interrupt multiplexing
- Diving into advanced peripheral IRQ management
- Demystifying per-CPU interrupts

Brief presentation of interrupts

On many platforms, a special device is responsible for managing IRQ lines. That device is the interrupt controller and it stands between the CPU and the interrupt lines it manages. The following is a diagram that shows the interactions that take place:

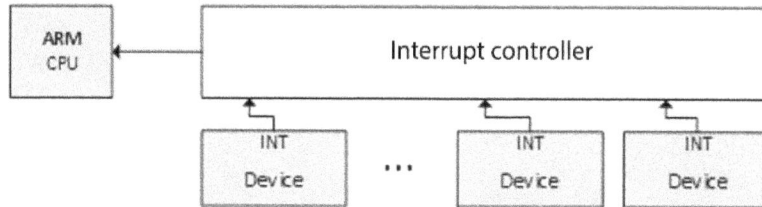

Figure 13.1 – Interrupt controller and IRQ lines

Not only can devices raise interrupts, but some processor operations can do that too. There are then two different kinds of interrupts:

- Synchronous interrupts, called **exceptions**, are produced by the CPU while processing instructions. These are **non-maskable interrupts** (**NMIs**) and result from a critical malfunction such as hardware failure. They are always processed by the CPU.

- Asynchronous interrupts, called **interrupts**, are issued by other hardware devices. These are normal and **maskable interrupts**. These are what we will discuss in the next sections of this chapter.

Before getting deeper into interrupt management in the Linux kernel, let's talk a bit more about exceptions.

Exceptions are consequences of programming errors, handled by the kernel, which sends a signal to the program and tries to recover from the error. These are classified into two categories, enumerated as follows:

- **Processor-detected exceptions**: Those the CPU generates in response to an anomalous condition, which are divided into three groups:

  - Faults, which can generally be corrected (bogus instruction).

  - Traps, which occur in the user process (invalid memory access, division by zero), are also a mechanism to switch to kernel mode in response to a system call. If the kernel code does cause a trap, it immediately panics.

  - Aborts – the serious errors.

- **Programmed exceptions**: These are requested by the programmer and handled like traps.

Now that we have introduced the different families of interrupts, let's learn how they are implemented from within the interrupt controller.

Understanding interrupt controllers and interrupt multiplexing

Having a single interrupt from the CPU is usually not enough. Most systems have tens or hundreds of them. Now comes interrupt controller, which allows them to be multiplexed. Very often, architecture or platform-specific implementations offer specific facilities, such as the following:

- Masking/unmasking individual interrupts

- Setting priorities

- SMP affinity

- Exotic features, such as wake-up interrupts

IRQ management and interrupt controller drivers both rely on the concept of the IRQ domain, which is built on top of the following structures:

- `struct irq_chip`: This is the interrupt controller data structure. This structure also implements a set of methods that allow to drive the interrupt controller and that are directly called by core IRQ code.

- `struct irqdomain`: This provides the following options:

  - A pointer to the interrupt controller's firmware node (`fwnode`)

  - A function for converting an IRQ's firmware description into an ID local to this interrupt controller (`hwirq`, also called hardware IRQ number)

  - A way to retrieve the Linux view (`virq`, also called virtual IRQ number) of an IRQ from `hwirq`

- `struct irq_desc`: This structure is Linux's view of an interrupt. It contains all the information about the interrupt as well as one-to-one mapping to the Linux interrupt number.

- `struct irq_action`: This structure is used to describe an IRQ handler.

- `struct irq_data`: This structure is embedded in the `struct irq_desc` structure and provides us with the following information:

  - The data that is relevant to the IRQ chip managing this interrupt.

  - Both `virq` and `hwirq`.

  - A pointer to `struct irq_chip` (the IRQ chip data structure). Note that most IRQ chip-related function calls are given `irq_data` as a parameter, from which you can obtain the corresponding `struct irq_desc`.

All the preceding data structures are part of the IRQ domain API. An interrupt controller is represented in the kernel by an instance of the `struct irq_chip` structure, which describes the actual hardware device, and some methods used by the IRQ core. The following code block shows its definition:

```
struct irq_chip {
    struct device    *parent_device;
    const char       *name;
    void    (*irq_enable)(struct irq_data *data);
    void    (*irq_disable)(struct irq_data *data);

    void    (*irq_ack)(struct irq_data *data);
    void    (*irq_mask)(struct irq_data *data);
    void    (*irq_unmask)(struct irq_data *data);
    void    (*irq_eoi)(struct irq_data *data);

    int     (*irq_set_affinity)(struct irq_data *data,
                const struct cpumask *dest, bool force);
    int     (*irq_retrigger)(struct irq_data *data);
    int     (*irq_set_type)(struct irq_data *data,
                        unsigned int flow_type);
    int     (*irq_set_wake)(struct irq_data *data,
                        unsigned int on);

    void    (*irq_bus_lock)(struct irq_data *data);
    void    (*irq_bus_sync_unlock)(struct irq_data *data);

    int     (*irq_get_irqchip_state)(struct irq_data *data,
                enum irqchip_irq_state which, bool *state);
```

```
   int     (*irq_set_irqchip_state)(struct irq_data *data,
              enum irqchip_irq_state which, bool state);
   void    (*ipi_send_single)(struct irq_data *data,
                                unsigned int cpu);
   void    (*ipi_send_mask)(struct irq_data *data,
                              const struct cpumask *dest);
   unsigned long     flags;
};
```

The following list explains the meanings of the elements in the structure:

- parent_device: This is a pointer to the parent of this IRQ chip.

- name: This is the name for the/proc/interrupts file.

- irq_enable: This hook enables the interrupt. If not set (if NULL), it defaults to chip->unmask.

- irq_disable: This disables the interrupt.

- irq_ack: This callback acknowledges an interrupt. It is unconditionally invoked by handle_edge_irq() and, therefore, must be defined (even an empty shell) for IRQ controller drivers that use handle_edge_irq() to handle interrupts. For such controllers, this callback is invoked at the start of the interrupt. Some controllers do not need this. Linux calls this function as soon as an interrupt is raised, long before it is serviced. This function is mapped to chip->disable() in some implementations so that if another interrupt request pokes on the line, it will not cause another interrupt until after the current interrupt request has been handled.

- irq_mask: This is the hook that masks an interrupt source in the hardware so that it cannot be raised anymore.

- irq_unmask: This hook unmasks an interrupt source.

- irq_eoi: Linux invokes this **end of interrupt** (**EOI**) hook right after an IRQ servicing completes. We use this function to reconfigure the controller as necessary in order to receive another interrupt request on that line. Some implementations map this function to chip->enable() to reverse operations done in chip->ack().

- irq_set_affinity: This sets the CPU affinity only on SMP machines. In such machines, this function is used to specify the CPU on which the interrupt will be handled. This function is unused in single-processor environments, as interrupts are always services on the same single CPU.

- `irq_retrigger`: This retriggers the interrupt in the hardware, which resends an IRQ to the CPU.

- `irq_set_type`: This sets the flow type, such as `IRQ_TYPE_LEVEL`, of an IRQ.

- `irq_set_wake`: This enables/disables the power management wake-on of an IRQ.

- `irq_bus_lock`: This function locks access to slow bus (I2C) chips. Locking a mutex here is sufficient.

- `irq_bus_sync_unlock`: This function syncs and unlocks slow bus (I2C) chips, and unlocks the mutex previously locked.

- `irq_get_irqchip_state` and `irq_set_irqchip_state`: These return or set the internal state of an interrupt, respectively.

- `ipi_send_single` and `ipi_send_mask`: These are used, respectively, to send **inter-processor interrupts** (**IPIs**) either to a single CPU or to a set of CPUs defined by a mask. IPIs are used on SMP systems to generate a CPU remote interrupt from the local CPU. We will discuss this later in the chapter, in the *Demystifying per-CPU interrupts* section.

Each interrupt controller is given a domain, which is to the controller what an address space is to a process (see *Chapter 10, Understanding the Linux Kernel Memory Allocation*). The interrupt controller domain is described in the kernel with a `struct irq_domain` structure. It manages mappings between hardware IRQ numbers and Linux IRQ numbers (that is, virtual IRQs). It is the hardware interrupt number translation object. The following code block shows its definition:

```
struct irq_domain {
    const char *name;
    const struct irq_domain_ops *ops;
    void *host_data;
    unsigned int flags;
    unsigned int mapcount;

    /* Optional data */
    struct fwnode_handle *fwnode;
    [...]
};
```

For the sake of readability, only elements that are relevant to us have been listed. The following list tells us their meanings:

- `name`: This is the name of the interrupt domain.

- `ops`: This is a pointer to the IRQ domain methods.

- `host_data`: This is a private data pointer for use by the owner. Not touched by the IRQ domain core code.

- `flags`: This hosts per-IRQ domain flags.

- `mapcount`: This is the number of mapped interrupts in this IRQ domain.

- Like all the remaining elements, `fwnode` is optional. It is a pointer to the **device tree** (**DT**) node associated with the IRQ domain. Used when decoding DT interrupt specifiers.

An interrupt controller driver creates and registers an IRQ domain by calling one of the `irq_domain_add_<mapping_method>()` functions, where `<mapping_method>` is the method by which `hwirq` should be mapped to Linux `virq`. These functions are described in the following list:

- `irq_domain_add_linear()`: This uses a fixed-size table indexed by the `hwirq` number. When an `hwirq` number is mapped, an `irq_desc` object is allocated for this `hwirq` and the IRQ number is stored in the table. This linear mapping is suitable for controllers or domains that have a fixed and small number of `hwirq` (~ < 256). The inconvenience of this mapping is the table size, being as large as the largest possible `hwirq` number. Therefore, the IRQ number lookup time is fixed, and IRQ descriptors are allocated for in-use IRQs only. Most drivers should use linear mapping. This function has the following prototype:

```
struct irq_domain *irq_domain_add_linear(
                struct device_node *of_node,
                unsigned int size,
                const struct irq_domain_ops *ops,
                void *host_data)
```

- `irq_domain_add_tree()`: With this mapping, the IRQ domain maintains the mapping between `virqs` (Linux IRQ numbers) and `hwirsq` (Hardware interrupt numbers) in a radix tree. An `irq_desc` object is allocated when an `hwirq` is mapped, and this hardware IRQ number is used as the radix tree's lookup key. If the `hwirq` number can be very large, then the treemap is a viable solution because it does not require allocating a table as large as the largest `hwirq` number. The drawback is that the `hwirq`-to-IRQ-number lookup is affected by the number of entries in the table. Very few drivers should need this mapping. There are fewer than 10 users of this API in the kernel. It has the prototype shown in the following code block:

```
struct irq_domain *irq_domain_add_tree(
                struct device_node *of_node,
                const struct irq_domain_ops *ops,
                void *host_data)
```

- `irq_domain_add_nomap()`: You will probably never use this method. Nonetheless, its entire description is available in `Documentation/IRQ-domain.txt`, in the kernel source tree. Its prototype is shown in the following code block:

```
struct irq_domain *irq_domain_add_nomap(
                struct device_node *of_node,
                unsigned int max_irq,
                const struct irq_domain_ops *ops,
                void *host_data)
```

In these functions, `of_node` is a pointer to the interrupt controller's DT node. `size` corresponds to the number of interrupts in the domain. `ops` represent map/unmap domain callbacks, and `host_data` is the controller's private data pointer.

When it is initially created, the IRQ domain is empty (no mapping). A mapping is created and added as and when the IRQ chip driver calls `irq_create_mapping()`, which has the following prototype:

```
unsigned int irq_create_mapping(struct irq_domain
                *domain, irq_hw_number_t hwirq)
```

In the preceding function, `domain` is the domain to which this hardware interrupt belongs, or `NULL` for the default domain; `hwirq` represents the hardware interrupt number in that domain space.

If a mapping for the `hwirq` number doesn't already exist in the IRQ domain, the function will allocate a new Linux IRQ descriptor (`struct irq_desc`) structure, returning a virtual interrupt number at the same time. Then, it will associate it with the `hwirq` number (by means of the `irq_domain_associate()` function, which in turn invokes the `irq_domain_ops.map` callback so that the driver can perform any required hardware setup). To understand this paragraph, we need to describe the IRQ domain operation data structure (`struct irq_domain_ops`), which is defined in the following code block:

```
struct irq_domain_ops {
    int (*map)(struct irq_domain *d, unsigned int virq,
          irq_hw_number_t hw);
    void (*unmap)(struct irq_domain *d,
                  unsigned int virq);
    int (*xlate)(struct irq_domain *d,
                  struct device_node *node,
                  const u32 *intspec,
                  unsigned int intsize,
                  unsigned long *out_hwirq,
                  unsigned int *out_type);
[...]
};
```

Elements in the data structure have been limited to the scope of this chapter. Nonetheless, the complete data structure can be found in `include/linux/irqdomain.h` in the kernel source. The following list tells us the meanings of the elements we have enumerated:

- map: This creates or updates mapping between a `virq` number and an `hwirq` number. This callback is invoked only once for a given mapping. It generally maps the `virq` number with a given handler using `irq_set_chip_and_handler()`, so that calling either `generic_handle_irq()` or `handle_nested_irq()` will trigger this handler. The function `irq_set_chip_and_handler()` is defined as in the following code block:

```
void irq_set_chip_and_handler(unsigned int irq,
                              struct irq_chip *chip,
                              irq_flow_handler_t handle)
```

In this function, `irq` is the Linux IRQ given as a parameter to the `map()` function, and `chip` is your IRQ chip. There are, however, dummy controllers that need almost nothing in their `irq_chip` structure. In this case, the driver passes `dummy_irq_chip`, defined in `kernel/irq/dummychip.c`, which is a kernel-predefined `irq_chip` structure defined for such controllers. `handle` determines the interrupt flow handler, the one that calls the real handler registered using `request_irq()`. Its value depends on the IRQ being edge- or level-triggered. In either case, `handle` should be set to `handle_edge_irq` or `handle_level_irq`. Both are kernel helper functions that do some operations before and after calling the real IRQ handler. An example is shown in this code block:

```
static int ativic32_irq_domain_map(
                  struct irq_domain *id,
                  unsigned int virq,
                  irq_hw_number_t hw)
{
[...]
    if (int_trigger_type & (BIT(hw))) {
        irq_set_chip_and_handler(virq,
                &ativic32_chip,
                handle_edge_irq);
        type = IRQ_TYPE_EDGE_RISING;
    } else {
        irq_set_chip_and_handler(virq,
                &ativic32_chip,
                handle_level_irq);
        type = IRQ_TYPE_LEVEL_HIGH;
    }
    irqd_set_trigger_type(irq_data, type);
    return 0;
}
```

- xlate: Given a DT node with an interrupt specifier, this hook decodes the hardware interrupt number in that specifier along with its Linux interrupt type value. Depending on the #interrupt-cells value specified in the DT controller node, the kernel provides generic translation functions:

  - irq_domain_xlate_twocell(): Generic translation function to be used for direct two-cell binding. It works with a device tree IRQ specifier with two-cell bindings where the cell values map directly to the hwirq number and Linux IRQ flags.

  - irq_domain_xlate_onecell(): Generic xlate for direct one-cell bindings.

  - Irq_domain_xlate_onetwocell(): Generic xlate for one- or two-cell bindings.

An example of domain operation is given in the following code block:

```
static struct irq_domain_ops mcp23016_irq_domain_ops = {
    .map    = mcp23016_irq_domain_map,
    .xlate  = irq_domain_xlate_twocell,
};
```

When an interrupt is received, the irq_find_mapping() function is used to find the Linux IRQ number from the hwirq number. Of course, the mapping must exist prior to being returned. A Linux IRQ number is always tied to a struct irq_desc structure, which is the structure by which Linux describes an IRQ and has the following definition:

```
struct irq_desc {
    struct irq_data         irq_data;
    unsigned int __percpu   *kstat_irqs;
    irq_flow_handler_t      handle_irq;
    struct irqaction        *action;
    unsigned int            irqs_unhandled;
    raw_spinlock_t          lock;
    struct cpumask          *percpu_enabled;
    atomic_t                threads_active;
    wait_queue_head_t       wait_for_threads;
#ifdef CONFIG_PM_SLEEP
    unsigned int            nr_actions;
    unsigned int            no_suspend_depth;
```

```
        unsigned int              force_resume_depth;
#endif
#ifdef CONFIG_PROC_FS
    struct proc_dir_entry    *dir;
#endif
    Int                parent_irq;
    struct module      *owner;
    const char         *name;
};
```

Some fields in this data structure are intentionally missing. For the remainder, the following list gives us their definitions:

- kstat_irqs: This is the per-CPU IRQ statistics since boot.
- handle_irq: This is the high-level IRQ events handler.
- action: This represents the list of the IRQ actions for this descriptor.
- irqs_unhandled: This is the stats field for spurious unhandled interrupts.
- lock: This represents locking for SMP.
- threads_active: This is the number of IRQ action threads currently running for this descriptor.
- wait_for_threads: This represents the wait queue for sync_irq to wait for threaded handlers.
- nr_actions: This is the number of installed actions on this descriptor.
- no_suspend_depth and force_resume_depth: This represents the number of irqaction instances on an IRQ descriptor that have IRQF_NO_SUSPEND or IRQF_FORCE_RESUME flags set.
- dir: This represents the /proc/irq/ procfs entry.
- name: This names the flow handler, visible in the /proc/interrupts output.

When registering an interrupt handler, this handler is added to the end of the `irq_desc.action` list associated with that interrupt line. For instance, each call to `request_irq()` (or the threaded version, `request_threaded_irq()`) creates and adds one `struct irqaction` structure to the end of the `irq_desc.action` list (knowing that `irq_desc` is the descriptor for this interrupt). For a shared interrupt, this field will contain as many `irqaction` objects as there are handlers registered. An IRQ action data structure has the following definition:

```
struct irqaction {
    irq_handler_t       handler;
    void                *dev_id;
    void __percpu       *percpu_dev_id;
    struct irqaction    *next;
    irq_handler_t       thread_fn;
    struct task_struct      *thread;
    unsigned int        irq;
    unsigned int        flags;
    unsigned long       thread_flags;
    unsigned long       thread_mask;
    const char          *name;
    struct proc_dir_entry   *dir;
};
```

The meanings of each element in this data structure are as follows:

- `handler`: This is the non-threaded (hard) interrupt handler function.
- `name`: This is the device name.
- `dev_id`: This is a cookie to identify the device.
- `percpu_dev_id`: This is a per-CPU cookie to identify the device.
- `next`: This is a pointer to the next IRQ action for shared interrupts.
- `irq`: This is the Linux interrupt number (`virq`).
- `flags`: This represents the IRQ flags (see `IRQF_*`).
- `thread_fn`: This is the threaded interrupt handler function for threaded interrupts.

- `thread`: This is a pointer to the thread structure in case of threaded interrupts.
- `thread_flags`: This represents the flags related to the thread.
- `thread_mask`: This is a bitmask for keeping track of thread activity.
- `dir`: This points to the `/proc/irq/NN/<name>/` entry.

The following is the definition of important fields in the `struct irq_data` structure, which is per-IRQ chip data passed down to chip functions:

```
struct irq_data {
    [...]
    unsigned int       irq;
    unsigned long           hwirq;
    struct irq_chip         *chip;
    struct irq_domain *domain;
    void               *chip_data;
};
```

The following list gives the meanings of elements in this data structure:

- `irq`: This is the interrupt number (Linux IRQ number).
- `hwirq`: This is the hardware interrupt number, local to the `irq_data.domain` interrupt domain.
- `chip`: This represents the low-level interrupt controller hardware access.
- `domain`: This represents the interrupt translation domain, responsible for mapping between the `hwirq` number and the Linux IRQ number.
- `chip_data`: This is platform-specific, per-chip private data for the chip methods, to allow shared chip implementations.

Now that we are familiar with the data structures of the IRQ framework, we can go a bit further and study how interrupts are requested and propagated all along the processing chain.

Diving into advanced peripheral IRQ management

In *Chapter 3*, *Dealing with Kernel Core Helpers*, we introduced peripheral IRQs, using `request_irq()` and `request_threaded_irq()`. With the former, you register a handler (top half) that will be executed in an atomic context, from which you can schedule a bottom half using one of the mechanisms discussed in that same chapter. On the other hand, with the `_threaded` variant, you can provide top and bottom halves to the function, so that the former will be run as the hard IRQ handler, which may decide to raise the second and threaded handler or not, which will be run in a kernel thread.

The problem with those approaches is that sometimes, drivers requesting an IRQ do not know about the nature of the interrupt controller that provides this IRQ line, especially when the interrupt controller is a discrete chip (typically a GPIO expander connected over SPI or I2C buses). Now comes the `request_any_context_irq()` function with which drivers requesting an IRQ know whether the handler will run in a thread context, and call `request_threaded_irq()` or `request_irq()` accordingly. This means that whether the IRQ associated with our device comes from an interrupt controller that may not sleep (memory-mapped one) or from one that can sleep (behind an I2C/SPI bus), there will be no need to change the code. Its prototype is shown in the following code block:

```
int request_any_context_irq(unsigned int irq,
                        irq_handler_t handler,
                        unsigned long flags,
                        const char * name,
                        void * dev_id);
```

Here are the meanings of each parameter in the function:

- `irq`: This represents the interrupt line to allocate.
- `handler`: This is the function to be called when the IRQ occurs. Depending on the context, this function might run as a hard IRQ or might be threaded.
- `flags`: This represents the interrupt type flags. It is the same as those in `request_irq()`.
- `name`: This will be used for debugging purposes to name the interrupt in `/proc/interrupts`.
- `dev_id`: This is a cookie passed back to the handler function.

`request_any_context_irq()` means that you can either get a hard IRQ or a threaded one. It works in the same way as the usual `request_irq()`, except that it checks whether the IRQ is configured as nested or not, and calls the right backend. In other words, it selects either a hard IRQ or threaded handling method depending on the context. This function returns a negative value on failure. On success, it returns either `IRQC_IS_HARDIRQ` or `IRQC_IS_NESTED`. A use case is shown in the following code block:

```
static irqreturn_t packt_btn_interrupt(int irq,
                                        void *dev_id)
{
    struct btn_data *priv = dev_id;

    input_report_key(priv->i_dev, BTN_0,
                    gpiod_get_value(priv->btn_gpiod) & 1);
    input_sync(priv->i_dev);
    return IRQ_HANDLED;
}


static int btn_probe(struct platform_device *pdev)
{
    struct gpio_desc *gpiod;
    int ret, irq;

    [...]
    gpiod = gpiod_get(&pdev->dev, "button", GPIOD_IN);
    if (IS_ERR(gpiod))
        return -ENODEV;

    priv->irq = gpiod_to_irq(priv->btn_gpiod);
    priv->btn_gpiod = gpiod;

    [...]

    ret = request_any_context_irq(
            priv->irq,
            packt_btn_interrupt,
```

```
                    (IRQF_TRIGGER_FALLING | IRQF_TRIGGER_RISING),
                "packt-input-button", priv);
    if (ret < 0) {
        dev_err(&pdev->dev,
            "Unable to request GPIO interrupt line\n");
        goto err_btn;
    }

    return ret;
}
```

The preceding code is an excerpt of the driver sample of an input device driver. The advantage of using `request_any_context_irq()` is that you do not need to care about what can be done in the IRQ handler, since the context in which the handler will run depends on the interrupt controller that provides the IRQ line. In our example, if the GPIO belongs to a controller sitting on an I2C or SPI bus, the handler will be threaded. Otherwise (memory mapped), the handler will run in a hard IRQ context.

Understanding IRQ and propagation

Let's consider the following diagram with a GPIO controller whose interrupt line is connected to a native GPIO on the SoC:

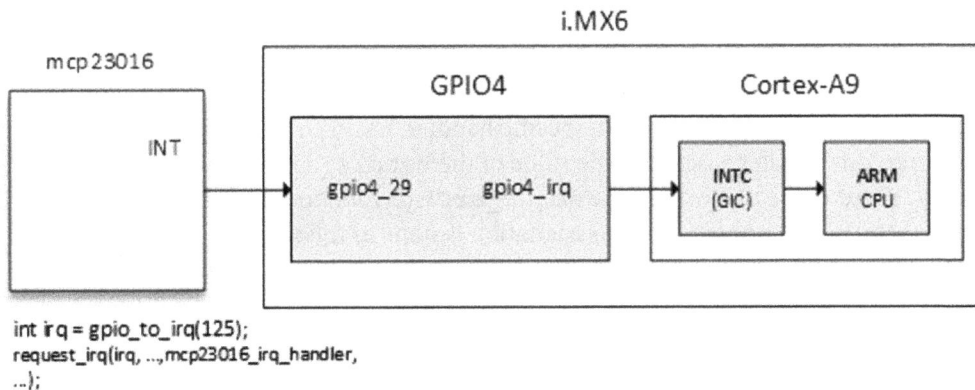

```
int irq = gpio_to_irq(125);
request_irq(irq, ...,mcp23016_irq_handler,
...);
```

Figure 13.2 – Interrupt propagation

IRQs are always processed based on the Linux IRQ number (not `hwirq`). The general function to request an IRQ on a Linux system is `request_threaded_irq()`. `request_irq()` is a wrapper on `request_threaded_irq()` which just don't provide the bottom half. The following code block shows its prototype:

```
int request_threaded_irq(unsigned int irq,
                 irq_handler_t handler,
                 irq_handler_t thread_fn,
                 unsigned long irqflags,
                 const char *devname, void *dev_id)
```

When called, the function extracts `struct irq_desc` associated with the IRQ using the `irq_to_desc()` macro. It then allocates a new `struct irqaction` structure and sets it up, filling parameters such as handler and flags. The following code block is an excerpt:

```
action->handler = handler;
action->thread_fn = thread_fn;
action->flags = irqflags;
action->name = devname;
action->dev_id = dev_id;
```

That same function finally inserts/registers the descriptor in the proper IRQ list by invoking the `__setup_irq()` (by means of `setup_irq()`) function, defined in `kernel/irq/manage.c`.

Now, when an IRQ is raised, the kernel executes some assembler code in order to save the current state and jumps to the arch-specific handler, `handle_arch_irq`. For ARM architectures, this handler is set with the value of the `handle_irq` field in `struct machine_desc` of the platform in the `setup_arch()` function implemented in `arch/arm/kernel/setup.c`. The assignation is done as follows:

```
handle_arch_irq = mdesc->handle_irq
```

For SoCs that use the ARM **Generic Interrupt Controller** (**GIC**), the `handle_irq` callback is set with `gic_handle_irq`, in either `drivers/irqchip/irq-gic.c` or `drivers/irqchip/irq-gic-v3.c`:

```
set_handle_irq(gic_handle_irq);
```

`gic_handle_irq()` calls `handle_domain_irq()`, which executes `generic_handle_irq()`, in turn calling `generic_handle_irq_desc()`, which ends by calling `desc->handle_irq()`. The whole chain can be seen in `arch/arm/kernel/irq.c`. Now, `handle_irq` is the actual call for the flow handler, which we registered as `mcp23016_irq_handler` in the diagram.

`gic_hande_irq()` is a GIC interrupt handler. `generic_handle_irq()` will execute the handler of the SoC's GPIO4 IRQ, which will look for GPIO pins that issued the interrupt, and call `generic_handle_irq_desc()`.

Chaining IRQs

This section describes how the interrupt handlers of a parent call its children's interrupt handlers, in turn calling their children's interrupt handlers, and so on. The kernel offers two approaches on how to call interrupt handlers for child devices in the IRQ handler of the parent (interrupt controller) device. These are the chained and nested methods.

Chained interrupts

This approach is used for SoC's internal GPIO controllers, which are memory-mapped and which do not put the caller to sleep when these are accessed. Chained means that those interrupts are just chains of function calls (for example, SoC's GPIO module interrupt handler is being called from the GIC interrupt handler, just as a function call). `generic_handle_irq()` is used for interrupts chaining. Child IRQ handlers are called from inside the parent's hard IRQ handler. This means that even from within the child interrupt handlers, we are still in an atomic context (HW interrupt), and the driver must not call functions that may sleep.

Nested interrupts

With this flow, function calls are nested, which means interrupt handlers are not invoked in the parent's handler. `handle_nested_irq()` is used for creating nested interrupt child IRQs. Handlers are called inside a new thread created for this purpose. This method is used by controllers that sit on slow buses such as SPI or I2C (such as GPIO expanders), and whose access may sleep (I2C and SPI access routines may sleep). Nested interrupt handlers that run in a process context can call any sleeping function.

Demystifying per-CPU interrupts

The most common ARM interrupt controller, GIC in the ARM multi-core processor, supports three types of interrupts:

- **CPU private interrupts**: These interrupts are private per CPU. If triggered, such a per-CPU interrupt will exclusively be serviced on the target CPU or CPU to which it is bound. Private interrupts can be split into two families:

  - **Private peripheral interrupts** (**PPIs**): These are private and can only be generated by hardware bound to the CPU.

  - **Software-generated interrupts** (**SGIs**): Unlike PPIs, these are generated by the software. Thanks to this, SGIs are usually used as interrupt IPIs for inter-core communication on multi-core systems, meaning that one CPU can generate an interrupt (by writing the appropriate message, made of the interrupt ID and the target CPU to the GIC controller) to (an)other CPU(s). This is what we will talk about in this section.

- **Shared peripheral interrupts** (**SPIs**) (not to be confused with the SPI bus): These are the classical interrupts that we have discussed so far. Such interrupts can route to any CPU.

In systems with an interrupt controller that supports private interrupts per core, some of the IRQ controller registers will be banked so that they're only accessible from one core (for example, a core will only be able to read/write its own interrupt configuration). Usually, to be able to do so, some interrupt controller registers are banked per CPU; a CPU can enable its local interrupt by writing to its banked registers.

The distributor block and the CPU interface block are logically partitioned in the GIC. Interacting with interrupt sources, the distributor block prioritizes interrupts and delivers them to the CPU interface block. The CPU interface block links to the system's processors and manages priority masking and preemption for the processors to which it is linked.

The GIC can support up to 8 CPU interfaces, each of which can handle up to 1,020 interrupts. Interrupt ID numbers 0–1019 are assigned by the GIC as follows:

- Interrupt numbers 0–31 are interrupts that are private to a CPU interface. These private interrupts are banked in the distributor block and split as follows:

  - SGIs use banked interrupt numbers 0–15.

  - PPIs use banked interrupt numbers 16–31. In SMP systems, for example, a per-CPU timer provided by clock event devices can generate such interrupts.

- SPIs use interrupt numbers 32–1,019.

- The remaining interrupts are reserved, that is, interrupt numbers 1020–1023.

Now that we are familiar with ARM GIC interrupt families, we can focus on the family we are interested in, that is, SGIs.

SGIs and IPIs

In ARM processors, there are 16 SGIs, numbered from 0 to 15, but the Linux kernel registers only a few of them: eight (from 0 to 7) to be precise. SGI8 to SGI15 are free for now. Registered SGIs are those defined in enum ipi_msg_type, which is defined as the following:

```
enum ipi_msg_type {
    IPI_WAKEUP,
    IPI_TIMER,
    IPI_RESCHEDULE,
    IPI_CALL_FUNC,
    IPI_CPU_STOP,
    IPI_IRQ_WORK,
    IPI_COMPLETION,
    NR_IPI,
[...]
    MAX_IPI
};
```

Their respective descriptions can be found in an array of strings, or ipi_types, defined in the following code block:

```
static const char *ipi_types[NR_IPI] = {
    [IPI_WAKEUP] = "CPU wakeup interrupts",
    [IPI_TIMER] = "Timer broadcast interrupts",
    [IPI_RESCHEDULE] = "Rescheduling interrupts",
    [IPI_CALL_FUNC]  = "Function call interrupts",
    [IPI_CPU_STOP]   = "CPU stop interrupts",
    [IPI_IRQ_WORK]   = "IRQ work interrupts",
    [IPI_COMPLETION] = "completion interrupts",
};
```

IPIs are registered in the `set_smp_ipi_range()` function, defined in the following code block:

```
void __init set_smp_ipi_range(int ipi_base, int n)
{
    int i;
    WARN_ON(n < MAX_IPI);
    nr_ipi = min(n, MAX_IPI);

    for (i = 0; i < nr_ipi; i++) {
        int err;

        err = request_percpu_irq(ipi_base + i,
                ipi_handler, "IPI", &irq_stat);
        WARN_ON(err);

        ipi_desc[i] = irq_to_desc(ipi_base + i);
        irq_set_status_flags(ipi_base + i, IRQ_HIDDEN);
    }

    ipi_irq_base = ipi_base;
    /* Setup the boot CPU immediately */
    ipi_setup(smp_processor_id());
}
```

In the preceding code block, each IPI is registered with `request_percpu_irq()` on a per-CPU basis. We can see that IPIs have the same handler, `ipi_handler()`, defined as follows:

```
static irqreturn_t ipi_handler(int irq, void *data)
{
    do_handle_IPI(irq - ipi_irq_base);
    return IRQ_HANDLED;
}
```

The underlying function executed in the handler is do_handle_IPI(), defined as follows:

```
static void do_handle_IPI(int ipinr)
{
    unsigned int cpu = smp_processor_id();

    if ((unsigned)ipinr < NR_IPI)
        trace_ipi_entry_rcuidle(ipi_types[ipinr]);

    switch (ipinr) {
    case IPI_WAKEUP:
        break;

#ifdef CONFIG_GENERIC_CLOCKEVENTS_BROADCAST
    case IPI_TIMER:
        tick_receive_broadcast();
        break;
#endif
    case IPI_RESCHEDULE:
        scheduler_ipi();
        break;

    case IPI_CPU_STOP:
        ipi_cpu_stop(cpu);
        break;
[...]
    default:
        pr_crit("CPU%u: Unknown IPI message 0x%x\n",
                cpu, ipinr);
        break;
    }

    if ((unsigned)ipinr < NR_IPI)
        trace_ipi_exit_rcuidle(ipi_types[ipinr]);
}
```

From the preceding function,

- `IPI_WAKEUP`: This is used to wake up and boot a secondary CPU. It is mostly issued by the boot CPU.

- `IPI_RESCHEDULE`: The Linux kernel uses rescheduling interrupts to tell another CPU core to schedule a thread. The scheduler on SMP systems does this to distribute the load over multiple CPU cores. As a general rule, it is ideal to have as many processes running on all the cores in lower power (lower clock frequencies) rather than have one busy core running at full speed while other cores are sleeping. When the scheduler needs to offload work from one core to another sleeping core, the scheduler sends a kernel IPI message to that sleeping core, causing it to wake up from its low-power sleep and begin running a process. These IPI events are reported by `powertop` as `Rescheduling Interrupts`.

- `IPI_TIMER`: This is the timer broadcast interrupt. This IPI emulates a timer interrupt on an idle CPU. It is sent by the broadcast clock event/tick device to CPUs represented in `tick_broadcast_mask`, which is the bitmap that represents the list of processors that are in a sleeping mode. Tick devices and broadcast masks are discussed in *Chapter 3, Dealing with Kernel Core Helpers.*

- `IPI_CPU_STOP`: When a kernel panic occurs on one CPU, other CPUs are instructed to dump their stack and to stop execution via the `IPI_CPU_STOP` IPI message. The target CPUs are not shut down or taken offline; instead, they stop execution and are placed in a low-power loop, in a **wait for event** (**WFE**) state.

- `IPI_CALL_FUNC`: This is used to run a function in another processor context.

- `IPI_IRQ_WORK`: This is used to run a work in a hardware IRQ context. The kernel offers a bunch of mechanisms to defer works to a later time, especially out of the hardware interrupt context. There might, however, be the occasional need to run a work in a hardware interrupt context and there is no hardware conveniently signaling interrupts at the time. To achieve that, an IPI is used to run the work in a hardware interrupt context. This is mainly used in code running from non-maskable interrupts, which needs to be able to interact with the rest of the system.

On a running system, you can look for available IPIs from the `/proc/interrupt` file, as shown in the following code block:

```
root@udoo-labcsmart:~# cat /proc/interrupts | grep IPI
IPI0:          0          0  CPU wakeup interrupts
IPI1:         29         22  Timer broadcast interrupts
IPI2:      84306     322774  Rescheduling interrupts
```

```
IPI3:          970        1264  Function call interruptsIPI4:
0             0  CPU stop interrupts
IPI5:      2505436     4064821  IRQ work interrupts
IPI6:            0           0  completion interrupts
root@udoo-labcsmart:~#
```

In the command output shown here, the first column is the IPI identifier and the last one is the description of the IPI. The columns in between are their respective numbers of executions on each CPU.

Summary

Now, IRQ multiplexing has no more secrets from you. We have discussed the most important element of IRQ management in Linux systems: the IRQ domain API. You have the basics to understand existing interrupt controller drivers, as well as their binding from within the DT. IRQ propagation has been discussed in order to explore what happens between the request and the handler invocation.

In the next chapter, we deal with a completely different topic: the Linux device model.

14
Introduction to the Linux Device Model

Until version 2.5, the Linux kernel had no way to describe and manage objects, and its code reusability was not as enhanced as it is now. In other words, there was no device topology, nor organization as we know it is in sysfs nowadays. There was no information on subsystem relationships, nor on how the system is put together. Then came the **Linux Device Model (LDM)**, which introduced the following features:

- The concept of classes. They are used to group devices of the same type or that expose the same functionalities (for example, mice and keyboards are both input devices).

- Communication with the user space through a virtual filesystem, allowing you to manage and enumerate devices and the properties they expose from user space.

- Object life cycle management using reference counting.

- A power management facility, allowing you to handle the order in which devices should shut down.

- The reusability of the code. Classes and frameworks expose interfaces, behaving like a contract that any driver that registers with them must respect.

- An **object-oriented (OO)**-like programming style and encapsulation in the kernel.

In this chapter, we will take advantage of LDM and export some properties to the user space through the sysfs filesystem. To do this, we will cover the following topics:

- Introduction to LDM data structures
- Getting deeper inside LDM
- Overview of the device model from sysfs

Introduction to LDM data structures

The Linux device model introduced device hierarchy. It is built on top of a few data structures. Among these is the bus, which is represented in the kernel as an instance of struct bus_type; the device driver, which is represented by a struct device_driver structure; and the device, which is the last element and is represented as an instance of the struct device structure. In this section, we will introduce all those structures and learn how they interact each with other.

The bus data structure

A bus is a channel link between devices and the processor. The hardware entity that manages the bus and exports its protocol to devices is called the bus controller. For example, the USB controller provides USB support, while the I2C controller provides I2C bus support. However, the bus controller, being a device on its own, must be registered like any device. It will be the parent of the devices that need to sit on this bus. In other words, every device sitting on the bus must have its parent field pointing to the bus device. A bus is represented in the kernel by the struct bus_type structure, which has the following definition:

```
struct bus_type {
    const char        *name;
    const char        *dev_name;
    struct device     *dev_root;
    const struct attribute_group **bus_groups;
    const struct attribute_group **dev_groups;
    const struct attribute_group **drv_groups;

    int (*match)(struct device *dev,
                    struct device_driver *drv);
    int (*probe)(struct device *dev);
    void (*sync_state)(struct device *dev);
```

```
    int (*remove)(struct device *dev);
    void (*shutdown)(struct device *dev);

    int (*suspend)(struct device *dev, pm_message_t state);
    int (*resume)(struct device *dev);
    const struct dev_pm_ops *pm;
[...]
};
```

Only the elements that are relevant to this book have been listed here; let's take a look at them in more detail:

- match is a callback that's invoked whenever a new device or driver is added to this bus. The callback must be smart enough and should return a nonzero value when there is a match between a device and a driver, both given as parameters. The main purpose of the match callback is to allow a bus to determine if a particular device can be handled by a given driver or the other logic if the given driver supports a given device. Most of the time, this verification is done by a simple string comparison (device and driver name, whether it's a table and device tree-compatible property, and so on). For enumerated devices (PCI, USB), verification is done by comparing the device IDs that are supported by the driver with the device ID of the given device, without sacrificing bus-specific functionality.

- probe is a callback that's invoked when a new device or driver is added to this bus after the match has occurred. This function is responsible for allocating the specific bus device structure and calling the given driver's probe function, which is supposed to manage the device (allocated earlier).

- remove is called when a device leaves this bus.

- suspend is a method that's called when a device on the bus needs to be put into sleep mode.

- resume is called when a device on the bus must be brought out of sleep mode.

- pm is the set of power management operations for this bus, which may also call driver-specific power management operations.

- `drv_groups` is a pointer to a list (array) of `struct attribute_group` elements, each of which has a pointer to a list (array) of `struct attribute` elements. It represents the default attributes of the device drivers on the bus. The attributes that are passed to this field will be given to every driver that's registered with the bus. Those attributes can be found in the driver' directory in `/sys/bus/<bus-name>/drivers/<driver-name>`.

- `dev_groups` represents the default attributes of the devices on the bus. Any attributes that are passed (through the list/array of `struct attribute_group` elements) to this field will be given to every device that's registered with the bus. Those attributes can be found in the device's directory in `/sys/bus/<bus-name>/devices/<device-name>`.

- `bus_group` holds the set (group) of default attributes that are added automatically when the bus is registered with the core.

Apart from defining `bus_type`, the bus driver must define a bus-specific driver structure that extends the generic `struct device_driver`, as well as a bus-specific device structure that extends the generic `struct device` structure. Both are part of the device model core. The bus driver must also allocate a bus-specific device structure for each physical device that's discovered when probing. It's also in charge of setting up the device's bus and parent fields, as well as registering them with the LDM core. These fields must point to the `bus_type` and the `bus_device` structures that are defined in the bus driver The LDM core uses them to build device hierarchy and initialize other fields.

Each bus internally manages two important lists: the list of devices that have been added and sitting on it, and the list of drivers that have been registered with it. Whenever you add/register or remove/unregister a device/driver with the bus, the corresponding list is updated with the new entry. The bus driver must provide helper functions to register/unregister device drivers that can handle devices on that bus, as well as helper functions to register/unregister devices sitting on the bus. These helper functions always wrap the generic functions that are provided by the LDM core, which are `driver_register()`, `device_register()`, `driver_unregister()`, and `device_unregister()`.

Now, let's start writing a new bus infrastructure called PACKT. PACKT is going to be our bus; the devices that will be sitting on this bus will be PACKT devices, while their drivers will be PACKT drivers. Let's start writing the helpers that will allow us to register the PACKT devices and drivers:

```
/*
 * Let's write and export symbols that people
 * writing drivers for packt devices must use.
 */
int packt_register_driver(struct packt_driver *driver)
{
    driver->driver.bus = &packt_bus_type;
    return driver_register(&driver->driver);
}
EXPORT_SYMBOL(packt_register_driver);

void packt_unregister_driver(struct packt_driver *driver)
{
    driver_unregister(&driver->driver);
}
EXPORT_SYMBOL(packt_unregister_driver);

int packt_register_device(struct packt_device *packt)
{
    packt->dev.bus = &packt_bus_type;
    return device_register(&packt->dev);
}
EXPORT_SYMBOL(packt_device_register);

void packt_unregister_device(struct packt_device *packt)
{
    device_unregister(&packt->dev);
}
EXPORT_SYMBOL(packt_unregister_device);
```

Now that we've created the registration helpers, let's write the only function that allows us to allocate a new PACKT device and register this device with the PACKT core:

```
struct packt_device * packt_device_alloc(const char *name,
                                                int id)
{
    struct packt_device    *packt_dev;
    int                    status;

    packt_dev = kzalloc(sizeof(*packt_dev), GFP_KERNEL);
    if (!packt_dev)
        return NULL;

    /* devices on the bus are children of the bus device */
    strcpy(packt_dev->name, name);
    packt_dev->dev.id = id;
    dev_dbg(&packt_dev->dev,
      "device [%s] registered with PACKT bus\n",
        packt_dev->name);

    return packt_dev;
}
EXPORT_SYMBOL_GPL(packt_device_alloc);
```

The packt_device_alloc() function allocates a bus-specific device structure that must be used to register a PACKT device with the bus. At this stage, we should expose the helpers that allow us to allocate a PACKT controller device and register it – that is, register a new PACKT bus. To do this, we must define the PACKT controller data structure, like so:

```
struct packt_controller {
    char name[48];
    struct device dev;     /* the controller device */
    struct list_head  list;
    int (*send_msg) (stuct packt_device *pdev,
                        const char *msg, int count);
    int (*recv_msg) (stuct packt_device *pdev,
                        char *dest, int count);
};
```

In the preceding data structure, `name` represents the controller's name, `dev` represents the underlying struct device associated with this controller, `list` is used to insert this controller into the system's global list of `PACKT` controllers, and `send_msg` and `recv_msg` are hooks that must be provided by the controller to access the `PACKT` devices that are sitting on it:

```c
/* system global list of controllers */
static LIST_HEAD(packt_controller_list);
struct packt_controller
    *packt_alloc_controller(struct device *dev)
{
    struct packt_controller *ctlr;
    if (!dev)
        return NULL;
    ctlr = kzalloc(sizeof(packt_controller), GFP_KERNEL);
    if (!ctlr)
        return NULL;
    device_initialize(&ctlr->dev);
    [...]
    return ctlr;
}
EXPORT_SYMBOL_GPL(packt_alloc_controller);
int packt_register_controller(
                        struct packt_controller *ctlr)
{
    /* must provide at least on hook */
if (!ctlr->send_msg && !ctlr->recv_msg){
        pr_err("Registering PACKT controller failure\n");
    }
    device_add(&ctlr->dev);
    [...] /* other sanity check */
    list_add_tail(&ctlr->list, &packt_controller_list);
}
EXPORT_SYMBOL_GPL(packt_register_controller);
```

In these two functions, we have demonstrated what the controller allocation and registration operations look like. The allocation method allocates memory and does some basic initialization, leaving room for the driver to do the rest. Note that after registering a controller, it will appear under /sys/devices in sysfs. Any devices that are added to this bus will appear under /sys/devices/packt-0/.

Bus registration

The bus controller is a device itself, and in most cases, buses are memory-mapped platform devices (even buses are, which support device enumeration). For example, the PCI controller is a platform device and so is its respective driver. We should use the bus_register(struct *bus_type) function to register a bus with the kernel. The PACKT bus structure looks as follows:

```c
/* This is our bus structure */
struct bus_type packt_bus_type = {
    .name     = "packt",
    .match    = packt_device_match,
    .probe    = packt_device_probe,
    .remove   = packt_device_remove,
    .shutdown = packt_device_shutdown,
};
```

Now that the basic bus operations have been defined, we need to register the PACKT bus framework and make it available for both the controller and slave drivers. The bus controller is a device itself; it must be registered with the kernel and will be used as the parent of the device that's sitting on the bus. This is done in the bus controller's probe or init function. In the case of the PACKT bus, the code would be as follows:

```c
static int __init packt_init(void)
{
    int status;
    status = bus_register(&packt_bus_type);
    if (status < 0)
        goto err0;

    status = class_register(&packt_master_class);
    if (status < 0)
        goto err1;
```

```
        return 0;

err1:
    bus_unregister(&packt_bus_type);
err0:
    return status;
}
postcore_initcall(packt_init);
```

When a device is registered by the bus controller driver, the `parent` member of the device must point to the bus controller device (this should be done by the device driver) and its `bus` property must point to the `PACKT` bus type (this is done by the core) to build the physical device tree. To register a `PACKT` device, you must call `packt_device_register()`, which should take as an argument a `PACKT` device allocated with `packt_device_alloc()`:

```
int packt_device_register(struct packt_device *packt)
{
    packt->dev.bus = &packt_bus_type;
    return device_register(&packt->dev);
}
EXPORT_SYMBOL(packt_device_register);
```

Now that we are done with bus registration, let's look at the driver's infrastructure and see how it is designed.

The driver data structure

A driver is a set of methods that allow us to drive a given device. A global device hierarchy allows all the system's devices to be represented in the same way. This makes it simple for the core to navigate the device tree and perform tasks such as properly ordered power management transitions.

Each device driver is represented as an instance of `struct device_driver`, which is defined like so:

```
struct device_driver {
    const char       *name;
    struct bus_type  *bus;
    struct module    *owner;
```

```
    const struct of_device_id    *of_match_table;
    const struct acpi_device_id  *acpi_match_table;

    int (*probe) (struct device *dev);
    int (*remove) (struct device *dev);
    void (*shutdown) (struct device *dev);
    int (*suspend) (struct device *dev,
                    pm_message_t state);
    int (*resume) (struct device *dev);
    const struct attribute_group **groups;

    const struct dev_pm_ops *pm;
};
```

Let's look at the elements in this data structure:

- name represents the driver's name. It can be used for matching, by comparing it with the device's name.

- bus represents the bus where this driver sits on. The bus driver must fill this field.

- module represents the module that owns this driver. In 99% of cases, you must set this field to THIS_MODULE.

- of_match_table is a pointer to the array of struct of_device_id. The struct of_device_id structure is used to perform open firmware matches through a special file called the device tree, which is passed to the kernel during the boot process:

```
struct of_device_id {
    char        compatible[128];
    const void *data;
};
```

- suspend and resume are power management callbacks that are invoked to put the device to sleep or wake it up from a sleep state, respectively. The remove callback is called when the device is physically removed from the system or when its reference count reaches 0. The remove callback is also called during system reboot.

- probe is the probe callback that runs when you're attempting to bind a driver to a device. The bus driver is in charge of calling the device driver's probe function.

- group is a pointer to a list (array) of struct attribute_group and is used as a default attribute for the driver. Prefer this method instead of creating an attribute separately.

Now that we are familiar with the driver data structure and all its elements, let's learn what APIs the kernel provides so that we can register it.

Driver registration

The low-level driver_register() function is used to register a device driver with the bus and it is added to the bus's driver list. When a device driver registers with the bus, the core travels through the list of devices on this same bus and calls the bus's match callback for each device that does not have a driver. It does this to find out if there are any devices that the driver can handle.

The following is the declaration of our driver infrastructure:

```
/*
 * Bus specific driver structure
 * You should provide your device's probe
 * and remove functions.
 */
struct packt_driver {
    int      (*probe)(struct packt_device *packt);
    int      (*remove)(struct packt_device *packt);
    void     (*shutdown)(struct packt_device *packt);
    struct device_driver driver;
    const struct i2c_device_id *id_table;
};
#define to_packt_driver(d) \
        container_of(d, struct packt_driver, driver)
#define to_packt_device(d) container_of(d, \
                        struct packt_device, dev)
```

In our example, there are two helper macros to get the PACKT device and the PACKT driver, given a generic struct device or struct driver.

Then comes the structure that's used to identify a PACKT device, which is defined as follows:

```
struct packt_device_id {
    char name[PACKT_NAME_SIZE];
    kernel_ulong_t driver_data;    /* Data private to the driver
*/
};
```

The device and the device driver are bound together when they match. Binding is the process of associating a device with a device driver, and it is performed by the bus framework.

Now, let's get back to registering our drivers with our PACKT bus. The drivers must use packt_register_driver(struct packt_driver *driver), which is a wrapper around driver_register(). The driver parameter must have been filled in before registering the PACKT driver. The LDM core provides helper functions for iterating over the list of drivers that have been registered with the bus:

```
int bus_for_each_drv(struct bus_type * bus,
                struct device_driver * start, void * data,
                int (*fn)(struct device_driver *, void *));
```

This helper iterates over the bus's list of drivers and calls the fn callback for each driver in the list.

The device data structure

struct device is the generic data structure that's used to describe and characterize each device on the system, whether it is physical or not. It contains details about the physical attributes of the device and provides proper linkage information to help build suitable device trees and reference counting:

```
struct device {
    struct device *parent;
    struct kobject kobj;
    const struct device_type *type;
    struct bus_type      *bus;
    struct device_driver *driver;
    void     *platform_data;
    void     *driver_data;
```

```
    struct device_node        *of_node;
    struct class *class;
    const struct attribute_group **groups;
    void     (*release)(struct device *dev);
[...]
};
```

The preceding data structure has been shortened for the sake of readability. That said, let's take a look at the elements that have been provided:

- parent represents the device's parent and is used to build the device tree hierarchy. When registered with a bus, the bus driver is responsible for setting this field with the bus device.

- bus represents the bus where this device sits. The bus driver must fill this field.

- type identifies the device's type.

- kobj is the kobject and handles reference counting and device model support.

- of_node is a pointer to the open firmware (device tree) node associated with the device. It is up to the bus driver to set this field.

- platform_data is a pointer to the platform data that's specific to the device. It's usually declared in a board-specific file during device provisioning.

- driver_data is a pointer to private data for the driver.

- class is a pointer to the class that this device belongs to.

- group is a pointer to a list (array) of struct attribute_group and is used as the default attributes for the device. You should use this instead of creating the attributes separately.

- release is a callback that's called when the device reference count reaches zero. The bus has the responsibility of setting up this field. The PACKT bus driver shows you how to do that.

Now that we have described the device's structure, let's learn how to make it part of the system by registering it.

Device registration

`device_register()` is a function that's provided by the LDM core to register a device with the bus. After this call, the bus's list of drivers is iterated over to find the driver that supports this device; then, this device is added to the bus's device list. `device_register()` internally calls `device_add()`:

```
int device_add(struct device *dev)
{
    [...]
    bus_probe_device(dev);
        if (parent)
                klist_add_tail(&dev->p->knode_parent,
                            &parent->p->klist_children);
    [...]
}
```

The helper function that's provided by the kernel to iterate over the bus's device list is `bus_for_each_dev`. It's defined as follows:

```
int bus_for_each_dev(struct bus_type * bus,
                    struct device * start, void * data,
                    int (*fn)(struct device *, void *));
```

Whenever a device is added, the core invokes the matching method of the bus driver (`bus_type->match`). If the matching function succeeds, the core will invoke the probing function of the bus driver (`bus_type->probe`), given that both the device and driver matched as parameters. Then, it is up to the bus driver to invoke the probing method of the device's driver (that is, `driver->probe`). For our PACKT bus driver, the function that's used to register a device is `packt_device_register(struct packt_device *packt)`, which internally calls `device_register()`. Here, the parameter is a PACKT device that's been allocated with `packt_device_alloc()`.

The bus-specific device data structure is then defined as follows:

```
/*
 * Bus specific device structure
 * This is what a PACKT device structure looks like
 */
struct packt_device {
    struct module        *owner;
```

```
    unsigned char            name[30];
    unsigned long            price;
    struct device            dev;
};
```

In the preceding code, `dev` is the underlying `struct device` structure for the device model, while `name` is the name of the device.

Now that we have defined the data structure, let's learn about the underlying mechanisms of LDM.

Getting deeper inside LDM

So far, we have discussed buses, drivers, and devices, which were used to build the system device topology. While this is true, the previous topics were the tip of the iceberg. Under the hood, LDM relies on the three lowest level data structures, which are `kobject`, `kobj_type`, and `kset`. These are used to link the objects.

Before we go any further, let's define some of the terms that will be used throughout this chapter:

- **sysfs**: sysfs is an in-memory virtual filesystem that shows the hierarchy of kernel objects, abstracted by instances of `struct kobject`.
- **Attribute**: An attribute (or sysfs attribute) appears as a file in sysfs. From within the kernel, it can be mapped to anything: a variable, a device property, a buffer, or anything useful to the driver that may need to be exported to the world.

In this section, we will learn how each of these structures is involved in the device model.

Understanding the kobject structure

`struct kobject`, which means kernel object (also abbreviated as **kobject** throughout this chapter) is the core data structure of the device model as it is the core of the concepts behind the sysfs. For each directory that's found in sysfs, there is a `struct kobject` wandering around somewhere within the kernel. Additionally, a kobject can export one or more attributes, which appear in that Kobject's sysfs directory as files. Now, let's get back to the code – `struct kobject` is defined in the kernel like so:

```
struct kobject {
    const char               *name;
    struct list_head         entry;
```

```
    struct kobject          *parent;
    struct kset             *kset;
    struct kobj_type        *ktype;
    struct sysfs_dirent     *sd;
    struct kref             kref;
[...]
};
```

In the preceding data structure, only the main elements have been listed. Let's take a look at them in more detail:

- name is the name of this kobject. It can be modified using the kobject_set_name(struct kobject *kobj, const char *name) function. It is used as the name of this kobject directory.

- parent is a pointer to another kobject, which is considered the parent of this kobject. It is used to build topologies and to describe the relationship between objects.

- sd points to a struct sysfs_dirent structure that represents the directory of this kobject in sysfs. name will be used as the name of this directory. If parent is set, then this directory will be a sub-directory in the parent's directory.

- kref provides reference counting for the kobject. It helps track whether an object is still in use or not and potentially releases it if it's not being used anymore. Alternatively, it can prevent it from being removed if it's still in use. When it's used from within a kernel object, its initial value is 1.

- ktype describes the kobject. Every kobject is given a set of default attributes when it is created. The ktype element, which belongs to the struct kobj_type structure, is used to specify these default attributes. Such a structure allows kernel objects to share common operations (sysfs_ops), whether those objects are functionally related or not.

- kset tells us which set (group) of objects this object belongs to.

Before a kobject can be used, it must be (exclusively) dynamically allocated and then initialized. To do so, drivers can use the `kzalloc()` (or `kmalloc()`) or `kobject_create()` function. With `kzalloc()`, the object is allocated and empty and must be initialized using another API, `kobject_init()`. With `kobject_create()`, allocation and initialization are implicit. These APIs are defined as follows:

```
void kobject_init(struct kobject *kobj,
                     struct kobj_type *ktype)
struct kobject *kobject_create(void)
```

In the preceding code, `kobject_init()` expects, in its first parameter, a kobject that has been initially allocated, via `kzalloc()` for example. The second parameter, `ktype`, is mandatory, so it can't be NULL; otherwise, the kernel will complain (`dump_stack()`). `kobject_create()`, on the other hand, expects nothing; it does allocation and initialization implicitly (it calls `kzalloc()` and `kobject_init()` internally). On success, it returns a freshly initialized kobject object.

Once a kobject has been initialized, the driver can use `kobject_add()` to link this object with the system, creating a sysfs directory entry for this kobject. Where this directory will be created depends on the parent element of the kobject being set or not. That said, if this parent is not set, it can be specified while adding the kobject to the system using `kobject_add()`, which is defined as follows:

```
int kobject_add(struct kobject *kobj, struct kobject *parent,
                     const char *fmt, ...);
```

In the preceding function, `kobj` is the kernel object to be added to the system and `parent` is its parent. The `kobject` directory will be created as a sub-directory of its parent. If `parent` is NULL, the directory will be created under `/sys/` directly.

Instead of using `kobject_init()` or `kobject_create()` and then `kobject_add()` individually, it is possible to use `kobject_init_and_add()`, which groups their actions. It's defined like so:

```
int kobject_init_and_add(struct kobject *kobj,
                     struct kobj_type *ktype,
                     struct kobject *parent,
                     const char *fmt, ...);
```

In the preceding interface, the object needs to be allocated first. There is another helper that will implicitly allocate, initialize, and add the kobject with the system. This function is `kobject_create_and_add()` and it's defined as follows:

```
struct kobject * kobject_create_and_add(const char *name,
                                        struct kobject *parent);
```

The preceding function takes the name of the kobject that will be used as the directory name, as well as a parent kobject whose created directory will be a sub-directory. Passing NULL as the second parameter will result in the directory being created under `/sys/` directly.

That said, some predefined kobjects in the kernel already represent some directories under `/sys/`. Let's look at a few of them:

- `kernel_kobj`: This kobject is responsible for the `/sys/kernel` directory.
- `mm_kobj`: This is responsible for `/sys/kernel/mm`.
- `fs_kobj`: This is the filesystem kobject and it's responsible for `/sys/fs`.
- `hypervisor_kobj`: This is responsible for `/sys/hypervisor`.
- `power_kobj`: This is a power management kobject that's used at the origin of `/sys/power`.
- `firmware_kobj`: This is the firmware kobject that owns the `/sys/firmware` directory.

Once you're done with a kobject, the driver should release it. The low-level function you can use to release a kobject is `kobject_release()`. However, this API does not consider other potential users of the kobject. It is raw and dummy. It is recommended to use `kobject_put()` instead, which will decrement the kobject's reference counter and then release this kobject if the new reference counter value is 0. Remember that when a kobject is initialized, the reference counter value is set to 1. Moreover, it is also recommended that users wrap the kobject's usage into the `kobject_get()` and `kobject_put()` functions, where `kobject_get()` will just increment the reference counter value.

These APIs have the following prototypes:

```
void kobject_put(struct kobject * kobj);
struct kobject *kobject_get(struct kobject *kobj);
```

In the preceding code, `kobject_get()` takes the kobject to increase the reference counter as a parameter and returns this same kobject once the parameter has been initialized. `kobject_put()` will decrement 1 from the reference counter and, if the new value is 0, `kobject_release()` will be automatically called on the object, which will release it.

The following code shows how to combine `kobject_create()`, `kobject_init()`, and `kobject_add()` to create and add a kernel object to the system:

```
/* Somewhere */
static struct kobject *mykobj;

[...]
mykobj = kobject_create();
if (!mykobj)
    return -ENOMEM;

kobject_init(mykobj, &my_ktype);
if (kobject_add(mykobj, NULL, "%s", "hello")) {
    pr_info("ldm: kobject_add() failed\n");
    kobject_put(mykobj);
    mykobj = NULL;
    return -1;
}
```

As we can see, we could use an all-in-one function, such as `kobject_create_and_add()`, which internally calls `kobject_create()` and `kobject_add()`. The following excerpt from `drivers/base/core.c` shows how to use it:

```
static struct kobject * class_kobj   = NULL;
static struct kobject * devices_kobj = NULL;

/* Create /sys/class */
class_kobj = kobject_create_and_add("class", NULL);
if (!class_kobj)
    return -ENOMEM;
[...]

/* Create /sys/devices */
```

```
devices_kobj = kobject_create_and_add("devices", NULL);
if (!devices_kobj)
    return -ENOMEM;
```

Keep in mind that for each `struct kobject`, the corresponding kobject directory can be found in `/sys/`, and the upper directory is pointed out by `kobj->parent`.

> **Note**
>
> Since it is mandatory to provide a `kobj_type` while initializing a kobject, all-in-one helpers use a default and kernel-provided `kobj_type`; that is, `dynamic_kobj_ktype`. Therefore, unless you have a good reason to initialize your `kobj_type` (most of the time, this will be if you wish to populate some default attributes), you should use `kobject_create*()` variants, which use the kernel-provided kobject type instead of `kobject_init*()`, which would require providing your own initialized `kobj_type`.

Understanding the kobj_type structure

A `struct kobj_type` structure, which means kernel object type, is a data structure that defines the behavior of a kobject element and controls what happens to this kobject when it is created or destroyed. Additionally, a `kobj_type` contains the default attributes of the kobject, as well as the hooks that allow it to operate on these attributes.

Because most devices of the same type have the same attributes, these attributes are isolated and stored in the `ktype` element. This allows them to be managed flexibly. Every kobject must have an associated `kobj_type` structure. Its data structure is defined as follows:

```
struct kobj_type {
    void (*release)(struct kobject *);
    const struct sysfs_ops sysfs_ops;
    struct attribute **default_attrs;
};
```

In the preceding data structure, `release` is a callback that's called on the release path of the kobject to give drivers a chance to release resources that have been allocated for the kobject. This callback is implicitly run when `kobject_put()` is about to free the kobject. `default_attrs` is an array of pointers to `attribute` structures. This field lists the attributes to be created for every kobject of this type, while `sysfs_ops` provides a set of methods that allow you to access those attributes.

The following code shows the definition of the `struct sysfs_ops` data structure in the kernel:

```
struct sysfs_ops {
    ssize_t (*show)(struct kobject *kobj,
                struct attribute *attr, char *buf);
    ssize_t (*store)(struct kobject *kobj,
                struct attribute *attr, const char *buf,
                size_t size);
};
```

In the preceding code, `show` is the callback that's invoked in response to a read operation of an attribute being exposed by this `kobj_type` – that is, whenever an attribute is read from the user space. `buf` is the output buffer. The buffer's size is fixed and is `PAGE_SIZE` in length. The data that must be exposed must be put inside `buf`, preferably using `scnprintf()`. Finally, if the callback succeeds, it must return the size (in bytes) of the data that was written into the buffer, or a negative error if it fails. Each attribute should contain/provide a single, human-readable value or property, according to the sysfs rules; if you have a lot of data to return, you should consider breaking it into numerous attributes.

`store` is called for writing purposes – that is, when users write something into an attribute. Its `buf` parameter is `PAGE_SIZE` at most but it can be smaller. It must return the size (in bytes) of the data that was read from the buffer on success or a negative error on failure (or if an unwanted value is received).

The `attr` pointer is passed as an argument into both methods and can be utilized to determine which attribute is being accessed. It is frequent for `show`/`store` methods to perform a series of tests you can perform on the attribute name to achieve that. Other implementations, on the other hand, wrap the attribute's structure in an enclosing data structure that provides the data that's required to return the attribute's value (`struct kobject_attribute`, `struct device_attribute`, `struct driver_attribute`, and `struct class_attribute` are some examples); in this case, the `container_of` macro is used to obtain a pointer to the embedding structure. This is the method that's used by `kobj_sysfs_ops`, which represents the operations that are provided by `dynamic_kobj_ktype`. Both methods are demonstrated in the examples provided with this book.

Understanding the kset structure

The purpose of `struct kset` is mainly to group related kernel objects together. `kset` stands for kernel object sets, which can be interpreted as a collection of kobjects. In other words, a `kset` gathers related kobjects into a single place, such as all *block devices*.

The `kset` data structure is defined in the kernel like so:

```
struct kset {
    struct list_head list;
    spinlock_t list_lock;
    struct kobject kobj;
};
```

All the elements in the data structure are quite self-explanatory. Simply put, `list` is a linked list of all kobjects in `kset`, `list_lock` is a spinlock that protects linked list access (while adding or removing kobject elements in `kset`), and `kobj` represents the base class kobject for the set. This kobject will be used as the default parent of kobjects to be added in the set with a `NULL` parent.

Each registered `kset` corresponds to a sysfs directory that's created on behalf of its `kobj` element. A `kset` can be created and added using the `kset_create_and_add()` function and removed with `kset_unregister()`. The following code shows the definitions for both:

```
struct kset * kset_create_and_add(const char *name,
                      const struct kset_uevent_ops *u,
                      struct kobject *parent_kobj);
void kset_unregister (struct kset * k);
```

In the preceding APIs, `name` is the name of `kset`, which is also used as the name of the directory that will be created for `kset`. The `u` parameter is a pointer to a `struct uevent_ops`, which represents a set of **user event** (**uevent**) operations that are called whenever a change is made to `kset` so that, for example, it can add new environment variables or filter out the uevents if so desired. This parameter can be (and most of the time is) `NULL`. Finally, `parent_kobj` is the parent kobject of `kset`.

Adding a kobject to the set is as simple as specifying its `.kset` field for the right `kset`:

```
static struct kobject foo_kobj, bar_kobj;
[...]

example_kset = kset_create_and_add("kset_example",
```

```
                        NULL, kernel_kobj);

  /* since we have a kset for this kobject,
   * we need to set it before calling into the kobject core.
   */
 foo_kobj.kset = example_kset;
 bar_kobj.kset = example_kset;

 retval = kobject_init_and_add(&foo_kobj, &foo_ktype,
                        NULL, "foo_name");
 retval = kobject_init_and_add(&bar_kobj, &bar_ktype,
                        NULL, "bar_name");
```

Once you're done with your `kset`, it can be released with `kset_unregister()`, after which it will be dynamically deallocated when it is no longer in use. The following code will release our example's `kset`:

```
 kset_unregister(example_kset);
```

Now that we are familiar with kobjects and type structures, let's learn how to deal with non-default sysfs attributes.

Working with non-default attributes

Attributes are sysfs files that are exported to the user space via kobjects. While default attributes might be enough most of the time, you can add other attributes. An attribute can be readable, writable, or both, from the user space.

An attribute definition looks as follows:

```
struct attribute {
        char           *name;
        struct module  *owner;
        umode_t        mode;
};
```

In the attribute data structure, `name` is the name of the attribute, which is also the name of the corresponding file entry. `owner` is the attribute owner – most of the time, this is `THIS_MODULE` – and `mode` specifies the read/write permissions for this attribute in **user-group-other (ugo)** format.

Default attributes are very convenient to use but are not flexible enough. Moreover, simple attributes cannot be read or written except by their `kobj_type` sysfs ops, which means that if there are too many attributes, the branches in the show/store functions will be messy. To address this, the kobject core provides a mechanism where each attribute is embedded in an enclosing and special data structure: `struct kobj_attribute`. This data structure exposes the wrapper routines for reading and writing.

`struct kobj_attribute` (defined in `include/linux/kobject.h`) looks as follows:

```
struct kobj_attribute {
  struct attribute attr;
  ssize_t (*show)(struct kobject *kobj,
                  struct kobj_attribute *attr, char *buf);
  ssize_t (*store)(struct kobject *kobj,
                  struct kobj_attribute *attr,
                  const char *buf, size_t count);
};
```

In this data structure, `attr` is the attribute representing the file to be created, `show` is a pointer to a function that will be called when the file is read from the user space, and `store` is a pointer to a function that will be called when the file is written, again from the user space.

Using the enclosing `kobj_attribute` structure makes developments more generic and extends attribute flexibility. This way, a pointer to `attr` is passed to either the `store` or `show` function and can be used not only to determine which attribute is being accessed but also to retrieve the enclosing data structure (that is, `kobj_attribute`) that this attribute's `show`/`store` method can be invoked from. To do so, you can use the `container_of` macro to obtain a pointer to the embedding structure.

The following is an excerpt from `lib/kobject.c` that demonstrates this generic mechanism in both the `show` and `store` methods of a kernel-provided sysfs operations element: `kobj_sysfs_ops`. This element is also the sysfs operations data structure (the `kobj_type->sysfs_ops` element) that's used by `dynamic_kobj_ktype`:

```
static ssize_t kobj_attr_show(struct kobject *kobj,
                      struct attribute *attr, char *buf)
{
    struct kobj_attribute *kattr;
    ssize_t ret = -EIO;
```

```
    kattr = container_of(attr, struct kobj_attribute, attr);
    if (kattr->show)
        ret = kattr->show(kobj, kattr, buf);
    return ret;
}


static ssize_t kobj_attr_store(struct kobject *kobj,
                               struct attribute *attr,
                               const char *buf, size_t count)
{
    struct kobj_attribute *kattr;
    ssize_t ret = -EIO;

    kattr = container_of(attr, struct kobj_attribute, attr);
    if (kattr->store)
        ret = kattr->store(kobj, kattr, buf, count);
    return ret;
}


const struct sysfs_ops kobj_sysfs_ops = {
    .show  = kobj_attr_show,
    .store = kobj_attr_store,
};
```

In the preceding code, the `container_of` macro does everything. This also reassures us that with this approach, we remain compatible with all the bus-, device-, class-, and driver-related kobject implementations, as we will see in the next section.

Let's go back to the APIs. You will probably always know which attributes you wish to expose in advance; thus, the attributes will almost always be declared statically. To help with this, the kernel provides the `__ATTR` macro for initializing `kobj_attribute`. This macro is defined as follows:

```
#define __ATTR(_name, _mode, _show, _store) {            \
    .attr = {.name = __stringify(_name),                 \
             .mode = VERIFY_OCTAL_PERMISSIONS(_mode) },\
    .show  = _show,                                      \
```

```
        .store = _store,                                    \
}
```

In the preceding macro definition, _name will be stringified and used as the attribute name, _mode represents the attribute mode, and _show and _store are pointers to the attribute's show and store methods, respectively.

The following is an example of two attribute's declarations, bar and foo (this example will be used as a base later in this section):

```
static struct kobj_attribute foo_attr =
    __ATTR(foo, 0660, attr_show, attr_store);
```

```
static struct kobj_attribute bar_attr =
    __ATTR(bar, 0660, attr_show, attr_store);
```

In the preceding example, we have two attributes with the 0660 permission. The first attribute is named foo, the second one is named bar, and both use the same show and store methods.

Now, we must create the underlying file. The low-level kernel APIs that are used to add/remove attributes from the sysfs filesystem are sysfs_create_file() and sysfs_remove_file(), respectively. They are defined as follows:

```
int sysfs_create_file(struct kobject * kobj,
                    const struct attribute * attr);
void sysfs_remove_file(struct kobject * kobj,
                    const struct attribute * attr);
```

sysfs_create_file() returns 0 on success or a negative error on failure. sysfs_remove_file() must be given the same parameters to remove the file attributes.

Let's use these APIs to add our bar and foo attributes to the system:

```
struct kobject *demo_kobj;
int err;

demo_kobj = kobject_create_and_add("demo", kernel_kobj);
if (!demo_kobj) {
    pr_err("demo: demo_kobj registration failed.\n");
    return -ENOMEM;
}
```

```
err = sysfs_create_file(demo_kobj, &foo_attr.attr);
if (err)
    pr_err("unable to create foo attribute\n");
err = sysfs_create_file(demo_kobj, &bar_attr.attr);
if (err){
    sysfs_remove_file(demo_kobj, &foo_attr.attr);
    pr_err("unable to create bar attribute\n");
}
```

Once the preceding code has been executed, the bar and foo files will be visible in sysfs, in the /sys/demo directory. In our example, we used the __ATTR macro to define our attributes. We had to specify the name, the mode, and the show/store methods. The kernel provides convenience macros for the most frequent cases to make specifying attributes and writing code more succinct, readable, and easier. These macros are as follows:

- __ATTR_RO(name): This assumes that name_show is the show callback's name and sets the mode to 0444.

- __ATTR_WO(name): This assumes that name_store is the store function's name and is restricted to mode 0200, which means root write access only.

- __ATTR_RW(name): This assumes that name_show and name_store are for the show and store callbacks' names, respectively, and sets the mode to 0644.

- __ATTR_NULL: This is used as a list terminator. It sets both names to NULL and is used as an end of list indicator (see kernel/workqueue.c).

All these macros only expect the name of the attribute as a parameter. The difference with these macros is that unlike __ATTR, whose store/show function names can be arbitrary, the attributes here are built under the assumption that the show and store methods are named <attribute_name>_show and <attribute_name>_store, respectively. The following code demonstrates this with the __ATTR_RW_MODE macro, which is defined as follows:

```
#define __ATTR_RW_MODE(_name, _mode) {                     \
    .attr = { .name = __stringify(_name),                  \
            .mode = VERIFY_OCTAL_PERMISSIONS(_mode) },\
    .show   = _name##_show,                                \
    .store  = _name##_store,                               \
}
```

As we can see, the .show and .store fields are set with their attribute names suffixed with _show and _store, respectively. Let's take a look at the following example:

```
static struct kobj_attribute attr_foo = __ATTR_RW(foo);
```

The preceding attribute declaration assumes that the show and store methods are defined as foo_show and foo_store, respectively.

> **Note**
>
> If you need to provide a single store/show operation pair for all the attributes, you should probably define these attributes with __ATTR. However, if processing the attributes requires providing a show/store pair per attribute, you can use the other attribute definition macros.

The following code shows the implementation of the show/store function for our previously defined foo and bar attributes:

```
static ssize_t attr_store(struct kobject *kobj,
                          struct kobj_attribute *attr,
                          const char *buf, size_t count)
{
    int value, ret;

    ret = kstrtoint(buf, 10, &value);
    if (ret < 0)
        return ret;

    if (strcmp(attr->attr.name, "foo") == 0)
        foo = value;
    else /* if (strcmp(attr->attr.name, "bar") == 0) */
        bar = value;

    return count;
}

static ssize_t attr_show(struct kobject *kobj,
                         struct kobj_attribute *attr,
                         char *buf)
```

```
{
    int value;

    if (strcmp(attr->attr.name, "foo") == 0)
        value = foo;
    else
        value = bar;
    return sprintf(buf, "%d\n", value);
}
```

In the preceding code, instead of providing a pair of show/store operations per attribute, we have used the same function pair for all the attributes, and we differentiated the attributes by their respective names. This is a common practice when you're using the generic `kobject_attribute` instead of framework-specific attributes. This is because they sometimes impose different show/store function names for each attribute since they do not rely on the `__ATTR` macro for defining attributes.

Working with binary attributes

So far, we have become familiar with the sysfs statement and mentioned that an attribute must store a single property/value in a human-readable text format, as well as that such an attribute has a `PAGE_SIZE` limit. However, there could be situations, although rare, which would require larger data to be exchanged in binary format, for example, all with random access. An example of such a situation is a device firmware transfer, where the user space would upload some binary data to be pushed to the hardware or PCI devices, exposing part or all of their configuration address spaces.

To cover those cases, the sysfs framework provides binary attributes. Note that these attributes are for sending/receiving binary data that is not interpreted/manipulated by the kernel at all. It should *only* be used as a pass-through to and from hardware, with no interpretation by the kernel. The only manipulations you can perform are some checks on the magic number and size, for example.

Now, let's get back to the code A binary attribute is represented using a `struct bin_attribute` and is defined as follows:

```
struct bin_attribute {
    struct attribute attr;
    size_t      size;
    void                *private;
    ssize_t (*read)(struct file *filp,
```

```
                struct kobject *kobj,
                struct bin_attribute *attr,
                char *buffer, loff_t off, size_t count);
    ssize_t (*write)(struct file *filp,
                struct kobject *kobj,
                struct bin_attribute *attr,
                const char *buffer,
                loff_t off, size_t count);
    int (*mmap)(struct file *filp, struct kobject *kobj,
                    struct bin_attribute *attr,
                    struct vm_area_struct *vma);
};
```

In the preceding code, `attr` is the underlying classic attribute for this binary attribute and holds the name, owner, and permissions for the binary attribute. `size` represents the maximum size of the binary attribute (or zero if there is no maximum limit). `private` is a field that can be used for any convenience. Most of the time, it is assigned the buffer of the binary attribute. The `read()`, `write()`, and `mmap()` functions are optional and work similarly to the normal `char` driver equivalents. In their parameters, `filp` is an opened file pointer instance that's associated with the attribute and `kobj` is the underlying `kobject` associated with this this binary attribute. `buffer` is the output or input buffer for read or write operations, respectively. `off` is the same offset argument that's found in all read or write methods for all types of files. It refers to the offset from the start of the file – that is, offset into the binary data. Finally, `count` is the number of bytes to read or write.

> **Note**
>
> Though binary attributes may not have size limitations, larger data is always requested/sent on a `PAGE_SIZE` chunk basis. This means that, for example, the `write()` function can be called multiple times for a single load. However, this split is handled by the kernel, which means it's transparent for the driver. The disadvantage is that sysfs has no way of signaling the end of a series of write operations, so code that implements a binary attribute must figure it out some other way.

For a binary attribute to be created, it must be allocated and initialized. Like classic attributes, there are two ways to allocate binary attributes – either statically or dynamically. For static allocation, the framework provides the low-level __BIN_ATTR macro, which is defined as follows:

```
#define __BIN_ATTR(_name, _mode, _read, _write, _size) {  \
    .attr = { .name = __stringify(_name), .mode = _mode }, \
    .read    = _read,                                       \
    .write   = _write,                                      \
    .size    = _size,                                       \
```

It works similarly to the __ATTR macro. In terms of parameters, _name is the binary attribute name, _mode represents its permissions, _read and _write are the read and write functions, respectively, and _size is the size of the binary attribute.

Like classic attributes, binary attributes have their own high-level helper macros to ease the process of defining them. Some of these macros are as follows:

```
BIN_ATTR_RO(name, size)
BIN_ATTR_WO(name, size)
BIN_ATTR_RW(name, size)
```

These macros declare a single instance of struct bin_attribute, whose corresponding variable is named bin_attribute_<name>, as shown in the following BIN_ATTR definition:

```
#define BIN_ATTR_RW(_name, _size)          \
struct bin_attribute bin_attr_##_name =    \
            __BIN_ATTR_RW(_name, _size)
```

Moreover, like classic attributes, these high-level macros expect the read/write methods to be named <attribute_name>_read and <attribute_name>_write, respectively, as shown in the following __BIN_ATTR_RW definition:

```
#define __BIN_ATTR_RW(_name, _size) \
    __BIN_ATTR(_name, 0644, _name##_read, _name##_write, \
              _size)
```

For dynamic allocation, a simple kzalloc() is enough. However, dynamically allocated binary attributes must be initialized using sysfs_bin_attr_init(), as shown here:

```
void sysfs_bin_attr_init(strict bin_attribute *bin_attr)
```

After this, the driver must set other properties, such as the underlying attribute's mode, name, and permission, and optionally the read/write/map functions.

Unlike classic attributes, which can be set up as default attributes, binary attributes must be created explicitly. This can be done using `sysfs_create_bin_file()`, as follows:

```
int sysfs_create_bin_file(struct kobject *kobj,
                          struct bin_attribute *attr);
```

This function returns 0 on success or a negative error on failure. Once you're done with a binary attribute, it can be removed with `sysfs_remove_bin_file()`, which is defined as follows:

```
int sysfs_remove_bin_file(struct kobject *kobj,
                          struct bin_attribute *attr);
```

The following is an excerpt (whose full version can be found in `drivers/i2c/i2c-slave-eeprom.c`) of a concrete example highlighting the use of a binary attribute that's been allocated and initialized dynamically:

```
struct eeprom_data {
[...]
    struct bin_attribute bin;
    u8 buffer[];
};


static int i2c_slave_eeprom_probe(
                         struct i2c_client *client)
{
    struct eeprom_data *eeprom;
    int ret;
    unsigned int size = FIELD_GET(I2C_SLAVE_BYTELEN,
                                  id->driver_data) + 1;


    eeprom = devm_kzalloc(&client->dev,
                sizeof(struct eeprom_data) + size,
                GFP_KERNEL);
    if (!eeprom)
        return -ENOMEM;
```

```
    [...]

    sysfs_bin_attr_init(&eeprom->bin);
    eeprom->bin.attr.name = "slave-eeprom";
    eeprom->bin.attr.mode = S_IRUSR | S_IWUSR;
    eeprom->bin.read = i2c_slave_eeprom_bin_read;
    eeprom->bin.write = i2c_slave_eeprom_bin_write;
    eeprom->bin.size = size;

    ret = sysfs_create_bin_file(&client->dev.kobj,
                                &eeprom->bin);
    if (ret)
        return ret;

    [...]
    return 0;
};
```

Upon unloading the path of the module or when the device leaves, the associated binary file is removed, as follows:

```
static int i2c_slave_eeprom_remove(struct i2c_client *client)
{
    struct eeprom_data *eeprom = i2c_get_clientdata(client);
    sysfs_remove_bin_file(&client->dev.kobj, &eeprom->bin);

[...]
    return 0;
}
```

Then, when it comes to implementing the read/write function, data can be moved back and forth using memcpy(), as shown here:

```
static ssize_t i2c_slave_eeprom_bin_read(struct file *filp,
        struct kobject *kobj, struct bin_attribute *attr,
        char *buf, loff_t off, size_t count)
{
    struct eeprom_data *eeprom;
```

```
    eeprom = dev_get_drvdata(kobj_to_dev(kobj));
[...]
    memcpy(buf, &eeprom->buffer[off], count);
[...]
    return count;
}


static ssize_t i2c_slave_eeprom_bin_write(
        struct file *filp, struct kobject *kobj,
        struct bin_attribute *attr,
        char *buf, loff_t off, size_t count)
{
    struct eeprom_data *eeprom;
    eeprom = dev_get_drvdata(kobj_to_dev(kobj));
[...]
    memcpy(&eeprom->buffer[off], buf, count);
[...]
    return count;
}
```

In the preceding excerpt, the offset (the off parameter) points to where the data should be read/written, and count determines the size of this data.

The concept of attribute group

So far, we have learned how to individually add (binary) attributes by calling the sysfs_create_file() or sysfs_create_bin_file() function. While this is enough if we have a few attributes to add, it may become painful as the number of attributes grows, either upon adding or removing them. The driver will have to loop over the attributes to create each of them or invoke sysfs_create_file() as many times as there are attributes. Here is where the attribute group comes in. It relies on the struct attribute_group structure, which is defined as follows:

```
struct attribute_group {
    const char          *name;
    umode_t             (*is_visible)(struct kobject *,
                            struct attribute *, int);
    umode_t             (*is_bin_visible)(struct kobject *,
                            struct bin_attribute *, int);
```

```
    struct attribute  **attrs;
    struct bin_attribute   **bin_attrs;
};
```

If it is unnamed, an attribute group will place all the attributes directly in the kobject's directory when defining a group of attributes. If, however, a name is supplied, a subdirectory will be created for the attributes, with the directory's name being the name of the attribute group. is_visible() is an optional callback that intends to return the permissions associated with a specific attribute in the group. It will be called repeatedly for each (non-binary) attribute in the group. This callback must then return the read/write permission of the attribute, or 0 if the attribute is not supposed to be accessed at all. is_bin_visible() is the counterpart of is_visible() for binary attributes. The returned value/permission will replace the static permissions that have been defined in struct attribute. The attrs element is a pointer to a NULL terminated list of attributes, while bin_attrs is its counterpart for binary attributes.

The kernel functions that are used to add/remove group attributes to/from the filesystem are as follows:

```
int sysfs_create_group(struct kobject *kobj,
                       const struct attribute_group *grp)
void sysfs_remove_group(struct kobject * kobj,
                        const struct attribute_group * grp)
```

Back to our demo example with standard attributes, the two bar and foo attributes can be embedded into a struct attribute_group. This will allow us adding these to the system in a single shot, using one function call as follows:

```
static struct kobj_attribute foo_attr =
    __ATTR(foo, 0660, attr_show, attr_store);

static struct kobj_attribute bar_attr =
    __ATTR(bar, 0660, attr_show, attr_store);

/* attrs is aa array of pointers to attributes */
static struct attribute *demo_attrs[] = {
    &bar_foo_attr.attr,
    &bar_attr.attr,
    NULL,
};
```

```
static struct attribute_group my_attr_group = {
    .attrs = demo_attrs,
    /*.bin_attrs = demo_bin_attrs,*/
};
```

Finally, to create the attributes in a single shot, we need to use `sysfs_create_group()`, as shown in the following code:

```
struct kobject *demo_kobj;
int err;

demo_kobj = kobject_create_and_add("demo", kernel_kobj);
if (!demo_kobj) {
    pr_err("demo: demo_kobj registration failed.\n");
    return -ENOMEM;
}
err = sysfs_create_group(demo_kobj, &foo_attr.attr);
```

Here, we have demonstrated the importance of creating a group of attributes and how easy it is to use their APIs. While we have been generic so far, in the next section, we'll learn how to create framework-specific attributes.

Creating symbolic links

Drivers can create/remove symbolic links on existing kobjects (directories) using `sysfs_{create|remove}_link()` functions, as shown here:

```
int sysfs_create_link(struct kobject * kobj,
                      struct kobject * target, char * name);
void sysfs_remove_link(struct kobject * kobj, char * name);
```

This allows an object to exist in more than one place or even create a shortcut. The `create` function will create a symbolic link called `name` that points to the remote `target` kobject's sysfs entry. The link will be created in the `kobj` kobject directory. A well-known example is devices appearing in both `/sys/bus` and `/sys/devices` since a bus controller is first a device on its own before exposing a bus. However, note that any symbolic links that are created will be persistent (unless the system is rebooted), even after target removal. Thus, the driver must consider that when the associated device leaves the system or when the module is unloaded.

Overview of the device model from sysfs

sysfs is a non-persistent virtual filesystem that provides a global view of the system and exposes the kernel objects hierarchy (topology) using their kobjects. Each kobject shows up as a directory. The files in these directories represent the kernel variables that are exported by the related kobject. These files are called attributes and can be read or written.

If any registered kobject creates a directory in sysfs, where the directory is created depends on the parent of this kobject (which is also a kobject, thus highlighting internal object hierarchies). In sysfs, top-level directories represent the common ancestors of object hierarchies or the subsystems that the objects belong to.

These top-level sysfs directories can be found in the /sys/ directory, as follows:

```
/sys$ tree -L 1
├── block
├── bus
├── class
├── dev
├── devices
├── firmware
├── fs
├── hypervisor
├── kernel
├── module
└── power
```

block contains a directory per block device on the system. Each of these contains subdirectories for partitions on the device. bus contains the registered bus on the system. dev contains the registered device nodes in a raw way (no hierarchy), with each being a symbolic link to the real device in the /sys/devices directory. The devices directory gives the real view of the topology of devices in the system. firmware shows a system-specific tree of low-level subsystems such as ACPI, EFI, and OF (device tree). fs lists the filesystems that are used on the system. kernel holds the kernel configuration options and status information. Finally, module is a list of loaded modules and power is the system power management control interface from the user space.

Each of these directories corresponds to a kobject, some of which are exported as kernel symbols. These are as follows:

- `kernel_kobj`, which corresponds to `/sys/kernel`.

- `power_kobj`, which corresponds to `/sys/power`.

- `firmware_kobj`, which corresponds to `/sys/firmware`. It's exported in the `drivers/base/firmware.c` source file.

- `hypervisor_kobj`, which corresponds to `/sys/hypervisor`. It's exported in the `drivers/base/hypervisor.c` source file.

- `fs_kobj`, which corresponds to `/sys/fs`. This is exported in the `fs/namespace.c` source file.

For the rest, `class/`, `dev/` and `devices/` are created at boot time by the `devices_init()` function in `drivers/base/core.c` in the kernel sources, `block/` is created in `block/genhd.c`, and `bus/` is created as a `kset` in `drivers/base/bus.c`.

Creating device-, driver-, bus- and class-related attributes

So far, we have learned how to create dedicated kobjects to populate attributes inside. However, the device, driver, bus, and class frameworks provide attribute abstractions and file creation, where the attributes that are created are directly tied to the respective framework in the appropriate kobject directory.

To do so, each framework provides a framework-specific attribute data structure that encloses the default attribute and allows us to provide a custom show/store callback. These are `struct device_attribute`, `struct driver_attribute`, `struct bus_atttribute`, and `struct class_attribute` for the device, driver, bus, and class frameworks, respectively. They are defined like `kboj_attribute` is but use different names. Let's look at their respective data structures:

- Devices have the following attribute data structure:

```
struct driver_attribute {
    struct attribute attr;
    ssize_t (*show)(struct device_driver *driver,
                char *buf);
    ssize_t (*store)(struct device_driver *driver,
                const char *buf, size_t count);
};
```

- Classes have the following attribute data structure:

```
struct class_attribute {
    struct attribute attr;
    ssize_t (*show)(struct class *class,
            struct class_attribute *attr, char *buf);
    ssize_t (*store)(struct class *class,
            struct class_attribute *attr,
            const char *buf, size_t count);
};
```

- The bus framework has the following attribute data structure:

```
struct bus_attribute {
    struct attribute  attr;
    ssize_t (*show)(struct bus_type *bus, char *buf);
    ssize_t (*store)(struct bus_type *bus,
            const char *buf, size_t count);
};
```

- Devices have the following attribute data structure:

```
struct device_attribute {
    struct attribute  attr;
    ssize_t (*show)(struct device *dev,
            struct device_attribute *attr,
            char *buf);
    ssize_t (*store)(struct device *dev,
                struct device_attribute *attr,
                const char *buf, size_t count);
};
```

The preceding device-specific data structure's show function takes an additional count parameter, whereas the others do not.

They can be dynamically allocated with `kzalloc()` and initialized by setting the fields of their inner attribute elements and providing the appropriate callback functions. However, each framework provides a set of macros to statically allocate, initialize, and assign a single instance of their respective attribute data structure. Let's look at these macros:

- The bus infrastructure provides the following macros:

```
BUS_ATTR_RW(_name)
BUS_ATTR_RO(_name)
BUS_ATTR_WO(_name)
```

  With these bus framework-specific macros, the resulting bus attribute variable is named `bus_attr_<_name>`. For example, the variable name that results from `BUS_ATTR_RW(foo)` will be `bus_attr_foo` and will be of the `struct bus_attribute` type.

- For drivers, the following macros are provided:

```
DRIVER_ATTR_RW(_name)
DRIVER_ATTR_RO(_name)
DRIVER_ATTR_WO(_name)
```

  These driver-specific attribute definition macros will name the resulting variable using the `driver_attr_<_name>` pattern. Therefore, the variable that results from `DRIVER_ATTR_RW(foo)` will be of the `struct driver_attribute` type and will be named `driver_attr_foo`.

- The class framework works with the following macros:

```
CLASS_ATTR_RW(_name)
CLASS_ATTR_RO(_name)
CLASS_ATTR_WO(_name)
```

  Using these class-specific macros, the resulting variable will be of the `struct class_atribute` type and will be named based on the `class_attr_<_name>` pattern. Thus, the resulting variable name of `CLASS_ATTR_RW(foo)` will be `class_attr_foo`.

- Finally, device-specific attributes can be statically allocated and initialized using the following macros:

```
DEVICE_ATTR(_name, _mode, _show, _store)
DEVICE_ATTR_RW(_name)
DEVICE_ATTR_RO(_name)
DEVICE_ATTR_WO(_name)
```

Device-specific attributes definition macros use their own pattern for variable names, which is dev_attr_<_name>. Thus, for example, DEVICE_ATTR_RO(foo) will result in a struct device_attribute object named dev_attr_foo.

Because all these macros are built on top of __ATTR_RW, __ATTR_RO, and __ATTR_WO, they statically allocate and initialize a single instance of the framework-specific attribute data structure and assume the show/store functions are named <attribute_name>_show and <attribute_name>_store, respectively (remember, this is because they do not rely on the __ATTR macro). There is an exception for DEVICE_ATTR(), which uses the show/store function as it was passed, without any suffix or prefix. This exception is because DEVICE_ATTR relies on __ATTR to define attributes.

As we have seen, all these framework-specific macros use a predefined prefix to name the resulting framework-specific attribute object variable. Let's take a look at the following class attribute:

```
static CLASS_ATTR_RW(foo);
```

This will create a static variable of the struct class_attribute type named class_attr_foo and will assume that its show and store functions are named foo_show and foo_store, respectively. This can be referenced in a group using its inner attribute element, as shown here:

```
static struct attribute *fake_class_attrs[] = {
    &class_attr_foo.attr,
    [...]
    NULL,
};
static struct attribute_group fake_attr_group = {
    .attrs = fake_class_attrs,
};
```

The most important thing when it comes to creating the respective files is that the driver can select the appropriate API from the following list:

```
int device_create_file(struct device *device,
            const struct device_attribute *entry);
int driver_create_file(struct device_driver *driver,
            const struct driver_attribute *attr);
int bus_create_file(struct bus_type *bus,
            struct bus_attribute *);
int class_create_file(struct class *class,
            const struct class_attribute *attr)
```

Here, the device, driver, bus, and class arguments are the respective device, driver, bus, and class entities that the attribute must be added to. Moreover, the attribute will be created in the directory that corresponds to the inner kobject of each entity, as shown in the following code:

```
int device_create_file(struct device *dev,
                    const struct device_attribute *attr)
{
    [...]
    error = sysfs_create_file(&dev->kobj, &attr->attr);
    [...]
}

int class_create_file(struct class *cls,
                    const struct class_attribute *attr)
{
    [...]
    error =
        sysfs_create_file(&cls->p->class_subsys.kobj,
                        &attr->attr);
    return error;
}

int bus_create_file(struct bus_type *bus,
                    struct bus_attribute *attr)
{
```

```
    [...]
    error =
        sysfs_create_file(&bus->p->subsys.kobj,
                            &attr->attr);
    [...]
}
```

To kill two birds with one stone, the preceding code also shows that `device_create_file()`, `bus_create_file()`, `driver_create_file()` and `class_create_file()` all make an internal call to `sysfs_create_file()`.

Once you're done with each respective attribute object, the appropriate removal method must be invoked. The following code shows the possible options:

```
void device_remove_file(struct device *device,
            const struct device_attribute *entry);
void driver_remove_file(struct device_driver *driver,
            const struct driver_attribute *attr);
void bus_remove_file(struct bus_type *,
            struct bus_attribute *);
void class_remove_file(struct class *class,
            const struct class_attribute *attr);
```

Each of these APIs expects the same arguments as those that are passed when the attributes are created.

Now that you know how the inner show/store functions of the `kobj_atribute` elements are invoked, it should be obvious to you how those framework-specific show/store functions are invoked as well.

Let's have a look at the device's implementation. The device framework has an internal `kobj_type` that implements device-specific show and store functions. These functions take in the inner attribute element as one of their arguments. After that, the `container_of` macro retrieves a pointer for the enclosing data structure (which is the framework-specific attribute data structure) that the framework-specific show and store functions are invoked from.

The following is an excerpt from `drivers/base/core.c` that shows the device-specific `sysfs_ops` implementation:

```
static ssize_t dev_attr_show(struct kobject *kobj,
                        struct attribute *attr,
```

```
                                 char *buf)
{
    struct device_attribute *dev_attr = to_dev_attr(attr);
    struct device *dev = kobj_to_dev(kobj);
    ssize_t ret = -EIO;

    if (dev_attr->show)
        ret = dev_attr->show(dev, dev_attr, buf);
    if (ret >= (ssize_t)PAGE_SIZE) {
        print_symbol("dev_attr_show:
                      %s returned bad count\n",
                    (unsigned long)dev_attr->show);
    }
    return ret;
}

static ssize_t dev_attr_store(struct kobject *kobj,
                      struct attribute *attr,
                      const char *buf, size_t count)
{
    struct device_attribute *dev_attr = to_dev_attr(attr);
    struct device *dev = kobj_to_dev(kobj);
    ssize_t ret = -EIO;

    if (dev_attr->store)
        ret = dev_attr->store(dev, dev_attr, buf, count);
    return ret;
}

static const struct sysfs_ops dev_sysfs_ops = {
    .show    = dev_attr_show,
    .store   = dev_attr_store,
};
```

Note that in the preceding code, `to_dev_attr()`, which is the macro that makes use of `container_of`, is defined as follows:

```
#define to_dev_attr(_attr) \
        container_of(_attr, struct device_attribute, attr)
```

The principle is the same for the bus (in `drivers/base/bus.c`), driver (in `drivers/base/bus.c`), and class (in `drivers/base/class.c`) attributes.

Making a sysfs attribute poll- and select-compatible

Though this is not a requirement for dealing with sysfs attributes, the main idea here is to allow the `poll()` or `select()` system calls to be used on a given attribute to passively wait for a change. This change could be firmware becoming available, an alarm notification, or information that the attribute value has changed. While the user would sleep on the file waiting for a change, the driver must invoke `sysfs_notify()` to release any sleeping user.

This notification API is defined as follows:

```
void sysfs_notify(struct kobject *kobj, const char *dir,
                  const char *attr)
```

If the `dir` parameter is not NULL, it is used to find a subdirectory from within the directory of `kobj`, which contains the attribute (presumably created by `sysfs_create_group`). This call will cause any polling process to wake up and process the event (which might be reading the new value, handling the alarm, and so on).

> **Note**
> There will be no notifications without this function call; therefore, any polling process will end up waiting indefinitely (unless a timeout was specified in the system call).

The following code, which shows the `store()` function of an attribute, is provided with this book:

```
static ssize_t store(struct kobject *kobj,
                     struct attribute *attr,
                     const char *buf, size_t len)
{
    struct d_attr *da = container_of(attr, struct d_attr,
```

```
                                        attr);

    sscanf(buf, "%d", &da->value);
    pr_info("sysfs_foo store %s = %d\n",
            a->attr.name, a->value);

    if (strcmp(a->attr.name, "foo") == 0){
        foo.value = a->value;
        sysfs_notify(mykobj, NULL, "foo");
    }
    else if(strcmp(a->attr.name, "bar") == 0){
        bar.value = a->value;
        sysfs_notify(mykobj, NULL, "bar");
    }
    return sizeof(int);
}
```

In the preceding code, it makes sense to call `sysfs_notify()` once the value has been updated so that the user code can read the accurate value.

The user code can directly pass the opened attribute file to `poll()` or `select()` without having to read the initial content of this attribute. Doing so is at the convenience of the developer. However, note that upon notification, `poll()` returns `POLLERR|POLLPRI` (as are flags, which users must request while invoking `poll()`), while `select()` returns the file descriptor, whether it is waiting for read, write, or exception events.

Summary

By completing this chapter, you should be familiar with LDM, its data structures (bus, class, device, and driver), and its low-level data structures, which are `kobject`, `kset`, `kobj_type`, and `attributes` (or a group of these). You should now know how objects are represented within the kernel (device topology) and be able to create an attribute (or group) that exposes your device or driver features and properties through sysfs.

In the next chapter, we will cover the **IIO (Industrial I/O)** framework, which heavily uses the power of sysfs.

Section 4 - Misc Kernel Subsystems for the Embedded World

This section will touch upon the IIO framework, pin control and GPIO systems, and other important feature-rich concepts. In this section, you will also gain an in-depth understanding of GPIOs mapped to interrupts and their processing mechanisms, describing every kernel data structure and API involved. Finally, you'll learn how to deal with input devices drivers. It goes without saying that user-space handlings of the enumerated subsystems will be introduced in this section as well.

The following chapters will be covered in this section:

15
Digging into the IIO Framework

Industrial input/output (IIO) is a kernel subsystem dedicated to **analog-to-digital converters** (**ADCs**) and **digital-to-analog converters** (**DACs**). With the growing numbers of sensors (measurement devices with analog-to-digital or digital-to-analog capabilities) with different code implementations, scattered across kernel sources, gathering them became necessary. That is what the IIO framework does, in a generic way. Jonathan Cameron and the Linux IIO community have been developing it since 2009. Accelerometers, gyroscopes, current/voltage measurement chips, light sensors, and pressure sensors all fall into the IIO family of devices.

The IIO model is based on device and channel architecture:

- The device represents the chip itself, the top level of the hierarchy.

- The channel represents a single acquisition line of the device. A device may have one or more channels. For example, an accelerometer is a device with three channels, one for each axis (*x*, *y*, and *z*).

The IIO chip is the physical and hardware sensor/converter. It is exposed to the user space as a character device (when a triggered buffer is supported) and a sysfs directory entry that will contain a set of files, some of which represent the channels.

These are the two ways to interact with an IIO device from user space:

- `/sys/bus/iio/iio:deviceX/`, a sysfs directory that represents the device along with its channels

- `/dev/iio:deviceX`, a character device that exports the device's events and data buffer

As a picture is worth a thousand words, the following is a figure showing an overview of the IIO framework:

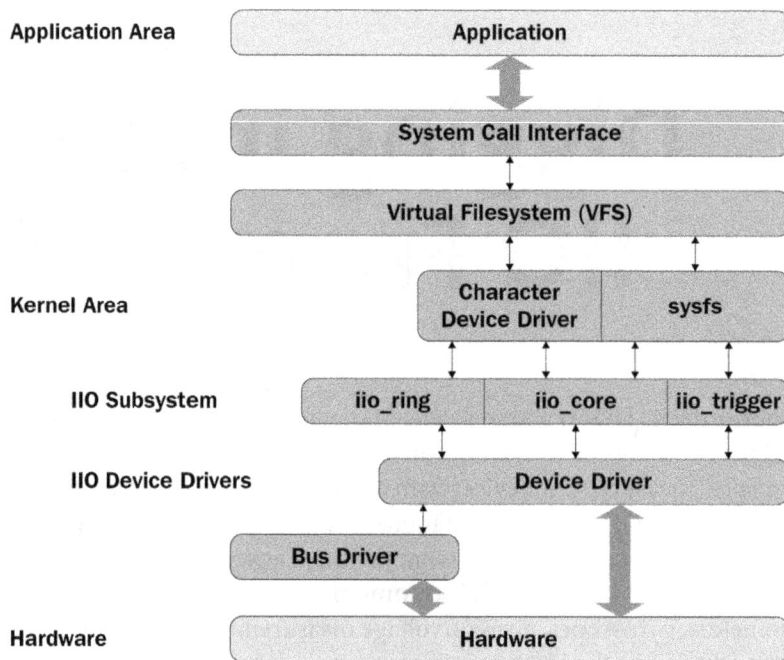

Figure 15.1 – IIO framework overview

The preceding figure shows how the IIO framework is organized between the kernel and the user space. The driver manages the hardware and reports processing to the IIO core, using a set of facilities and APIs exposed by the IIO core. The IIO subsystem then abstracts the whole underlying mechanism to user space by means of the sysfs interface and the character device, on top of which users can execute system calls.

IIO APIs are spread over several header files, as follows:

```
/* mandatory, the core */
#include <linux/iio/iio.h>
/* mandatory since sysfs is used */
```

```
#include <linux/iio/sysfs.h>
/* Optional. Advanced feature, to manage iio events */
#include <linux/iio/events.h>
/* mandatory for triggered buffers */
#include <linux/iio/buffer.h>
/* rarely used. Only if the driver implements a trigger */
#include <linux/iio/trigger.h>
```

In this chapter, we will describe and handle every concept of the IIO framework, such as walking through its data structure (devices, channels, and so on), dealing with triggered buffer support and continuous capture, along with its sysfs interface, exploring existing IIO triggers, learning how to capture data in either one-shot mode or continuous mode, and listing tools that can help the developer in testing their devices.

In other words, we will cover the following topics in this chapter:

- Introduction to IIO data structures
- Integrating IIO triggered buffer support
- Accessing IIO data
- Dealing with the in-kernel IIO consumer interface
- Walking through user-space IIO tools

Introduction to IIO data structures

The IIO framework is made of a few data structures among which is one representing the IIO device, another one describing this device, and the last one enumerating the channels exposed by the device. An IIO device is represented in the kernel as an instance of `struct iio_dev` and described by a `struct iio_info` structure. All the important IIO structures are defined in `include/linux/iio/iio.h`.

Understanding the struct iio_dev structure

The `struct iio_dev` structure represents the IIO device, describing the device and its driver. It tells us how many channels are available on the device and what modes the device can operate in (one-shot or triggered buffer, for example). Moreover, this data structure exposes some hooks to be provided by the driver.

This data structure has the following definition:

```
struct iio_dev {
    [...]
    int                 modes;
    int                 currentmode;
    struct device   dev;
    struct iio_buffer           *buffer;
    int                         scan_bytes;
    const unsigned long         *available_scan_masks;
    const unsigned long         *active_scan_mask;
    bool                        scan_timestamp;
    struct iio_trigger          *trig;
    struct iio_poll_func        *pollfunc;
    struct iio_chan_spec const  *channels;
    int                         num_channels;
    const char                  *name;
    const struct iio_info       *info;
    const struct iio_buffer_setup_ops   *setup_ops;
    struct cdev                 chrdev;
};
```

For the sake of readability, only relevant elements for us have been listed in the preceding excerpt. The complete structure definition lies in `include/linux/iio/iio.h`. The following are the meanings of the elements in the data structure:

- `modes` represents the different modes supported by the device. Possible modes are as follows:

  - `INDIO_DIRECT_MODE`: This says the device provides sysfs-type interfaces.

  - `INDIO_BUFFER_TRIGGERED`: This says that the device supports hardware triggers associated with a buffer. This flag mode is automatically set when you set up a triggered buffer using the `iio_triggered_buffer_setup()` function.

  - `INDIO_BUFFER_SOFTWARE`: In continuous conversions, the buffering will be implemented in software, by the kernel itself. The kernel will push data into the internal FIFO with a possible interrupt at a specified watermark.

- INDIO_BUFFER_HARDWARE: This means the device has a hardware buffer. In continuous conversions, the buffering can be handled by the device. This means that the data stream can be obtained directly from the hardware backend.

- INDIO_ALL_BUFFER_MODES: A union of the preceding three.

- INDIO_EVENT_TRIGGERED: Conversion can be triggered by some sort of event, such as a threshold voltage reached on an ADC, but no interrupt or timer trigger. This flag is intended to be used for comparator-equipped chips with no other way to trigger conversion.

- INDIO_HARDWARE_TRIGGERED: Can be triggered by hardware events, such as IRQ or clock events.

- INDIO_ALL_TRIGGERED_MODES union of INDIO_BUFFER_TRIGGERED, INDIO_EVENT_TRIGGERED, and INDIO_HARDWARE_TRIGGERED.

- currentmode: This represents the mode used by the device.

- dev: This represents the struct device (according to Linux Device Model) the IIO device is tied to.

- buffer: This is your data buffer, pushed to the user space when using triggered buffer mode. It is automatically allocated and associated with your device when you enable triggered buffer support using the iio_triggered_buffer_setup function.

- scan_bytes: This is the number of bytes captured to be fed to the buffer. When using a trigger buffer from the user space, the buffer should be at least indio->scan_bytes bytes large.

- available_scan_masks: This is an optional array of allowed bitmasks. When using a triggered buffer, you can enable channels to be captured and fed into the IIO buffer. If you do not want to allow some channels to be enabled, you should fill this array with only allowed ones. An example of an accelerometer (with X, Y, and Z channels) is as follows:

```
/*
 * Bitmasks 0x7 (0b111) and 0 (0b000) are allowed.
 * It means one can enable none or all of them.
 * You can't for example enable only channel X and Y
 */
static const unsigned long my_scan_masks[] = {0x7, 0};
indio_dev->available_scan_masks = my_scan_masks;
```

- `active_scan_mask`: This is a bitmask of enabled channels. Only the data from those channels should be pushed into the buffer. For example, for an eight-channel ADC converter, if you only enable the first (index 0), the third (index 2), and the last (index 7) channels, the bitmask would be `0b10000101 (0x85)`. `active_scan_mask` will be set to `0x85`. The driver can then use the `for_each_set_bit` macro to walk through each set bit, fetch the data from the corresponding channels, and fill the buffer.

- `scan_timestamp`: This tells whether to push the capture timestamp into the buffer or not. If `true`, the timestamp will be pushed as the last element of the buffer. The timestamp is 8 bytes (64 bits) large.

- `trig`: This is the current device trigger (when buffer mode is supported).

- `pollfunc`: This is the function run on the trigger being received.

- `channels`: This represents the table channel specification structure, to describe every channel the device has.

- `num_channels`: This represents the number of channels specified in `channels`.

- `name`: This represents the device name.

- `info`: Callbacks and constant information from the driver.

- `setup_ops`: A set of callback functions to call before and after the buffer is enabled/disabled. This structure is defined in `include/linux/iio/iio.h`, as follows:

```
struct iio_buffer_setup_ops {
    int (* preenable) (struct iio_dev *);
    int (* postenable) (struct iio_dev *);
    int (* predisable) (struct iio_dev *);
    int (* postdisable) (struct iio_dev *);
    bool (* validate_scan_mask) (
                    struct iio_dev *indio_dev,
                    const unsigned long *scan_mask);
};
```

Note that each callback in this data structure is optional.

- `chrdev`: Associated character device created by the IIO core, with `iio_buffer_fileops` as the file operation table.

Now that we are familiar with the IIO device structure, the next step is to allocate memory for it. The appropriate function to achieve that is devm_iio_device_alloc(), which is the managed version for iio_device_alloc() and has the following definition:

```
struct iio_dev *devm_iio_device_alloc(struct device *dev,
                                      int sizeof_priv)
```

It is recommended to use the managed version in a new driver as the devres core takes care of freeing the memory when it is no longer needed. In the preceding function prototype, dev is the device to allocate iio_dev for and sizeof_priv is the extra memory space to allocate for any private data structure. The function returns NULL if the allocation fails.

After the IIO device memory has been allocated, the next step is to initialize different fields. Once done, the device must be registered with the IIO subsystem using the devm_iio_device_register() function, the prototype of which is the following:

```
int devm_iio_device_register(struct device *dev,
                            struct iio_dev *indio_dev);
```

This function is the managed version of iio_device_register() and takes care of unregistering the IIO device on driver detach. In its parameters, dev is the same device as the one for which the IIO device has been allocated, and indio_dev is the IIO device previously initialized. The device will be ready to accept requests from the user space after this function succeeds (returns 0). The following is an example showing how to register an IIO device:

```
static int ad7476_probe(struct spi_device *spi)
{
    struct ad7476_state *st;
    struct iio_dev *indio_dev;
    int ret;

    indio_dev = devm_iio_device_alloc(&spi->dev,
                                      sizeof(*st));
    if (!indio_dev)
        return -ENOMEM;

    /* st is given the address of reserved memory for
     * private data
```

```
*/
st = iio_priv(indio_dev);
[...]


/* iio device setup */
indio_dev->name = spi_get_device_id(spi)->name;
indio_dev->modes = INDIO_DIRECT_MODE;
indio_dev->num_channels = 2;
[...]


return devm_iio_device_register(&spi->dev, indio_dev);
}
```

If an error occurs, devm_iio_device_register() will return a negative error code.
The reverse operation for the non-managed variant (usually done in the release function)
is iio_device_unregister(), which has the following declaration:

```
void iio_device_unregister(struct iio_dev *indio_dev)
```

However, managed registration takes care of unregistering the device on driver detach
or when the device leaves the system. Moreover, because we used a managed allocation
variant, there is no need to free the memory as this will be internal to the core.

You might have also noticed we used a new function in the excerpt, iio_priv().
This accessor returns the address of the private data allocated with the IIO device. It is
recommended to use this function instead of doing a direct dereference. As an example,
given an IIO device, the corresponding private data can be retrieved as follows:

```
struct my_private_data *the_data = iio_priv(indio_dev);
```

The IIO device is useless on its own. Now that we are done with the main IIO device data
structure, we have to add a set of hooks allowing us to interact with the device.

Understanding the struct iio_info structure

The struct iio_info structure is used to declare the hooks used by the IIO core to
read/write channel/attribute values. The following is part of its declaration:

```
struct iio_info {
    const struct attribute_group  *attrs;
```

```
    int (*read_raw)(struct iio_dev *indio_dev,
            struct iio_chan_spec const *chan,
            int *val, int *val2, long mask);

    int (*write_raw)(struct iio_dev *indio_dev,
            struct iio_chan_spec const *chan,
            int val, int val2, long mask);
    [...]
};
```

Again, the full definition of this data structure can be found in `/include/linux/iio/iio.h`. For the enumerated elements in the preceding structure excerpt, the following are their meanings:

- `attrs` represents the device attributes exposed to user space.

- `read_raw` is the callback invoked when a user reads a device sysfs file attribute. The `mask` parameter is a bitmask allowing us to know which type of value is requested. The `chan` parameter lets us know the channel concerned. `*val` and `*val2` are output parameters that must contain the elements making up the returned value. They must be set with raw values read from the device.

  The return value of this callback is kind of standardized and indicates how `*val` and `*val2` must be handled by the IIO core to compute the real value. Possible return values are the following:

  - `IIO_VAL_INT`: The output value is an integer. In this case, the driver must set `*val` only.

  - `IIO_VAL_INT_PLUS_MICRO`: The output value is made of an integer part and a micro part. The driver must set `*val` with the integer value, while `*val2` must be set with the micro value.

  - `IIO_VAL_INT_PLUS_NANO`: This is the same as the micro, but `*val2` must be set with the nano value.

  - `IIO_VAL_INT_PLUS_MICRO_DB`: The output values are in **dB**. `*val` must be set with the integer part and `*val2` must set with the micro part, if any.

  - `IIO_VAL_INT_MULTIPLE`: `val` is considered as an array of integers and `*val2` is the number of entries in the array. They must be set accordingly then. The maximum size of `val` is `INDIO_MAX_RAW_ELEMENTS`, defined as 4.

- IIO_VAL_FRACTIONAL: The final value is fractional. The driver must set *val with the numerator and *val2 with the denominator.

- IIO_VAL_FRACTIONAL_LOG2: The final value is a logarithmic fractional. The IIO core expects the denominator (*val2) to be specified as the *log2* of the actual denominator. For example, for ADCs and DACs, this will usually be the number of significant bits. *val is a normal integer denominator.

- IIO_VAL_CHAR: The IIO core expects *val to be a character. This is, most of the time, used with the IIO_CHAN_INFO_THERMOCOUPLE_TYPE mask, in which case the driver must return the type of thermocouple.

  All the preceding does not change the fact that, in case of an error, the callback must return a negative error code, for example, -EINVAL. I recommend you have a look at how the final value is processed in iio_convert_raw_to_processed_unlocked() in the drivers/iio/inkern.c source file.

- write_raw is the callback used to write a value to the device. You can use it, for example, to set the sampling frequency or change the scale.

An example of setting up the struct iio_info structure is the following:

```
static const struct iio_info iio_dummy_info = {
    .read_raw = &iio_dummy_read_raw,
    .write_raw = &iio_dummy_write_raw,
    [...]
};


/*
 * Provide device type specific interface functions and
 * constant data.
 */
indio_dev->info = &iio_dummy_info;
```

You must not confuse this struct iio_info with the user-space iio_info tool, which is part of the libiio package.

The concept of IIO channels

In IIO terminology, a channel represents a single acquisition line of a sensor. This means each data mesurement entity a sensor can provide/sense is called a **channel**. For example, an accelerometer will have three channels (X, Y, and Z), since each axis represents a single acquisition line. `struct iio_chan_spec` is the structure that represents and describes a single channel in the kernel, as follows:

```c
struct iio_chan_spec {
    enum iio_chan_type      type;
    int                channel;
    int                channel2;
    unsigned long      address;
    int                scan_index;
    struct {
        char sign;
        u8    realbits;
        u8    storagebits;
        u8    shift;
        u8    repeat;
        enum iio_endian endianness;
    } scan_type;
    long               info_mask_separate;
    long               info_mask_shared_by_type;
    long               info_mask_shared_by_dir;
    long               info_mask_shared_by_all;
    const struct iio_event_spec *event_spec;
    unsigned int       num_event_specs;
    const struct iio_chan_spec_ext_info *ext_info;
    const char         *extend_name;
    const char         *datasheet_name;
    unsigned           modified:1;
    unsigned           indexed:1;
    unsigned           output:1;
    unsigned           differential:1;
};
```

The following are the meanings of elements in the data structure:

- `type` specifies which type of measurement the channel makes. In the case of voltage measurement, it should be `IIO_VOLTAGE`. For a light sensor, it is `IIO_LIGHT`. For an accelerometer, `IIO_ACCEL` is used. All available types are defined in `include/uapi/linux/iio/types.h`, as `enum iio_chan_type`. To write a driver for a given converter, you have to look into that file to see the type each of your converter channels falls into.

- `channel` specifies the channel index when `.indexed` is set to 1.

- `channel2` specifies the channel modifier when `.modified` is set to 1.

- The `scan_index` and `scan_type` fields are used to identify elements from a buffer, when using buffer triggers. `scan_index` sets the position of the captured channel inside the buffer. Channels are placed in the buffer ordered by `scan_index`, from the lowest index (placed first) to the highest index. Setting `.scan_index` to `-1` will prevent the channel from buffered capture (no entry in the `scan_elements` directory). Elements in this substructure have the folowing meanings:

  - `sign`: `s` or `u` specifies symbols (signed (complement of 2) or unsigned).

  - `realbits`: The number of valid data bits.

  - `storagebits`: The number of digits occupied by this channel in the buffer. That is to say, a value can really be encoded with 12 bits, but it occupies 16 bits (storage bits) in the buffer. Therefore, the data must be moved four times to the right to get the actual value. This parameter depends on the device and you should refer to its datasheet.

  - `shift`: Represents the number of times data values should be right-shifted before masking out unused bits. This parameter is not always required. If the number of valid bits equals the number of storage bits, the shift will be 0. This parameter can also be found in the device datasheet.

  - `repeat`: The number of times real/storage bits repeat.

  - `endianness`: Represents the data endianness. It is of the `enum iio_endian` type and should be set with one of `IIO_CPU`, `IIO_LE`, or `IIO_BE`, which mean, the native CPU endianness, little endian, or big endian respectively.

- The `modified` field specifies whether a modifier is to be applied to this channel attribute name or not. In that case, the modifier is set in `.channel2`. (For example, `IIO_MOD_X`, `IIO_MOD_Y`, and `IIO_MOD_Z` are modifiers for axial-sensors about the X, Y, and Z axis). The available modifier list is defined in the kernel IIO header as `enum iio_modifier`. Modifiers only mangle the channel attribute name in sysfs, not the value.

- `indexed` specifies whether the channel attribute name has an index or not. If yes, the index is specified in the `.channel` field.

- `info_mask_separate` marks the attribute as being specific to this channel.

- `info_mask_shared_by_type` marks the attribute as being shared by all channels of the same type. The information exported is shared by all channels of the same type.

- `info_mask_shared_by_dir` marks the attribute as being shared by all channels of the same direction. The information exported is shared by all channels of the same direction.

- `info_mask_shared_by_all` marks the attribute as being shared by all channels, whatever their type or their direction may be. The information exported is shared by all channels.

`iio_chan_spec.info_mask_*` elements are masks used to specify channel sysfs attributes exposed to user space depending on their shared information. Therefore, masks must be set by ORing one or more bitmasks, all of which are defined in `include/linux/iio/types.h`, as follows:

```
enum iio_chan_info_enum {
    IIO_CHAN_INFO_RAW = 0,
    IIO_CHAN_INFO_PROCESSED,
    IIO_CHAN_INFO_SCALE,
    IIO_CHAN_INFO_OFFSET,
    IIO_CHAN_INFO_CALIBSCALE,
    [...]
    IIO_CHAN_INFO_SAMP_FREQ,
    IIO_CHAN_INFO_FREQUENCY,
    IIO_CHAN_INFO_PHASE,
    IIO_CHAN_INFO_HARDWAREGAIN,
    IIO_CHAN_INFO_HYSTERESIS,
    [...]
};
```

The following is an example of specifying a mask for a given channel:

```
iio_chan->info_mask_separate = BIT(IIO_CHAN_INFO_RAW) |
                    BIT(IIO_CHAN_INFO_PROCESSED);
```

This means raw and processed attributes are specific to the channel.

> **Note**
>
> While not specified in the preceding `struct iio_chan_spec` structure description, the term *attribute* refers to a *sysfs attribute*. This applies across the whole chapter.

Having described the channel data structure, let's decipher the mystery about channel attribute naming, which respects a specific convention.

Channel attribute naming convention

An attribute's name is automatically generated by the IIO core following a predefined pattern, `{direction}_{type}{index}_{modifier}_{info_mask}`. The following are descriptions of each field in the pattern:

- `{direction}` corresponds to the attribute direction, according to the `struct iio_direction` structure in `drivers/iio/industrialio-core.c`:

```
static const char * const iio_direction[] = {
    [0] = "in",
    [1] = "out",
};
```

  Do note that an input channel is a channel that can generate samples (such channels are handled in the read method, for instance, an ADC channel). On the other hand, an output channel is a channel that can receive samples (such channels are handled in the write method, for instance, a DAC channel).

- `{type}` corresponds to the channel type string, according to the constant `iio_chan_type_name_spec` char array (indexed by the channel type of type enum `iio_chan_type`) defined in `drivers/iio/industrialio-core.c`, as follows:

```
static const char * const iio_chan_type_name_spec[] = {
    [IIO_VOLTAGE] = "voltage",
    [IIO_CURRENT] = "current",
```

```
        [IIO_POWER] = "power",
        [IIO_ACCEL] = "accel",
        [...]
        [IIO_UVINDEX] = "uvindex",
        [IIO_ELECTRICALCONDUCTIVITY] =
                    "electricalconductivity",
        [IIO_COUNT] = "count",
        [IIO_INDEX] = "index",
        [IIO_GRAVITY] = "gravity",
    };
```

- {index} depends on the channel .indexed field being set or not. If set, the index will be taken from the .channel field in order to replace the {index} pattern.

- The {modifier} pattern depends on the channel .modified field being set or not. If set, the modifier will be taken from the .channel2 field, and the {modifier} field in the pattern will be replaced according to the char array struct iio_modifier_names structure:

```
    static const char * const iio_modifier_names[] = {
        [IIO_MOD_X] = "x",
        [IIO_MOD_Y] = "y",
        [IIO_MOD_Z] = "z",
        [IIO_MOD_X_AND_Y] = "x&y",
        [IIO_MOD_X_AND_Z] = "x&z",
        [IIO_MOD_Y_AND_Z] = "y&z",
        [...]
        [IIO_MOD_CO2] = "co2",
        [IIO_MOD_VOC] = "voc",
    };
```

- {info_mask} depends on the channel info mask, private or shared, indexing the value in the iio_chan_info_postfix char array, defined as the following:

```
    /* relies on pairs of these shared then separate */
    static const char * const iio_chan_info_postfix[] = {
        [IIO_CHAN_INFO_RAW] = "raw",
        [IIO_CHAN_INFO_PROCESSED] = "input",
        [IIO_CHAN_INFO_SCALE] = "scale",
```

```
            [IIO_CHAN_INFO_CALIBBIAS] = "calibbias",
            [...]
            [IIO_CHAN_INFO_SAMP_FREQ] = "sampling_frequency",
            [IIO_CHAN_INFO_FREQUENCY] = "frequency",
            [...]
    };
```

Channel naming convention should have no more secrets for us now. Now that we are familiar with the naming, let's learn how to precisely identify channels.

> **Note**
>
> In this naming pattern, if an element is not present, then the directly preceding underscore will be omitted. For example, if the modifier is not specified, the pattern becomes {direction}_{type}{index}_{info_mask} instead of {direction}_{type}{index}__{info_mask}.

Distinguishing channels

You may face some difficulties when there are multiple data channels of the same type. The dilemma would be *how to precisely identify each of them*. There are two solutions for that: **indexes** and **modifiers**.

Channel identification using an index

Given an ADC device with one channel line, indexing is not needed. Its channel definition would be as follows:

```
static const struct iio_chan_spec adc_channels[] = {
    {
            .type = IIO_VOLTAGE,
            .info_mask_separate = BIT(IIO_CHAN_INFO_RAW),
    },
}
```

Given the preceding excerpt, the attribute name will be `in_voltage_raw`, and its absolute sysfs path will be `/sys/bus/iio/iio:deviceX/in_voltage_raw`.

Now let's say the ADC has four or even eight channels. How do we identify each of them? The solution is to use indexes. Setting the `.indexed` field to `1` will modify the channel attribute name with the `.channel` value, replacing `{index}` in the naming pattern:

```
static const struct iio_chan_spec adc_channels[] = {
    {
        .type = IIO_VOLTAGE,
        .indexed = 1,
        .channel = 0,
        .info_mask_separate = BIT(IIO_CHAN_INFO_RAW),
    },
    {
        .type = IIO_VOLTAGE,
        .indexed = 1,
        .channel = 1,
        .info_mask_separate = BIT(IIO_CHAN_INFO_RAW),
    },
    {
        .type = IIO_VOLTAGE,
        .indexed = 1,
        .channel = 2,
        .info_mask_separate = BIT(IIO_CHAN_INFO_RAW),
    },
    {
        .type = IIO_VOLTAGE,
        .indexed = 1,
        .channel = 3,
        .info_mask_separate = BIT(IIO_CHAN_INFO_RAW),
    },
}
```

The following are the full sysfs paths of the resulting channel attributes:

```
/sys/bus/iio/iio:deviceX/in_voltage0_raw
/sys/bus/iio/iio:deviceX/in_voltage1_raw
```

```
/sys/bus/iio/iio:deviceX/in_voltage2_raw
/sys/bus/iio/iio:deviceX/in_voltage3_raw
```

As we can see, even if they all have the same type, they are differentiated by their index.

Channel identification using a modifier

To highlight the concept of modifiers, let's consider a light sensor with two channels – one for infrared light and the other for both infrared and visible light. Without an index or a modifier, an attribute name would be `in_intensity_raw`. Using indexes here can be error-prone because it makes no sense to have `in_intensity0_ir_raw` and `in_intensity1_ir_raw` as it would mean they are channels of the same type. Using a modifier will help us to have meaningful attribute names. The channel definition could look as follows:

```
static const struct iio_chan_spec mylight_channels[] = {
    {
        .type = IIO_INTENSITY,
        .modified = 1,
        .channel2 = IIO_MOD_LIGHT_IR,
        .info_mask_separate = BIT(IIO_CHAN_INFO_RAW),
        .info_mask_shared = BIT(IIO_CHAN_INFO_SAMP_FREQ),
    },
    {
        .type = IIO_INTENSITY,
        .modified = 1,
        .channel2 = IIO_MOD_LIGHT_BOTH,
        .info_mask_separate = BIT(IIO_CHAN_INFO_RAW),
        .info_mask_shared = BIT(IIO_CHAN_INFO_SAMP_FREQ),
    },
    {
        .type = IIO_LIGHT,
        .info_mask_separate = BIT(IIO_CHAN_INFO_PROCESSED),
        .info_mask_shared = BIT(IIO_CHAN_INFO_SAMP_FREQ),
    },
}
```

The resulting attributes would be as follows:

- `/sys/bus/iio/iio:deviceX/in_intensity_ir_raw` for the channel measuring IR intensity
- `/sys/bus/iio/iio:deviceX/in_intensity_both_raw` for the channel measuring both
- `/sys/bus/iio/iio:deviceX/in_illuminance_input` for the processed data
- `/sys/bus/iio/iio:deviceX/sampling_frequency` for the sampling frequency, shared by all

This is valid with an accelerometer too, as we will see in a later case study. For now, let's summarize what we have discussed so far by implementing a dummy IIO driver.

Putting it all together – writing a dummy IIO driver

Let's summarize what we have seen so far with a simple dummy driver, which will expose four voltage channels. We will not care about the `read()` or `write()` functions for the moment.

First, let's define the headers we'll need for the development:

```
#include <linux/init.h>
#include <linux/module.h>
#include <linux/kernel.h>
#include <linux/platform_device.h>
#include <linux/interrupt.h>
#include <linux/of.h>
#include <linux/iio/iio.h>
```

Then, because channel description is a generic and repetitive operation, let's define a macro that will populate the channel description for us, as follows:

```
#define FAKE_VOLTAGE_CHANNEL(num)                    \
    {                                                \
        .type = IIO_VOLTAGE,                         \
        .indexed = 1,                                \
        .channel = (num),                            \
        .address = (num),                            \
```

```
        .info_mask_separate = BIT(IIO_CHAN_INFO_RAW),          \
        .info_mask_shared_by_type = BIT(IIO_CHAN_INFO_SCALE) \
  }
```

After the channel population macro has been defined, let's define our driver state data structure, as follows:

```
struct my_private_data {
    int foo;
    int bar;
    struct mutex lock;
};
```

The data structure defined previously is useless. It is there just to show the concept. Then, since we do not need read or write operations in this dummy driver example, let's create empty read and write functions that just return 0 (meaning that everything went successfully):

```
static int fake_read_raw(struct iio_dev *indio_dev,
    struct iio_chan_spec const *channel, int *val,
    int *val2, long mask)
{
    return 0;
}
static int fake_write_raw(struct iio_dev *indio_dev,
                    struct iio_chan_spec const *chan,
                    int val, int val2, long mask)
{
    return 0;
}
```

We can now declare our IIO channels using the macro we defined earlier. Moreover, we can set up our `iio_info` data structure as follows, assigned at the same time as the fake read and write operations:

```
static const struct iio_chan_spec fake_channels[] = {
    FAKE_VOLTAGE_CHANNEL(0),
    FAKE_VOLTAGE_CHANNEL(1),
    FAKE_VOLTAGE_CHANNEL(2),
```

```
      FAKE_VOLTAGE_CHANNEL(3),
};


static const struct iio_info fake_iio_info = {
    .read_raw  = fake_read_raw,
    .write_raw = fake_write_raw,
    .driver_module = THIS_MODULE,
};
```

Now that all the necessary IIO data structures have been set up, we can switch to platform driver-related data structures and implementing its methods, as follows:

```
static const struct of_device_id iio_dummy_ids[] = {
    { .compatible = "packt,iio-dummy-random", },
    { /* sentinel */ }
};


static int my_pdrv_probe (struct platform_device *pdev)
{
    struct iio_dev *indio_dev;
    struct my_private_data *data;

    indio_dev = devm_iio_device_alloc(&pdev->dev,
                                      sizeof(*data));
    if (!indio_dev) {
        dev_err(&pdev->dev, "iio allocation failed!\n");
        return -ENOMEM;
    }

    data = iio_priv(indio_dev);
    mutex_init(&data->lock);
    indio_dev->dev.parent = &pdev->dev;
    indio_dev->info = &fake_iio_info;
    indio_dev->name = KBUILD_MODNAME;
    indio_dev->modes = INDIO_DIRECT_MODE;
    indio_dev->channels = fake_channels;
    indio_dev->num_channels = ARRAY_SIZE(fake_channels);
```

```
    indio_dev->available_scan_masks = 0xF;

    devm_iio_device_register(&pdev->dev, indio_dev);
    platform_set_drvdata(pdev, indio_dev);
    return 0;
}
```

In the preceding probing method, we have exclusively used resource-managed APIs for allocation and registering. This significantly simplifies the code and gets rid of the driver's remove method. The driver declaration and registering would then look like the following:

```
static struct platform_driver my_iio_pdrv = {
    .probe      = my_pdrv_probe,
    .driver     = {
        .name       = "iio-dummy-random",
        .of_match_table = of_match_ptr(iio_dummy_ids),
        .owner      = THIS_MODULE,
    },
};
module_platform_driver(my_iio_pdrv);
MODULE_AUTHOR("John Madieu <john.madieu@labcsmart.com>");
MODULE_LICENSE("GPL");
```

After loading the preceding module, you will have the following output while listing available IIO devices on the system:

```
~# ls -l /sys/bus/iio/devices/
lrwxrwxrwx    1 root      root               0 Jul 31 20:26
iio:device0 -> ../../../devices/platform/iio-dummy-random.0/
iio:device0
lrwxrwxrwx    1 root      root               0 Jul 31 20:23 iio_
sysfs_trigger -> ../../../devices/iio_sysfs_trigger

~# ls /sys/bus/iio/devices/iio\:device0/
dev                         in_voltage2_raw         name
uevent
in_voltage0_raw         in_voltage3_raw         power
in_voltage1_raw         in_voltage_scale        subsystem
```

```
~# cat /sys/bus/iio/devices/iio:device0/name
iio_dummy_random
```

> **Note**
>
> A very complete IIO driver that can be used for learning purposes or a development model is the IIO simple dummy driver, in `drivers/iio/dummy/iio_simple_dummy.c`. It can be made available on the target by enabling the `IIO_SIMPLE_DUMMY` kernel config option.

Now that we have addressed the basic IIO concept, we can go a step further by implementing buffer support and the concept of triggers.

Integrating IIO triggered buffer support

It might be useful to be able to capture data based on some external signals or events (triggers) in data acquisition applications. These triggers might be the following:

- A data ready signal

- An IRQ line connected to some external system (GPIO or whatever)

- On processor periodic interrupt (a timer, for example)

- User space reading/writing a specific file in sysfs

IIO device drivers are completely decorrelated from the triggers, whose drivers are implemented in `drivers/iio/trigger/`. A trigger may initialize data capture on one or many devices. These triggers are used to fill buffers, exposed to user space through the character device created during the registration of the IIO device.

You can develop your own trigger driver, but it is out of the scope of this book. We will try to focus on existing ones only. These are as follows:

- `iio-trig-interrupt`: This allows using IRQs as IIO triggers. In old kernel versions (prior to v3.11), it used to be `iio-trig-gpio`. To support this trigger mode, you should enable `CONFIG_IIO_INTERRUPT_TRIGGER` in the kernel config. If built as a module, the module will be called `iio-trig-interrupt`.

- `iio-trig-hrtimer`: Provides a frequency-based IIO trigger using high-resolution timers as an interrupt source (since kernel v4.5). In an older kernel version, it used to be `iio-trig-rtc`. To support this trigger mode in the kernel, the `IIO_HRTIMER_TRIGGER` config option must be enabled. If built as a module, the module will be called `iio-trig-hrtimer`.

- `iio-trig-sysfs`: This allows us to use the `SYSFS` entry to trigger data capture. `CONFIG_IIO_SYSFS_TRIGGER` is the kernel option to add the support of this trigger mode.

- `iio-trig-bfin-timer`: This allows us to use a Blackfin timer as an IIO trigger (still in staging).

IIO exposes an API so that we can do the following:

- Declare any given number of triggers.

- Choose which channels will have their data pushed into a buffer.

If your IIO device provides the support of a trigger buffer, you must set `iio_dev.pollfunc`, which is executed when the trigger fires. This handler has the responsibility of finding enabled channels through `indio_dev->active_scan_mask`, retrieving their data, and feeding them into `indio_dev->buffer` using the `iio_push_to_buffers_with_timestamp` function. Therefore, buffers and triggers are tightly connected in the IIO subsystem.

The IIO core provides a set of helper functions to set up triggered buffers, which you can find in `drivers/iio/industrialio-triggered-buffer.c`. The following are the steps to support a triggered buffer from within your driver:

1. Fill an `iio_buffer_setup_ops` structure if needed:

    ```
    const struct iio_buffer_setup_ops sensor_buffer_setup_ops
    = {
        .preenable    = my_sensor_buffer_preenable,
        .postenable   = my_sensor_buffer_postenable,
        .postdisable  = my_sensor_buffer_postdisable,
        .predisable   = my_sensor_buffer_predisable,
    };
    ```

2. Write the top half associated with the trigger. In 99% of cases, you just have to feed the timestamp associated with the capture:

    ```
    irqreturn_t sensor_iio_pollfunc(int irq, void *p)
    {
        pf->timestamp = iio_get_time_ns(
                        (struct indio_dev *)p);
        return IRQ_WAKE_THREAD;
    }
    ```

We then return a special value so the kernel knows it must schedule the bottom half, which will run in a threaded context.

3. Write the trigger bottom half, which will fetch data from each enabled channel and feed it into the buffer:

```
irqreturn_t sensor_trigger_handler(int irq, void *p)
{
    u16 buf[8];
    int bit, i = 0;
    struct iio_poll_func *pf = p;
    struct iio_dev *indio_dev = pf->indio_dev;

    /* one can use lock here to protect the buffer */
    /* mutex_lock(&my_mutex); */

    /* read data for each active channel */
    for_each_set_bit(bit, indio_dev->active_scan_mask,
                    indio_dev->masklength)
        buf[i++] = sensor_get_data(bit);

    /*
     * If iio_dev.scan_timestamp = true, the capture
     * timestamp will be pushed and stored too,
     * as the last element in the sample data buffer
     * before pushing it to the device buffers.
     */
    iio_push_to_buffers_with_timestamp(indio_dev, buf,
                                       timestamp);

    /* Please unlock any lock */
    /* mutex_unlock(&my_mutex); */

    /* Notify trigger */
    iio_trigger_notify_done(indio_dev->trig);
    return IRQ_HANDLED;
}
```

4. Finally, in the probe function, you have to set up the buffer itself, prior to registering the device:

```
iio_triggered_buffer_setup(
    indio_dev, sensor_iio_pollfunc,
    sensor_trigger_handler,
    sensor_buffer_setup_ops);
```

The magic function here is `iio_triggered_buffer_setup()`. It will also give the `INDIO_BUFFER_TRIGGERED` capability to the device, meaning that a polled ring buffer is possible.

When a trigger is assigned (from user space) to the device, the driver has no way of knowing when the capture will be fired. This is the reason why, while continuous buffered capture is active, you should prevent (by returning an error) the driver from handling sysfs per-channel data capture (performed by the `read_raw()` hook) in order to avoid undetermined behavior, since both the trigger handler and the `read_raw()` hook will try to access the device at the same time. The function used to check whether buffered mode is currently enabled is `iio_buffer_enabled()`. The hook will look as follows:

```
static int my_read_raw(struct iio_dev *indio_dev,
                const struct iio_chan_spec *chan,
                int *val, int *val2, long mask)
{
    [...]
    switch (mask) {
    case IIO_CHAN_INFO_RAW:
        if (iio_buffer_enabled(indio_dev))
            return -EBUSY;
    [...]
}
```

The `iio_buffer_enabled()` function simply tests whether the device's current mode corresponds to one of the IIO buffered modes. This function is defined as the following in `include/linux/iio/iio.h`:

```
static bool iio_buffer_enabled(struct iio_dev *indio_dev)
{
    return indio_dev->currentmode
        & (INDIO_BUFFER_TRIGGERED | INDIO_BUFFER_HARDWARE |
            INDIO_BUFFER_SOFTWARE);
}
```

Let's now describe some important things used in the preceding code:

- `iio_buffer_setup_ops` provides buffer setup functions to be called at a fixed step of the buffer configuration sequence (before/after enable/disable). If not specified, the default `iio_triggered_buffer_setup_ops` will be given to your device by the IIO core.

- `sensor_iio_pollfunc` is the trigger's top half. As with every top half, it runs in an interrupt context and must do as little processing as possible. In 99% of cases, recording the timestamp associated with the capture will be enough. Once again, you can use the default IIO `iio_pollfunc_store_time()` function.

- `sensor_trigger_handler` is the bottom half, which runs in a kernel thread, allowing you to do any processing, even acquiring a mutex or sleeping. The heavy processing should take place here. Most of the job here consists of reading data from the device and storing this data in the internal buffer together with the timestamp that has been recorded in the top half and pushing these to the IIO device buffer.

> **Note**
> A triggered buffer involves a trigger. It tells the driver when to read the sample from the device and put it into the buffer. A triggered buffer is not mandatory for writing an IIO device driver. You can use a single-shot capture through sysfs too, by reading the raw attribute of the channel, which will only perform a single conversion (for the channel attribute being read). Buffer mode allows continuous conversions, thus capturing more than one channel in a single shot.

Now that we are comfortable with all the in-kernel aspects of triggered buffers, let's introduce their setup in user space using the sysfs interface.

IIO trigger and sysfs (user space)

At runtime, there are two sysfs directories from where triggers can be managed:

- `/sys/bus/iio/devices/trigger<Y>/`: This directory is created once an IIO trigger is registered with the IIO core. In this path, `<Y>` corresponds to a trigger with an index. There is at least a `name` attribute in that directory, which is the trigger name that can be later used for association with a device.

- `/sys/bus/iio/devices/iio:deviceX/trigger/*`: This directory will be automatically created if your device supports a triggered buffer. A trigger can be associated with our device by writing the trigger's name in the `current_trigger` file in this directory.

Having enumerated the trigger-related sysfs directories, let's start by describing how the sysfs trigger interface works.

The sysfs trigger interface

A sysfs trigger is enabled in the kernel with the `CONFIG_IIO_SYSFS_TRIGGER=y` config option, which will result in the `/sys/bus/iio/devices/iio_sysfs_trigger/` folder being automatically created, which can be used for sysfs trigger management. There will be two files in the directory, `add_trigger` and `remove_trigger`. Its driver is `drivers/iio/trigger/iio-trig-sysfs.c`. The following are descriptions of each of these attributes:

- `add_trigger`: Used to create a new sysfs trigger. You can create a new trigger by writing a positive value (which will be used as a trigger ID) into that file. It will create the new sysfs trigger, accessible at `/sys/bus/iio/devices/triggerX`, where X is the trigger number. For example, `echo 2 > add_trigger` will create a new sysfs trigger, accessible at `/sys/bus/iio/devices/trigger2`. An invalid argument message will be returned if a trigger with the supplied ID already exists in the system. The sysfs trigger name pattern is `sysfstrig{ID}`. The `echo 2 > add_trigger` command will create the `/sys/bus/iio/devices/trigger2` trigger, whose name is `sysfstrig2`, and you can check it with `cat /sys/bus/iio/devices/trigger2/name`. Each sysfs trigger contains at list one file: `trigger_now`. Writing 1 into that file will instruct all devices with the corresponding trigger name in their `current_trigger` to start the capture and push data into their respective buffers. Each device buffer must have its size set and must be enabled (`echo 1 > /sys/bus/iio/devices/iio:deviceX/buffer/enable`).

- `remove_trigger`: Used to remove a trigger. The following command will be sufficient to remove the previously created trigger:

```
echo 2 > remove_trigger
```

As you can see, the value used in `add_trigger` while creating the trigger must be the same value you use when removing the trigger.

> **Note**
>
> You should note that the driver will only capture data when the associated trigger is triggered. Thus, when using the sysfs trigger, the data will only be captured at the time when 1 is written into the `trigger_now` attribute. Thus, to implement continuous data capture, you should run `echo 1 > trigger_now` as many times as you need a sample count, in a loop, for example. This is because a single call of `echo 1 > trigger_now` is equivalent to a single trigging and thus will perform only one capture, which will be pushed in the buffer. With interrupt-based triggers, data is captured and pushed in the buffer anytime an interrupt occurs.

Now we are done with the trigger setup, this trigger must be assigned to a device so that it can trigger data capture on this device, as we will see in the next section.

Tying a device to a trigger

Associating a device with a given trigger consists of writing the name of the trigger to the `current_trigger` file available under the device's trigger directory. For example, let's say we need to tie a device with the trigger that has index 2:

```
# set trigger2 as current trigger for device0
echo sysfstrig2 > /sys/bus/iio/devices/iio:device0/trigger/
current_trigger
```

To detach the trigger from the device, you should write an empty string to the `current_trigger` file of the device trigger directory, as follows:

```
echo "" > iio:device0/trigger/current_trigger
```

We will see later in the chapter (in the *Capturing data using a sysfs trigger* section) a practical example dealing with sysfs triggers for data capture.

Interrupt trigger interface

Say we have the following sample:

```
static struct resource iio_irq_trigger_resources[] = {
    [0] = {
        .start = IRQ_NR_FOR_YOUR_IRQ,
        .flags = IORESOURCE_IRQ | IORESOURCE_IRQ_LOWEDGE,
    },
};

static struct platform_device iio_irq_trigger = {
    .name = "iio_interrupt_trigger",
    .num_resources = ARRAY_SIZE(iio_irq_trigger_resources),
    .resource = iio_irq_trigger_resources,
};

platform_device_register(&iio_irq_trigger);
```

In this sample, we declare our IRQ- (that the IIO interrupt trigger will register using `request_irq()`) based trigger as a platform device. It will result in the IRQ trigger standalone module (whose source file is `drivers/iio/trigger/iio-trig-interrupt.c`) being loaded. After the probing succeeds, there will be a directory corresponding to the trigger. IRQ trigger names have the form `irqtrigX`, where X corresponds to the IRQ we just passed. This name is the one you will see in `/proc/interrupt`:

```
$ cd /sys/bus/iio/devices/trigger0/
$ cat name
    irqtrig85
```

As we have done with other triggers, you just have to assign that trigger to your device, by writing its name into your device's `current_trigger` file:

```
echo "irqtrig85" > /sys/bus/iio/devices/iio:device0/trigger/
current_trigger
```

Now, every time the interrupt fires, device data will be captured.

The IRQ trigger driver is implemented in `drivers/iio/trigger/iio-trig-interrupt.c`. Since the driver requires a resource, we can use a device tree without any code change, with the only condition to respect the `compatible` property, as follows:

```
mylabel: my_trigger@0{
    compatible = "iio_interrupt_trigger";
    interrupt-parent = <&gpio4>;
    interrupts = <30 0x0>;
};
```

The example assumes the IRQ line is `GPIO#30`, which belongs to the `gpio4` GPIO controller node. This consists of using a GPIO as an interrupt source, so that whenever the GPIO changes to a given state, the interrupt is raised, thus triggering the capture.

The hrtimer trigger interface

`hrtimer trigger` is implemented in `drivers/iio/trigger/iio-trig-hrtimer.c` and relies on the `configfs` filesystem (see `Documentation/iio/iio_configfs.txt` in kernel sources), which can be enabled via the `CONFIG_IIO_CONFIGFS` config option and mounted on our system (usually under the `/config` directory):

```
$ mkdir /config
$ mount -t configfs none /config
```

Now, loading the `iio-trig-hrtimer` module will create IIO groups accessible under `/config/iio`, allowing users to create `hrtimer` triggers under `/config/iio/triggers/hrtimer`. The following is an example:

```
# create a hrtimer trigger
$ mkdir /config/iio/triggers/hrtimer/my_trigger_name
# remove the trigger
$ rmdir /config/iio/triggers/hrtimer/my_trigger_name
```

Each `hrtimer` trigger contains a single `sampling_frequency` attribute in the trigger directory. A full and working example is provided later in the chapter in the *Data capture using an hrtimer trigger* section.

IIO buffers

An IIO buffer offers continuous data capture, where more than one data channel can be read at once. The buffer is accessible from the user space via the `/dev/iio:device` character device node. From within the trigger handler, the function used to fill the buffer is `iio_push_to_buffers_with_timestamp()`. In order to allocate and set up a trigger buffer for a device, drivers must use `iio_triggered_buffer_setup()`.

IIO buffer sysfs interface

An IIO buffer has an associated attributes directory under `/sys/bus/iio/iio:deviceX/buffer/*`. The following are some of the existing attributes:

- `length`: The capacity of the buffer. It represents the total number of data samples that can be stored by the buffer. It is the number of scans contained by the buffer.

- `enable`: Activate the buffer capture and start the buffer capture up.

- `watermark`: This attribute has been available since kernel version v4.2. It is a positive number that specifies how many scan elements a blocking read should wait for. If using the `poll()` system call, for example, it will block until the watermark is reached. It makes sense only if the watermark is greater than the requested amount of reads. It does not affect non-blocking reads. A maximum delay guarantee can be achieved by blocking on `poll()` with a timeout and reading the available samples after the timeout expires.

Now that we have enumerated and described the attributes present in the IIO buffer directory, let's discuss how to set up the IIO buffer.

IIO buffer setup

A channel whose data is to be read and pushed into the buffer is called a **scan element**. Its configurations are accessible from the user space via the `/sys/bus/iio/iio:deviceX/scan_elements/*` directory, containing the following attributes:

- `*_en`: This is a suffix for the attribute name, used to enable the channel. If, and only if, the value of its attribute is non-zero, then a triggered capture will contain data samples for this channel. For example, `in_voltage0_en` and `in_voltage1_en` are attributes that enable `in_voltage0` and `in_voltage1`. Therefore, if the value of `in_voltage1_en` is non-zero, then the output of a triggered capture on the underlying IIO device will include the `in_voltage1` channel value.

- `type`: Describes the scan element data storage within the buffer and hence the form in which it is read from user space. For example, `in_voltage0_type` is an example of a channel type. The format respects the following pattern: `[be|le]:[s|u]bits/storagebitsXrepeat[>>shift]`. The following are the meanings of each field in the following format:

  - `be` or `le` specifies the endianness (big or little).

  - `s` or `u` specifies the sign, either signed (two's complement) or unsigned.

  - `bits` is the number of valid data bits.

  - `storagebits` is the number of bits this channel occupies in the buffer. That said, a value may really be coded on 12 bits (`bits`) but occupies 16 bits (`storagebits`) in the buffer. You must, therefore, shift the data four times to the right to obtain the actual value. This parameter depends on the device, and you should refer to its datasheet.

  - `shift` represents the number of times you must shift the data value prior to masking out unused bits. This parameter is not always needed. If the number of valid bits is equal to the number of storage bits, the shift will be 0. You can also find this parameter in the device datasheet.

  - The `repeat` element specifies the number of times `bits/storagebits` is repeated. The repeat value is omitted when the repeat element is 0 or 1.

The best way to explain this section is by providing an excerpt of the kernel docs, which you can find here: `https://www.kernel.org/doc/html/latest/driver-api/iio/buffers.html`. Let's consider a driver for a 3-axis accelerometer with 12-bit resolution where data is stored in two 8-bit (thus 16 bits) registers, as follows:

```
 7   6   5   4   3   2   1   0
+---+---+---+---+---+---+---+---+
|D3 |D2 |D1 |D0 | X | X | X | X |  (LOW byte, address 0x06)
+---+---+---+---+---+---+---+---+


 7   6   5   4   3   2   1   0
+---+---+---+---+---+---+---+---+
|D11|D10|D9 |D8 |D7 |D6 |D5 |D4 |(HIGH byte, address 0x07)
+---+---+---+---+---+---+---+---+
```

According to the preceding description, each axis will have the following scan element:

```
$ cat  /sys/bus/iio/devices/iio:device0/scan_elements/in_
accel_y_type
le:s12/16>>4
```

You should interpret this as being little-endian signed data, 16 bits in size, which needs to be shifted right by 4 bits before masking out the 12 valid bits of data.

The element of `struct iio_chan_spec` responsible for determining how a channel's value should be stored in a buffer is `scant_type`:

```
struct iio_chan_spec {
    [...]
    struct {
        char sign; /* either u or s as explained above */
        u8 realbits;
        u8 storagebits;
        u8 shift;
        u8 repeat;
        enum iio_endian endianness;
    } scan_type;
    [...]
};
```

This structure absolutely matches `[be|le]:[s|u]bits/ storagebitsXrepeat[>>shift]`, which was the pattern described previously. Let's have a look at each part of the structure:

- `sign` represents the sign of the data and matches `[s|u]` in the pattern.
- `realbits` corresponds to `bits` in the pattern.
- `storagebits` matches `storagebits` in the pattern.
- `shift` corresponds to `shift` in the pattern, as well as `repeat`.
- `iio_indian` represents the endianness and matches `[be|le]` in the pattern.

At this point, we should be able to implement the IIO channel structure that corresponds to the type explained previously:

```
struct struct iio_chan_spec accel_channels[] = {
    {
```

```
            .type = IIO_ACCEL,
            .modified = 1,
            .channel2 = IIO_MOD_X,
            /* other stuff here */
            .scan_index = 0,
            .scan_type = {
                .sign = 's',
                .realbits = 12,
                .storagebits = 16,
                .shift = 4,
                .endianness = IIO_LE,
                },
        }
    /* similar for Y (with channel2 = IIO_MOD_Y,
     * scan_index = 1) and Z (with channel2
     * = IIO_MOD_Z, scan_index = 2) axis
     */
}
```

Buffer and trigger support are the last concepts in our learning process of the IIO framework. Now that we are familiar with that, we can put everything together and summarize the knowledge we have acquired with a concrete, lite example.

Putting it all together

Let's have a closer look at the BMA220 digital triaxial acceleration sensor from Bosch. This is an SPI/I2C-compatible device, with 8-bit-sized registers, along with an on-chip motion-triggered interrupt controller, which senses tilt, motion, and shock vibration. Its datasheet is available here: http://www.mouser.fr/pdfdocs/ BSTBMA220DS00308.PDF. Its driver is available thanks to the CONFIG_BMA200 kernel config option. Let's walk through it.

We first declare our channels using struct iio_chan_spec. If the triggered buffer will be used, then we need to fill in the scan_index and scan_type fields. The following code excerpt shows the declaration of our channels:

```
#define BMA220_DATA_SHIFT       2
#define BMA220_DEVICE_NAME      "bma220"
#define BMA220_SCALE_AVAILABLE  "0.623 1.248 2.491 4.983"
```

```
#define BMA220_ACCEL_CHANNEL(index, reg, axis) {     \
    .type = IIO_ACCEL,                               \
    .address = reg,                                  \
    .modified = 1,                                   \
    .channel2 = IIO_MOD_##axis,                      \
    .info_mask_separate = BIT(IIO_CHAN_INFO_RAW), \
    .info_mask_shared_by_type = BIT(IIO_CHAN_INFO_SCALE),\
    .scan_index = index,                             \
    .scan_type = {                                   \
        .sign = 's',                                 \
        .realbits = 6,                               \
        .storagebits = 8,                            \
        .shift = BMA220_DATA_SHIFT,                  \
        .endianness = IIO_CPU,                       \
    },                                               \
}


static const struct iio_chan_spec bma220_channels[] = {
    BMA220_ACCEL_CHANNEL(0, BMA220_REG_ACCEL_X, X),
    BMA220_ACCEL_CHANNEL(1, BMA220_REG_ACCEL_Y, Y),
    BMA220_ACCEL_CHANNEL(2, BMA220_REG_ACCEL_Z, Z),
};
```

.info_mask_separate = BIT(IIO_CHAN_INFO_RAW) means there will be a
*_raw sysfs entry (attribute) for each channel, and .info_mask_shared_by_type
= BIT(IIO_CHAN_INFO_SCALE) says that there is only a *_scale sysfs entry for all
channels of the same type:

```
jma@jma:~$ ls -l /sys/bus/iio/devices/iio:device0/
(...)
# without modifier, a channel name would have in_accel_raw
(bad)
-rw-r--r-- 1 root root 4096 jul 20 14:13 in_accel_scale
-rw-r--r-- 1 root root 4096 jul 20 14:13 in_accel_x_raw
-rw-r--r-- 1 root root 4096 jul 20 14:13 in_accel_y_raw
-rw-r--r-- 1 root root 4096 jul 20 14:13 in_accel_z_raw
(...)
```

Reading `in_accel_scale` calls the `read_raw()` hook with the mask set to `IIO_CHAN_INFO_SCALE`. Reading `in_accel_x_raw` calls the `read_raw()` hook with the mask set to `IIO_CHAN_INFO_RAW`. The real value is then `raw_value x scale`.

What `.scan_type` says is that each channel's return value is signed, 8 bits in size (will occupy 8 bits in the buffer), but the useful payload only occupies 6 bits, and data must be right-shifted twice prior to masking out unused bits. Any scan element type will look as follows:

```
$ cat /sys/bus/iio/devices/iio:device0/scan_elements/in_
accel_x_type
le:s6/8>>2
```

The following is our `pullfunc` (actually, it is the bottom half), which reads a sample from the device and pushes read values into the buffer (`iio_push_to_buffers_with_timestamp()`). Once done, we inform the core (`iio_trigger_notify_done()`):

```c
static irqreturn_t bma220_trigger_handler(int irq, void *p)
{
    int ret;
    struct iio_poll_func *pf = p;
    struct iio_dev *indio_dev = pf->indio_dev;
    struct bma220_data *data = iio_priv(indio_dev);
    struct spi_device *spi = data->spi_device;

    mutex_lock(&data->lock);
    data->tx_buf[0] =
                BMA220_REG_ACCEL_X | BMA220_READ_MASK;
    ret = spi_write_then_read(spi, data->tx_buf,
                    1, data->buffer,
            ARRAY_SIZE(bma220_channels) - 1);
    if (ret < 0)
        goto err;

    iio_push_to_buffers_with_timestamp(indio_dev,
                        data->buffer, pf->timestamp);
err:
    mutex_unlock(&data->lock);
```

```
    iio_trigger_notify_done(indio_dev->trig);

    return IRQ_HANDLED;
}
```

The following is the read function. It is a hook called every time you read a sysfs entry of the device:

```
static int bma220_read_raw(struct iio_dev *indio_dev,
            struct iio_chan_spec const *chan,
            int *val, int *val2, long mask)
{
    int ret;
    u8 range_idx;
    struct bma220_data *data = iio_priv(indio_dev);

    switch (mask) {
     case IIO_CHAN_INFO_RAW:
            /* do not process single-channel read
             * if buffer mode is enabled
             */
            if (iio_buffer_enabled(indio_dev))
                    return -EBUSY;
            /* Else we read the channel */
            ret = bma220_read_reg(data->spi_device,
                                    chan->address);
            if (ret < 0)
                    return -EINVAL;
            *val = sign_extend32(ret >> BMA220_DATA_SHIFT,
                                    5);
            return IIO_VAL_INT;
     case IIO_CHAN_INFO_SCALE:
            ret = bma220_read_reg(data->spi_device,
                                    BMA220_REG_RANGE);
            if (ret < 0)
                    return ret;
            range_idx = ret & BMA220_RANGE_MASK;
```

```
                *val = bma220_scale_table[range_idx][0];
                *val2 = bma220_scale_table[range_idx][1];
                return IIO_VAL_INT_PLUS_MICRO;
    }
    return -EINVAL;
}
```

When you read a *raw sysfs file, the hook is called given IIO_CHAN_INFO_RAW in the mask parameter and the corresponding channel in the *chan parameter. *val and *val2 are actually output parameters that must be set with the raw value (read from the device). Any read performed on the *scale sysfs file will call the hook with IIO_CHAN_INFO_SCALE in the mask parameter, and so on for each attribute mask.

The same principle applies in the write function, used to write a value to the device. There is an 80% chance your driver does not require a write operation. In the following example, the write hook lets the user change the device's scale, though other parameters can be changed, such as sampling frequency or digital-to-analog raw value:

```
static int bma220_write_raw(struct iio_dev *indio_dev,
                struct iio_chan_spec const *chan,
                int val, int val2, long mask)
{
    int i;
    int ret;
    int index = -1;
    struct bma220_data *data = iio_priv(indio_dev);

    switch (mask) {
    case IIO_CHAN_INFO_SCALE:
     for (i = 0; i < ARRAY_SIZE(bma220_scale_table); i++)
      if (val == bma220_scale_table[i][0] &&
            val2 == bma220_scale_table[i][1]) {
                index = i;
                break;
            }
        if (index < 0)
          return -EINVAL;
```

```
        mutex_lock(&data->lock);
        data->tx_buf[0] = BMA220_REG_RANGE;
        data->tx_buf[1] = index;
        ret = spi_write(data->spi_device, data->tx_buf,
                sizeof(data->tx_buf));
        if (ret < 0)
            dev_err(&data->spi_device->dev,
                "failed to set measurement range\n");
        mutex_unlock(&data->lock);

        return 0;
    }
    return -EINVAL;
}
```

This function is called whenever you write a value to the device, and only supports scaling value change. An example of usage in user space could be `echo $desired_scale > /sys/bus/iio/devices/iio:devices0/in_accel_scale`.

Now it comes time to fill a `struct iio_info` structure to be given to our `iio_device`:

```
static const struct iio_info bma220_info = {
    .driver_module    = THIS_MODULE,
    .read_raw         = bma220_read_raw,
    .write_raw        = bma220_write_raw,
      /* Only if your needed */
};
```

In the `probe` function, we allocate and set up a `struct iio_dev` iio device. Memory for private data is reserved too:

```
/*
 * We only provide two mask possibilities,
 * allowing to select none or all channels.
 */
static const unsigned long bma220_accel_scan_masks[] = {
    BIT(AXIS_X) | BIT(AXIS_Y) | BIT(AXIS_Z),
    0
```

```
};

static int bma220_probe(struct spi_device *spi)
{
    int ret;
    struct iio_dev *indio_dev;
    struct bma220_data *data;
    indio_dev = devm_iio_device_alloc(&spi->dev,
                                        sizeof(*data));
    if (!indio_dev) {
        dev_err(&spi->dev, "iio allocation failed!\n");
        return -ENOMEM;
    }
    data = iio_priv(indio_dev);
    data->spi_device = spi;
    spi_set_drvdata(spi, indio_dev);
    mutex_init(&data->lock);

    indio_dev->dev.parent = &spi->dev;
    indio_dev->info = &bma220_info;
    indio_dev->name = BMA220_DEVICE_NAME;
    indio_dev->modes = INDIO_DIRECT_MODE;
    indio_dev->channels = bma220_channels;
    indio_dev->num_channels = ARRAY_SIZE(bma220_channels);
    indio_dev->available_scan_masks =
                                    bma220_accel_scan_masks;
    ret = bma220_init(data->spi_device);
    if (ret < 0)
        return ret;
    /* this will enable trigger buffer
     * support for the device */
    ret = iio_triggered_buffer_setup(indio_dev,
                            iio_pollfunc_store_time,
                            bma220_trigger_handler, NULL);
    if (ret < 0) {
        dev_err(&spi->dev,
```

```
                            "iio triggered buffer setup failed\n");
        goto err_suspend;
    }


    ret = devm_iio_device_register(&spi->dev, indio_dev);
    if (ret < 0) {
        dev_err(&spi->dev, "iio_device_register
                            failed\n");
        iio_triggered_buffer_cleanup(indio_dev);
        goto err_suspend;
    }


    return 0;

err_suspend:
    return bma220_deinit(spi);
}
```

You can enable this driver by means of the CONFIG_BMA220 kernel option. That says, *this is available only from v4.8 in the kernel.* The closest device you can use on older kernel versions is BMA180, which you can enable using the CONFIG_BMA180 option.

> **Note**
> To enable buffered capture in the IIO simple dummy driver, you must enable
> the IIO_SIMPLE_DUMMY_BUFFER kernel config option.

Now that we are familiar with IIO buffers, we will learn how to access the data coming from IIO devices and resulting from channel acquisitions.

Accessing IIO data

You may have guessed, there are only two ways to access data with the IIO framework: one-shot capture through sysfs channels or continuous mode (triggered buffer) via an IIO character device.

Single-shot capture

Single-shot data capture is done through the sysfs interface. By reading the sysfs entry that corresponds to a channel, you'll capture only the data specific to that channel. Say we have a temperature sensor with two channels: one for the ambient temperature and the other for the thermocouple temperature:

```
# cd /sys/bus/iio/devices/iio:device0
# cat in_voltage3_raw
6646
# cat in_voltage_scale
0.305175781
```

The processed value is obtained by multiplying the scale by the raw value:

*Voltage value: 6646 * 0.305175781 = 2028.19824053*

The device datasheet says the process value is given in mV. In our case, it corresponds to *2.02819 V*.

Accessing the data buffer

To get a triggered acquisition working, trigger support must have been implemented in your driver. Then, to acquire data from within the user space, you must create a trigger, assign it, enable the ADC channels, set the dimension of the buffer, and enable it. The code for this is given in the following section.

Capturing data using a sysfs trigger

Data capture using sysfs triggers consists of sending a set of commands and a few sysfs files. Let's go through what you should do to achieve that:

1. **Creating the trigger**: Before the trigger can be assigned to any device, it should be created:

   ```
   echo 0 > /sys/devices/iio_sysfs_trigger/add_trigger
   ```

   In the preceding command, 0 corresponds to the index we need to assign to the trigger. After this command, the trigger directory will be available under /sys/bus/iio/devices/ as trigger0. The trigger's full patch will be /sys/bus/iio/devices/trigger0.

2. **Assigning the trigger to the device**: A trigger is uniquely identified by its name, which you can use in order to tie the device to the trigger. Since we used 0 as the index, the trigger will be named `sysfstrig0`:

```
echo sysfstrig0 >
/sys/bus/iio/devices/iio:device0/trigger/current_trigger
```

We could have used this command too:

```
cat /sys/bus/iio/devices/trigger0/name > /sys/bus/iio/
devices/iio:device0/trigger/current_trigger.
```

However, if the value you have written does not correspond to an existing trigger name, nothing will happen. To make sure the trigger has been defined successfully, you can use the following command:

```
cat /sys/bus/iio/devices/iio:device0/trigger/current_
trigger
```

3. **Enabling some scan elements**: This step consists of choosing which channels should have their data value pushed into the buffer. You should pay attention to `available_scan_masks` in the driver:

```
echo 1 > /sys/bus/iio/devices/iio:device0/scan_elements/
in_voltage4_en
echo 1 > /sys/bus/iio/devices/iio:device0/scan_elements/
in_voltage5_en
echo 1 > /sys/bus/iio/devices/iio:device0/scan_elements/
in_voltage6_en
echo 1 > /sys/bus/iio/devices/iio:device0/scan_elements/
in_voltage7_en
```

4. **Setting up the buffer size**: Here, you should set the number of sample sets that may be held by the buffer:

```
echo 100 > /sys/bus/iio/devices/iio:device0/buffer/length
```

5. **Enabling the buffer**: This step consists of marking the buffer as being ready to receive pushed data:

```
echo 1 > /sys/bus/iio/devices/iio:device0/buffer/enable
```

To stop the capture, we'll have to write 0 in the same file.

6. **Firing the trigger**: Launch acquisition. This must be done as many times as data sample counts are needed in the buffer, in a loop, for example:

```
echo 1 > /sys/bus/iio/devices/trigger0/trigger_now
```

Now that acquisition is done, you can do the following.

7. Disable the buffer:

```
echo 0 > /sys/bus/iio/devices/iio:device0/buffer/enable
```

8. Detach the trigger:

```
echo "" > /sys/bus/iio/devices/iio:device0/trigger/
current_trigger
```

9. Dump the contents of our IIO character device:

```
cat /dev/iio\:device0 | xxd -
```

Now that we have learned how to use sysfs triggers, it will be easier to deal with hrtimer-based ones as they kind of use the same theoretical principle.

Data capture using an hrtimer trigger

hrtimers are high-resolution kernel timers with up to nanosecond granularity when the hardware allows it. As with sysfs-based triggers, data capture using hrtimer triggers requires a few commands for their setup. These commands can be split into the following steps:

1. Create the hrtimer-based trigger:

```
mkdir /sys/kernel/config/iio/triggers/hrtimer/trigger0
```

The preceding command will create a trigger named `trigger0`. This name will be used to assign this trigger to a device.

2. Define the sampling frequency:

```
echo 50 > /sys/bus/iio/devices/trigger0/sampling_
frequency
```

There is no configurable attribute in the `config` directory for the `hrtimer` trigger type. It introduces the `sampling_frequency` attribute to trigger directory. That attribute sets the polling frequency in Hz, with mHz precision. In the preceding example, we have defined a polling at 50 Hz (every 20 ms).

3. Link the trigger with the IIO device:

    ```
    echo trigger0 > /sys/bus/iio/devices/iio:device0/trigger/
    current_trigger
    ```

4. Choose on which channels data must be captured and pushed into the buffer:

    ```
    # echo 1 > /sys/bus/iio/devices/iio:device0/scan_
    elements/in_voltage4_en
    ```
    ```
    # echo 1 > /sys/bus/iio/devices/iio:device0/scan_
    elements/in_voltage5_en
    ```
    ```
    # echo 1 > /sys/bus/iio/devices/iio:device0/scan_
    elements/in_voltage6_en
    ```
    ```
    # echo 1 > /sys/bus/iio/devices/iio:device0/scan_
    elements/in_voltage7_en
    ```

5. Start the hrtimer capture, which will perform periodic data capture at the frequency
    we defined earlier and on channels that have been enabled previously:

    ```
    echo 1 > /sys/bus/iio/devices/iio:device0/buffer/enable
    ```

6. Finally, data can be dumped using `cat /dev/iio\:device0 | xxd -`.
    Because the trigger is an hrtimer, data will be captured and pushed at every hrtimer
    period interval.

7. To disable this periodic capture, the command to use is the following:

    ```
    echo 0 > /sys/bus/iio/devices/iio:device0/buffer/enable
    ```

8. Then, to remove this hrtimer trigger, the following command must be used:

    ```
    rmdir /sys/kernel/config/iio/triggers/hrtimer/trigger0
    ```

We can notice how easy it is to set up either a simple sysfs trigger or an hrtimer-based one.
They both consist of a few commands to set up and start the capture. However, captured
data would be meaningless or even dangerously misleading if not interpreted as it should,
which we'll discuss in the next section.

Interpreting the data

Now that everything has been set up, we can dump the data using the following
command:

```
# cat /dev/iio:device0 | xxd -
0000000: 0188 1a30 0000 0000 8312 68a8 c24f 5a14    ...0......h..
OZ.
```

```
0000010: 0188 1a30 0000 0000 192d 98a9 c24f 5a14  ...0.....-...
oz.
[...]
```

The preceding command will dump raw data that would need more processing to obtain the real data. In order to be able to understand the data output and process it, we need to look at the channel type, as follows:

```
$ cat /sys/bus/iio/devices/iio:device0/scan_elements/in_
voltage_type
be:s14/16>>2
```

In the preceding, be:s14/16>>2 means big-endian (be:) signed data (s) stored on 16 bits but whose real number of bits is 14. Moreover, it also means that the data must be shifted to the right two times (>>2) to obtain the real value. This means, for example, to obtain the voltage value in the first sample (0x188), this value must be right-shifted twice in order to mask unused bits: *0 x 188 >> 2 = 0 x 62 = 98*. Now, the real value is *98 * 250 = 24500 = 24.5 V*. If there were an offset attribute, the real value would be (raw + offset) * scale.

We are now familiar with IIO data access (from user space) and we are also done with the IIO producer interface in the kernel. It is not just the user space that can consume data from the IIO channel. There is an in-kernel interface as well, which we will discuss in the next section.

Dealing with the in-kernel IIO consumer interface

So far, we have dealt with the user-space consumer interface since data was consumed in user space. There are situations where a driver will require a dedicated IIO channel. An example is a battery charger that needs to measure the battery voltage as well. This measurement can be achieved using a dedicated IIO channel.

IIO channel attribution is done in the device tree. From the producer side, only one thing must be done: specifying the #io-channel-cells property according to the number of channels of the IIO device. Typically, it is 0 for nodes with a single IIO output and 1 for nodes with multiple IIO outputs. The following is an example:

```
adc: max1139@35 {
    compatible = "maxim,max1139";
    reg = <0x35>;
```

```
    #io-channel-cells = <1>;
};
```

On the consumer side, there are a few properties to provide. These are the following:

- io-channels: This is the only mandatory property. It represents the list of phandle (reference or pointer to a device tree node) and IIO specifier pairs, one pair for each IIO input to the device. Do note that if the IIO provider's #io-channel-cells property is 0, then only the phandle portion should be specified when referring to it in the consumer node. This is the case for single-channel IIO devices, for example, a temperature sensor. Otherwise, both the phandle and the channel index must be specified.

- io-channel-names: This is an optional but recommended property that is a list of IIO channel name strings. These names must be sorted in the same order as their corresponding channels, which are enumerated in the io-channels property. Consumer drivers should use these names to match IIO input names with IIO specifiers. This eases the channel identification in the driver.

Take the following example:

```
device {
    io-channels = <&adc 1>, <&ref 0>;
    io-channel-names = "vcc", "vdd";
};
```

The preceding node describes a device with two IIO resources, named vcc and vdd, respectively. The vcc channel originates from the &adc device output 1, while the vdd channel comes from the &ref device output 0.

Another example consuming several channels of the same ADC is the following:

```
some_consumer {
    compatible = "some-consumer";
    io-channels = <&adc 10>, <&adc 11>;
    io-channel-names = "adc1", "adc2";
};
```

Now that we are familiar with IIO binding and channel hogging, we can see how to play with those channels using the kernel IIO consumer API.

Consumer kernel API

The kernel IIO consumer interface relies on a few functions and data structures. The following is the main API:

```
struct iio_channel *devm_iio_channel_get(
        struct device *dev, const char *consumer_channel);
struct iio_channel * devm_iio_channel_get_all(
                                        struct device *dev);
int iio_get_channel_type(struct iio_channel *channel,
                    enum iio_chan_type *type);
int iio_read_channel_processed(struct iio_channel *chan,
                        int *val);
int iio_read_channel_raw(struct iio_channel *chan,
                    int *val);
```

The following are descriptions of each API:

- `devm_iio_channel_get()`: Used to get a single channel. `dev` is the pointer to the consumer device, and `consumer_channel` is the channel name as specified in the `io-channel-names` property. On success, it returns a pointer to a valid IIO channel, or a pointer to a negative error number if it is not able to get the IIO channel.

- `devm_iio_channel_get_all()`: Used to look up IIO channels. It returns a pointer to a negative error number if it is not able to get the IIO channel; otherwise, it returns an array of `iio_channel` structures terminated with 0 null `iio_dev` pointer. Say we have the following consumer node:

```
iio-hwmon {
    compatible = "iio-hwmon";
    io-channels = <&adc 0>, <&adc 1>, <&adc 2>,
    <&adc 3>, <&adc 4>, <&adc 5>,
    <&adc 6>, <&adc 7>, <&adc 8>,
    <&adc 9>;
};
```

The following code is an example of using `devm_iio_channel_get_all()` to get the IIO channels. This code also shows how to check for the last valid channel (the one with the null `iio_dev` pointer):

```
struct iio_channel *channels;
struct device *dev = &pdev->dev;
int num_adc_channels;

channels = devm_iio_channel_get_all(dev);
if (IS_ERR(channels)) {
    if (PTR_ERR(channels) == -ENODEV)
        return -EPROBE_DEFER;
        return PTR_ERR(channels);
}

num_adc_channels = 0;
/* count how many attributes we have */
while (channels[num_adc_channels].indio_dev)
        num_adc_channels++;

if (num_adc_channels !=
    EXPECTED_ADC_CHAN_COUNT) {
        dev_err(dev,
            "Inadequate ADC channels specified\n");
        return -EINVAL;
}
```

- `iio_get_channel_type()`: Returns the type of a channel, such as `IIO_VOLTAGE` or `IIO_TEMP`. This function fills enum `iio_chan_type` of the channel in the `type` output parameter. On error, the function returns a negative error number; otherwise, it returns 0.

- `iio_read_channel_processed()`: Reads the channel processed value in the correct unit, for example, in micro-volts for voltage and milli-degrees for temperature. `val` is the processed value read back. This function returns 0 on success or a negative value otherwise.

- `iio_read_channel_raw()`: Used to read a raw value from the channel. In this case, the consumer may need scale (`iio_read_channel_scale()`) and offset (`iio_read_channel_offset()`) in order to compute the processed value. `val` is the raw value read back.

In the preceding APIs, `struct iio_channel` represents an IIO channel from the consumer point of view. It has the following declaration:

```
struct iio_channel {
    struct iio_dev *indio_dev;
    const struct iio_chan_spec *channel;
    void *data;
};
```

In the preceding code, `iio_dev` is the IIO device to which the channel belongs, and `channel` is the underlying channel spec as seen by the provider.

Writing user-space IIO applications

After the long journey through the kernel-side implementation, it might be interesting to have a look at the other side, the user space. IIO support in user space can be handled through sysfs or using **libiio**, a library that has been specially developed for this purpose and follows the kernel-side evolutions. This library abstracts the hardware's low-level details and provides an easy and comprehensive programming interface that can also be used for complex projects.

In this section, we will be using version 0.21 of the library, whose documentation can be found here: `https://analogdevicesinc.github.io/libiio/v0.21/libiio/index.html`.

`libiio` can run on the following:

- A target, that is, the embedded system running Linux that includes IIO drivers for devices that are physically connected to the system, such as ADCs and DACs.

- A remote computer connected to the embedded system through a network, USB, or serial connection. This remote computer may be a PC running a Linux distribution, Windows, macOS, or OpenBSD/NetBSD. This remote PC communicates with the embedded system via the `iiod` server, which is a daemon running on the target.

The following diagram summarizes the architecture:

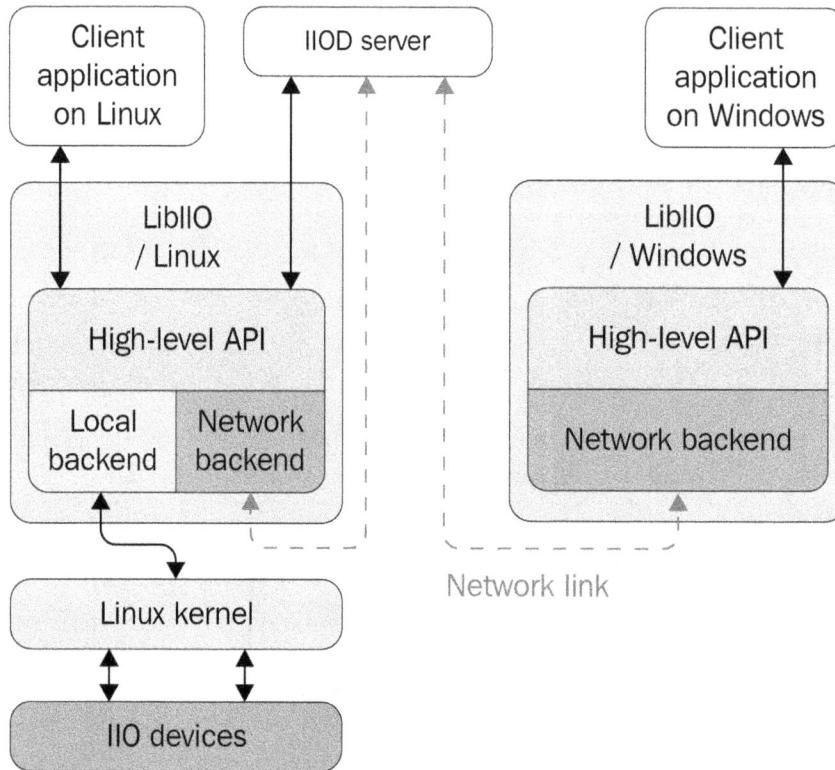

Figure 15.2 – libiio overview

libiio is built around five concepts, each of which corresponds to a data structure, altogether making almost all the API. These concepts are the following:

- **The backend**: This represents the connectivity (or the communication channel) between your application and the target on which the IIO devices to interact with are connected. This backend (thus connectivity) can be via USB, network, serial, or local. Independently from the hardware connectivity available, supported backends are library compile-time defined.

- **The context**: A context is a library instance that represents a collection of IIO devices, which in most cases correspond to a global view of the IIO devices on a running target. In this way, a context gathers all the IIO devices the target contains, as well as their channels and their attributes. For instance, when looking for an IIO device, code must create a context and request the target IIO device from this context.

Because applications may run remotely to the target board, the context will need a communication channel with that target. This is where the backend intervenes. Therefore, a context must be backed by a backend, which represents the connectivity between the target and the machine running the application. However, remotely running applications are not always aware of the target environment; thus, the library allows look up for available backends, allowing, among other things, dynamic behavior. This lookup is referred to as IIO context scanning. That said, applications may not bother with scanning if running locally to the target.

A context is represented with an instance of `struct iio_context`. A context object may contain zero or more devices. However, a device object is associated with only one context.

- **The device**: This is the IIO device. It is represented with `struct iio_device`, which is the user-space (`libiio` actually) counterpart of the in-kernel `struct iio_dev`. A device object may contain zero or more channels, while a channel is associated with only one device.

- **The buffer**: A buffer allows continuous data capture and chunk- (or slot-, instead of channel-) based reading. Such an object is represented with an instance of `struct iio_buffer`. A device may be associated with one buffer object, and a buffer is associated with only one device.

- **The channel**: A channel is an acquisition line, represented by an instance of `struct iio_channel`. A device may contain zero or more channels, and a channel is associated with only one device.

After becoming familiar with these concepts, we can split IIO application development into the following steps:

1. Creating a context, after having (optionally) scanned for available backends to create this context with.

2. Iterating over all devices, or looking for and picking the one of interest. Eventually getting/setting the device parameters via its attributes.

3. Walking through the device channels and enabling channels of interest (or disabling the ones we are not interested in). Eventually getting/setting the channel parameters via their attributes.

4. If a device needs a trigger, then associating a trigger with the given device. This trigger must have been created before creating the context.

5. Creating a buffer and associating this buffer with the device, and then starting streaming.

6. Starting the capture and reading the data.

Scanning and creating an IIO context

When creating a context, the library will identify the IIO devices (including triggers) that can be used and identify the channels for each device; then, it will identify all device- and channel-specific attributes and also identify attributes shared by all channels; finally, the library will create a context where all those entities are placed.

A context can be created using one of the following APIs:

```
iio_create_local_context()
iio_create_network_context()
iio_create_context_from_uri()
iio_context_clone(const struct iio_context *ctx)
```

Each of these functions returns a valid context object on success and NULL otherwise, with errno set appropriately. That said, while they all return the same values, their arguments may vary, as described in the following:

- iio_create_local_context(): Used to create a local context:

  ```
  struct iio_context * local_ctx;
  local_ctx = iio_create_local_context();
  ```

  Note that the local backend interfaces the Linux kernel through the sysfs virtual filesystem.

- iio_create_network_context(): Creates a network context. It takes as a parameter a string representing the IPv4 or IPv6 network address of the remote target:

  ```
  struct iio_context * network_ctx;
  network_ctx =
       iio_create_network_context("192.168.100.15");
  ```

- USB context can be created using an URI-based API, `iio_create_context_from_uri()`. The argument is a string identifying the USB device using the following pattern – `usb:[device:port:instance]`:

```
struct iio_context * usb_ctx;
usb_ctx = iio_create_context_from_uri("usb:3.80.5");
```

- A serial context, like a USB context, uses a URI-based API. However, its URI must match the following pattern – `serial:[port][,baud][,config]`:

```
struct iio_context * serial_ctx;
serial_ctx = iio_create_context_from_uri(
                  "serial:/dev/ttyUSB0,115200,8n1");
```

- `iio_create_context_from_uri()` is a URI-based API, taking as a parameter a valid URI (starting with the backend to use). For local context, the URI must be `"local:"`. For a URI-based network context, the URI pattern must match `"ip:<ipaddr>"`, where `<ipaddr>` is the IPv4 or IPv6 of the remote target. More information on URI-based contexts can be found here: `https://analogdevicesinc.github.io/libiio/v0.21/libiio/group__Context.html#gafdcee40508700fa395370b6c636e16fe`.

- `iio_context_clone()` duplicates the context given as a parameter and returns the new clone. This function is not supported on `usb:` contexts, since `libusb` can only claim the interface once.

Before creating a context, the user might be interested in scanning the available contexts (that is, looking for available backends). To find what IIO contexts are available, the user code must do the following:

- Invoke `iio_create_scan_context()` to create an instance of `iio_scan_context object`. The first argument to this function is a string that is used as a filter (`usb:`, `ip:`, `local:`, `serial:`, or a mix, such as `usb:ip`, where the default (`NULL`) means any backend that is compiled in).

- Call `iio_scan_context_get_info_list()` given the previous `iio_scan_context` object as parameter. This will return an array `iio_context_info` object from the `iio_scan_context` object. Each `iio_context_info` object can be examined with `iio_context_info_get_description()` and `iio_context_info_get_uri()` to determine which URI you want to attach to.

- Once done, the `info` object array and the `scan` object must be released with `iio_context_info_list_free()` and `iio_scan_context_destroy()`, respectively.

The following is a demonstration of scanning available contexts and creating one:

```c
int i;
ssize_t nb_ctx;
const char *uri;
struct iio_context *ctx = NULL;

#ifdef CHECK_REMOTE
struct iio_context_info **info;
struct iio_scan_context *scan_ctx =
                iio_create_scan_context("usb:ip:", 0);

if (!scan_ctx) {
    printf("Unable to create scan context!\n");
    return NULL;
}

nb_ctx = iio_scan_context_get_info_list(scan_ctx, &info);
if (nb_ctx < 0) {
    printf("Unable to scan!\n");
    iio_scan_context_destroy(scan_ctx);
    return NULL;
}

for (i = 0; i < nb_ctx; i++) {
    uri = iio_context_info_get_uri(info[0]);
    if (strcmp ("usb:", uri) == 0) {
        ctx = iio_create_context_from_uri(uri);
        break;
    }
    if (strcmp ("ip:", uri) == 0) {
        ctx =
            iio_create_context_from_uri("ip:192.168.3.18");
        break;
    }
}
iio_context_info_list_free(info);
```

```
iio_scan_context_destroy(scan_ctx);
#endif

if (!ctx) {
    printf("creating local context\n");
    ctx = iio_create_local_context();

    if (!ctx) {
        printf("unable to create local context\n");
        goto err_free_info_list;
    }
}
return ctx;
```

In the preceding code, if the CHECK_REMOTE macro is defined, the code will first scan for available contexts (that is, backends) by filtering USB and network ones. The code first looks for the USB context before looking for a network context. If none is available, it falls back to a local context.

In addition, you can get some context-related information using the following APIs:

```
int iio_context_get_version (
        const struct iio_context * ctx,
        unsigned int *major, unsigned int *minor,
        char git_tag[8])
const char * iio_context_get_name(
                        const struct iio_context *ctx)
const char * iio_context_get_description(
                        const struct iio_context *ctx)
```

In the preceding APIs, iio_context_get_version() returns the version of the backend in use into major, minor, and git_tag output arguments, and iio_context_get_name() returns a pointer to a static NULL-terminated string corresponding to the backend name, which can be local, xml, or network when the context has been created with the local, XML, and network backends, respectively.

The following is a demonstration:

```
unsigned int major, minor;
char git_tag[8];
struct iio_context *ctx;
[...] /* the context must be created */
iio_context_get_version(ctx, &major, &minor, git_tag);
printf("Backend version: %u.%u (git tag: %s)\n",
            major, minor, git_tag);
printf("Backend description string: %s\n",
            iio_context_get_description(ctx));
```

Now that the context has been created, and we are able to read its information, the user might be interested in walking through it, that is, navigating the entities this context is made of, for instance, getting the number of IIO devices or getting an instance of a given device.

> **Note**
>
> A context is a punctual and fixed view of IIO entities on the target. For instance, if a user creates an IIO trigger device after having created the context, this trigger device won't be accessible from this context. Because there is no context synchronization API, the proper way to do things would be to destroy and re-create things or to create the needed dynamic IIO elements at the beginning of the program before creating the context.

Walking through and managing IIO devices

The following are APIs to navigate through the devices in an IIO context:

```
unsigned int iio_context_get_devices_count(
                            const struct iio_context *ctx)
struct iio_device * iio_context_get_device(
        const struct iio_context *ctx, unsigned int index)
struct iio_device * iio_context_find_device(
        const struct iio_context *ctx, const char *name)
```

From a context, iio_context_get_devices_count() returns the number of IIO devices in this context.

`iio_context_get_device()` returns a handle for an IIO device specified by its index (or ID). This ID corresponds to `<X>` in `/sys/bus/iio/devices/iio:device<X>/`. For example, the ID of the `/sys/bus/iio/devices/iio:device1` device is 1. If the index is invalid, `NULL` is returned. Alternatively, given a device object, its ID can be retrieved with `iio_device_get_id()`.

`iio_context_find_device()` looks for an IIO device by its name. This name must correspond to the name specified in `iio_indev->name` specified in the driver. You can obtain this name either by using a dedicated `iio_device_get_name()` API or by reading the `name` attribute in this device's sysfs directory:

```
root:/sys/bus/iio/devices/iio:device1> cat name
ad9361-phy
```

The following is an example of going through all devices and printing their names and IDs:

```
struct iio_context * local_ctx;
local_ctx = iio_create_local_context();
int i;
for (i = 0; i < iio_context_get_devices_count(local_ctx);
     ++i) {
   struct iio_device *dev =
           iio_context_get_device(local_ctx, i);
   const char *name = iio_device_get_name(dev);
   printf("\t%s: %s\r\n", iio_device_get_id(dev), name );
}
iio_context_destroy(ctx);
```

The preceding code example iterates over IIO devices present in the context (a local context) and prints their names and IDs.

Walking through and managing IIO channels

The main channel management APIs are the following:

```
unsigned int iio_device_get_channels_count(
                        const struct iio_device *dev)
struct iio_channel* iio_device_get_channel(
        const struct iio_device *dev, unsigned int index)
struct iio_channel* iio_device_find_channel(
                        const struct iio_device *dev,
                        const char *name, bool output)
```

We can get the number of available channels from an `iio_device` object thanks to `iio_device_get_channels_count()`. Then, each `iio_channel` object can be accessed with `iio_device_get_channel()`, specifying the index of this channel. For example, on a three-axis (*x*, *y*, *z*) accelerometer, `iio_device_get_channel(iio_device, 0)` will correspond to getting channel 0, that is, `accel_x`. On an eight-channel ADC converter, `iio_device_get_channel(iio_device, 0)` will correspond to getting channel 0, that is, `voltage0`.

Alternatively, it is possible to look up a channel by its name using `iio_device_find_channel()`, which expects in arguments the channel name and a Boolean, which tells you whether the channel is an output or not. If you remember, in the *Channel attribute naming convention* section, we saw that attribute names respect the following pattern: `{direction}_{type}{index}_{modifier}_{info_mask}`. The subset in this pattern that needs to be used with `iio_device_find_channel()` is `{type}{index}_{modifier}`. Then, depending on the value of the Boolean parameter, the final name will be obtained by adding either `in_` or `out_` as a prefix. For instance, to obtain channel X of the accelerometer, we would use `iio_device_find_channel(iio_device, "accel_x", 0)`. For the first channel of the analog-to-digital converter, we would use `iio_device_find_channel(iio_device, "voltage0", 0)`.

The following is an example of going through all devices and all channels of each device:

```
struct iio_context * local_ctx;
struct iio_channel *chan;
local_ctx = iio_create_local_context();
int i, j;
for (i = 0; i < iio_context_get_devices_count(local_ctx);
        ++i) {
    struct iio_device *dev =
            iio_context_get_device(local_ctx, i);
    printf("Device %d\n", i);
    for (j = 0; j < iio_device_get_channels_count(dev);
            ++j) {
        chan = iio_device_get_channel(dev, j);
        const char *name = iio_channel_get_name(ch) ? :
                            iio_channel_get_id(ch);
        printf("\tchannel %d: %s\n", j, name);
    }
}
```

The preceding code creates a local context and walks through all the devices in this context. Then, for each device, it iterates over channels and prints their name.

Additionally, there are miscellaneous APIs allowing us to obtain channel properties. These are the following:

```
bool iio_channel_is_output(const struct iio_channel *chn);
const char* iio_channel_get_id(
                        const struct iio_channel *chn);
enum iio_modifier iio_channel_get_modifier(
                        const struct iio_channel *chn);
enum iio_chan_type iio_channel_get_type(
                        const struct iio_channel *chn);
const char* iio_channel_get_name(
                        const struct iio_channel *chn);
```

In the preceding APIs, the first one checks whether the IIO channel is output or not, and the others mainly return each of the elements the name pattern is made of.

Working with a trigger

In `libiio`, a trigger is assimilated to a device, as both are represented by `struct iio_device`. The trigger must be created before creating the context, else this trigger won't be seen/available from that context.

In order to do so, you must create the trigger yourself, as we saw in the *IIO trigger and sysfs (user space)* section. Then, to find this trigger from a context, as it is assimilated to a device, you can use one of the device-related lookup APIs that we described in the *Walking through and managing IIO devices* section. In this section, let's use `iio_context_find_device()`, which as you'll recall is defined as the following:

```
struct iio_device* iio_context_find_device(
        const struct iio_context *ctx, const char *name)
```

This function looks for a device by its name in the given context. This is the reason why the trigger must have been created before creating the context. In parameters, `ctx` is the context from where to look for the trigger and `name` is the name of the trigger, as you would have written it to the `current_trigger` sysfs file.

Once the trigger found, it must be assigned to a device using `iio_device_set_trigger()`, defined as the following:

```
int iio_device_set_trigger(const struct iio_device *dev,
                           const struct iio_device *trig)
```

This function associates the trigger, `trig`, to the device, `dev`, and returns `0` on success or a negative `errno` code on failure. If the `trig` parameter is `NULL`, then any trigger associated with the given device will be disassociated. In other words, to disassociate a trigger from the device, you should call `iio_device_set_trigger(dev, NULL)`.

Let's see how trigger lookup and association work in a little example:

```
struct iio_context *ctx;
struct iio_device *trigger, *dev;
[...]

ctx = iio_create_local_context();
/* at least 2 iio_device must exist:
 * a trigger and a device */
if (!(iio_context_get_devices_count(ctx) > 1))
    return -1;
```

```
trigger = iio_context_find_device(ctx, "hrtimer-1");
if (!trigger) {
    printf("no trigger found\n");
    return -1;
}

dev = iio_context_find_device(ctx, "iio-device-dummy");
if (!dev) {
    printf("unable to find the IIO device\n");
    return -1;
}
printf("Enabling IIO buffer trigger\n");
iio_device_set_trigger(dev, trigger);
[...]

/* When done with the trigger */
iio_device_set_trigger(dev, NULL);
```

In the preceding example, we first create a local context, and we make sure this context contains at least two devices. Then, from this context, we look for a trigger named hrtimer-1 and a device named iio-device-dummy. Once both are found, we associate the trigger to the device. Finally, when done with the trigger, it is disassociated from the device.

Creating a buffer and reading data samples

Note that channels we are interested in need to be enabled before creating the buffer. To do so, you can use the following APIs:

```
void iio_channel_enable(struct iio_channel * chn)
bool iio_channel_is_enabled(struct iio_channel * chn)
```

The first function enables the channel so that its data will be captured and pushed in the buffer. The second one is a helper checking whether a channel has already been enabled or not.

In order to disable a channel, you can use `iio_channel_disable()`, defined as the following:

```
void iio_channel_disable(struct iio_channel * chn)
```

Now that we are able to enable the channels, we need their data to be captured. We can create a buffer using `iio_device_create_buffer()`, defined as the following:

```
struct iio_buffer * iio_device_create_buffer(
        const struct iio_device *dev,
        size_t samples_count, bool cyclic)
```

This function configures and enables a buffer. In the preceding function, `samples_count` is the total number of data samples that can be stored by the buffer, whatever the number of enabled channels. It corresponds to the `length` attribute described in the *IIO buffer sysfs interface* section. `cyclic`, if `true`, enables cyclic mode. This mode makes sense for output devices only (such as DACs). However, in this section, we deal with input devices only (that is, ADCs).

Once you are done with a buffer, you can call `iio_buffer_destroy()` on this buffer, which disables it (thus stopping the capture) and frees the data structure. This API is defined as the following:

```
void  iio_buffer_destroy(struct iio_buffer *buf)
```

Do note that capturing starts as soon as the buffer is created, that is, after `iio_device_create_buffer()` has succeeded. However, samples are only pushed into the kernel buffers. In order to fetch samples from the kernel buffer to the user-space buffer, we need to use `iio_buffer_refill()`. While `iio_device_create_buffer()` has to be called only once to create the buffer and start the in-kernel continuous capture, `iio_buffer_refill()` must be called every time we need to fetch samples from the kernel buffer. It could be used in the processing loop, for example. The following is its definition:

```
ssize_t iio_buffer_refill (struct iio_buffer *buf)
```

With `iio_device_create_buffer()`, with the low-speed interface, the kernel allocates a single underlying buffer block (whose size equals `samples_count * nb_buffers * sample_size`) to handle the captures and immediately starts feeding samples inside. This default block count is 4 by default, and can be changed with `iio_device_set_kernel_buffers_count()`, defined as the following:

```
int iio_device_set_kernel_buffers_count(
                const struct iio_device *dev,
                unsigned int nb_buffers)
```

In high-speed mode, the kernel allocates `nb_buffers` buffer blocks, managed with the FIFO concept of an input queue (empty buffers) and output queue (buffers containing samples) in a way that, upon creation, all the buffers are filled with samples and put in the outgoing queue. When `iio_buffer_refill()` is called, the first buffer's data in the output queue is pushed (or mapped) to user space and this buffer is put back in the input queue waiting to be filled again. At the next call to `iio_buffer_refill()`, the second one is used, and so on, over and over. It must be noted that small buffers result in less latency but more overhead, while large buffers result in less overhead but more latency. The application must make tradeoffs between latency and management overhead. When cyclic mode is `true`, only a single buffer will be created, whatever the number of blocks specified.

In order to read the data samples, the following APIs can be used:

```
void iio_buffer_destroy(struct iio_buffer *buf)
void* iio_buffer_end(const struct iio_buffer *cbuf)
void* iio_buffer_start(const struct iio_buffer *buf)
ptrdiff_t iio_buffer_step(const struct iio_buffer *buf)


void* iio_buffer_first(const struct iio_buffer *buf,
                        const struct iio_channel *chn)
ssize_t iio_buffer_foreach_sample(struct iio_buffer *buf,
        ssize_t(*callback)(const struct iio_channel *chn,
                        void *src, size_t bytes, void *d),
        void *data)
```

The following are the meanings and usages of each API listed:

- `iio_buffer_end()` returns a pointer corresponding to the user-space address that immediately follows the last sample present in the buffer.

- `iio_buffer_start()` returns the address of the user-space buffer. Do, however, note that this address might change after `iio_buffer_refill()` (especially with a high-speed interface, where several buffer blocks are used).

- `iio_buffer_step()` returns the spacing between sample sets in the buffer. That is, it returns the difference between the addresses of two consecutive samples of one same channel.

- `iio_buffer_first()` returns the address of the first sample for a channel or the address of the end of the buffer if no sample for the given channel is present in the buffer.

- `iio_buffer_foreach_sample()` iterates over each sample in a buffer and calls a supplied callback for each sample found.

The preceding list of APIs can be split into three families, depending on how the data samples are read.

Buffer pointer reading

In this read method, `iio_buffer_first()` is coupled with `iio_buffer_step()` and `iio_buffer_end()` in order to iterate on all the samples of a given channel present in the buffer. This can be achieved in the following manner:

```
for (void *ptr = iio_buffer_first(buffer, chan);
            ptr < iio_buffer_end(buffer);
            ptr += iio_buffer_step(buffer)) {
[...]
}
```

In the preceding example, from within the loop, `ptr` will point to one sample of the channel we're interested in, that is, `chan`.

The following is an example:

```
const struct iio_data_format *fmt;
unsigned int i, repeat;
struct iio_channel *channels[8] = {0};
ptrdiff_t p_inc;
```

```
char *p_dat;
[...]


IIOC_DBG("Enter buffer refill loop.\n");
while (true) {
    nbytes = iio_buffer_refill(buf);
    p_inc = iio_buffer_step(buf);
    p_end = iio_buffer_end(buf);
    for (i = 0; i < channel_count; ++i) {
        fmt = iio_channel_get_data_format(channels[i]);
        repeat = fmt->repeat ? : 1;
        for (p_dat = iio_buffer_first(rxbuf, channels[i]);
                    p_dat < p_end; p_dat += p_inc) {
            for (j = 0; j < repeat; ++j) {
                if (fmt->length/8 == sizeof(int16_t))
                    printf("Read 16bit value: " "%" PRIi16,
                            ((int16_t *)p_dat)[j]);
                else if (fmt->length/8 == sizeof(int64_t))
                    printf("Read 64bit value: " "%" PRIi64,
                            ((int64_t *)p_dat)[j]);
            }
        }
    }
    printf("\n");
}
```

The preceding code reads the channel data format to check whether the value is repeated or not. This repeat corresponds to iio_chan_spec.scan_type.repeat. Then, assuming the code could work with two variants of a converter (the first one coding data on 16 bits and the second coding data on 64 bits), a check for the data length is performed to print in the appropriate format. This length corresponds to iio_chan_spec.scan_type.storagebits. Do note that PRIi16 and PRIi64 are the integer printf formats for int16_t and int64_t, respectively.

Callback-based sample reading

In callback-based sample reading, `iio_buffer_foreach_sample()` is at the heart of the reading logic. It has the following definition:

```
ssize_t iio_buffer_foreach_sample(struct iio_buffer *buf,
            ssize_t(*)(const struct iio_channel *chn,
                void *src, size_t bytes, void *d) callback,
            void *data)
```

This function calls the supplied callback for each sample found in a buffer. `data` is user data, which, if set, will be passed to the callback in the last argument. This function iterates over samples, and each sample is read and passed to a callback, along with the channel from where this sample originates. This callback has the following definition:

```
ssize_t sample_cb(const struct iio_channel *chn,
            void *src, size_t bytes, __notused void *d)
```

The callback receives four arguments, as follows:

- A pointer to the `iio_channel` structure that produced the sample
- A pointer to the sample itself
- The length of the sample in bytes, that is, the storage bits divided by 8, `iio_chan_spec.scan_type.storagebits/8`
- The user-specified pointer optionally passed to `iio_buffer_foreach_sample()`

This method may be used to read from (in the case of input devices) or write to (in the case of output devices) the buffer. The main difference from the previous method is that the callback function is invoked for each sample of the buffer, not ordered by channels, but in the order that they appear in the buffer.

The following is an example of this kind of callback implementation:

```
static ssize_t sample_cb(const struct iio_channel *chn,
            void *src, size_t bytes, __notused void *d)
{
    const struct iio_data_format *fmt =
                    iio_channel_get_data_format(chn);
    unsigned int j, repeat = fmt->repeat ? : 1;
    printf("%s ", iio_channel_get_id(chn));
```

```
    for (j = 0; j < repeat; ++j) {
        if (bytes == sizeof(int16_t))
            printf("Read 16bit value: " "%" PRIi16,
                   ((int16_t *)src)[j]);
        else if (bytes == sizeof(int64_t))
            printf("Read 64bit value: " "%" PRIi64,
                   ((int64_t *)src)[j]);
    }
    return bytes * repeat;
}
```

Then, in the main code, we loop and iterate over samples in the buffer, as follows:

```
int ret;
[...]
IIOC_DBG("Enter buffer refill loop.\n");
while (true) {
    nbytes = iio_buffer_refill(buf);
    ret = iio_buffer_foreach_sample(buf, sample_cb, NULL);
    if (ret < 0) {
        char text[256];
        iio_strerror(-ret, buf, sizeof(text));
        printf("%s (%d) while processing buffer\n",
               text, ret);
    }
    printf("\n");
}
```

The preceding code, instead of playing with samples directly, delegates the job to a callback.

High-level channel (raw) reading

The last method in this read series is to use one of the higher-level functions provided by the `iio_channel` class. These are `iio_channel_read_raw()`, `iio_channel_write_raw()`, `iio_channel_read()`, and `iio_channel_write()`, all defined as the following:

```
size_t iio_channel_read_raw(const struct iio_channel *chn,
        struct iio_buffer *buffer, void *dst, size_t len)
size_t iio_channel_read(onst struct iio_channel *chn,
        struct iio_buffer *buffer, void *dst, size_t len)
size_t iio_channel_write_raw(const struct iio_channel *chn,
        struct iio_buffer * buffer, const void *src,
        size_t len)
size_t iio_channel_write(const struct iio_channel *chn,
        struct iio_buffer *buffer, const void *src,
        size_t len)
```

The former two will basically copy the first N samples of a channel (chan) to a user-specified buffer (dst), which must have been allocated beforehand (N depending on the size of this buffer and a sample's storage size, that is, `iio_chan_spec.scan_type.storagebits / 8`). The difference between the two is that the _raw variant won't convert the samples and the user buffer will contain raw data, while the other variant will convert each sample so that the user buffer will contain processed values. These functions kind of demultiplex (since they target one channel's samples among several ones) samples of a given channel.

On the other hand, `iio_channel_write_raw()` and `iio_channel_write()` will copy the sample data from the user-specified buffer to the device, by targeting a given channel. These functions multiplex the samples as they gather samples targeting one channel among many. The difference between the two is that the _raw variant will copy data as is and the other will convert the data into hardware format before sending it to the device.

Let's try to use the preceding APIs to read data from a device:

```
#define CBUF_LENGTH 2048 /* the number of sample we need */
[...]
const struct iio_data_format *fmt;
unsigned int i, repeat;
struct iio_channel *chan[8] = {0};
```

```
[...]
IIOC_DBG("Enter buffer refill loop.\n");
while (true) {
    nbytes = iio_buffer_refill(buf);
    for (i = 0; i < channel_count; ++i) {
        uint8_t *c_buf;
        size_t sample, bytes;
        fmt = iio_channel_get_data_format(chan[i]);
        repeat = fmt->repeat ? : 1;
        size_t sample_size = fmt->length / 8 * repeat;

        c_buf = malloc(sample_size * CBUF_LENGTH);
        if (!c_buf) {
            printf("No memory space for c_buf\n");
            return -1;
        }

        if (buffer_read_method == CHANNEL_READ_RAW)
            bytes = iio_channel_read_raw(chan[i], buf,
                    c_buf, sample_size * CBUF_LENGTH);
        else
            bytes = iio_channel_read(chan[i], buf, c_buf,
                    sample_size * CBUF_LENGTH);

        printf("%s ", iio_channel_get_id(chan[i]));
        for (sample = 0; sample < bytes / sample_size;
                ++sample) {
            for (j = 0; j < repeat; ++j) {
                if (fmt->length / 8 == sizeof(int16_t))
                    printf("%" PRIi16 " ",
                            ((int16_t *)buf)[sample+j]);
                else if (fmt->length / 8 == sizeof(int64_t))
                    printf("%" PRId64 " ",
                            ((int64_t *)buf)[sample+j]);
            }
        }
```

```
        free(c_buf);
    }
    printf("\n");
}
```

In the preceding example, we first fetch data samples from the kernel using `iio_buffer_refill()`. Then, for each channel, we obtain the data format of this channel using `iio_channel_get_data_format()`, from which we grab the size of a sample for this channel. After that, we use this sample's size to compute the user buffer size to allocate for receiving this channel's samples. Obtaining a channel's sample size allows us to precisely determine the size of the user buffer to allocate.

Walking through user-space IIO tools

Though we have already gone through the steps required to capture IIO data, it might be tedious and confusing since each step must be performed manually. There are some useful tools you can use to ease and speed up your app development dealing with IIO devices. These are all from the `libiio` package, developed by Analog Devices, Inc. to interface IIO devices, available here: `https://github.com/analogdevicesinc/libiio`.

User-space applications can easily use the `libiio` library, which under the hood is a wrapper that relies on the following interfaces:

- `/sys/bus/iio/devices`, the IIO sysfs interface, which is mainly used for configuration/settings
- The `/dev/iio/deviceX` character device, for data/acquisitions

The preceding are exactly what we have manually dealt with so far. The tool's source code can be found under the library's `tests` directory: `https://github.com/analogdevicesinc/libiio/tree/master/tests` offers tools such as the following:

- The `iiod` server daemon, acting as a network backend to serve any application over a network link
- `iio_info` to dump attributes
- `iio_readdev` to read or scan from a device

We ended this chapter by enumerating tools, which can ease prototyping or device/driver testing. Links pointing to either sources, documentation, or examples of usage of these tools have been mentioned.

Summary

After reading this chapter, you are familiar with the IIO framework and vocabulary. You know what channels, devices, and triggers are. You can even play with your IIO device from the user space, through sysfs or a character device. The time to write your own IIO driver has come. There are a lot of available existing drivers that don't support trigger buffers. You can try to add this feature to one of them.

In the next chapter, we will play with the GPIO subsystem, which is a basic concept that has been introduced in this chapter as well.

16

Getting the Most Out of the Pin Controller and GPIO Subsystems

System-on-chips (**SoCs**) are becoming more and more complex and feature-rich. These features are mostly exposed through electrical lines originating from the SoC and are called pins. Most of these pins are routed to or multiplexed with several functional blocks (for instance, UART, SPI, RGMI, **General-Purpose Input Output** (**GPIO**), and so on), and the underlying device responsible for configuring these pins and switching between operating modes (switching between functional blocks) is called the **pin controller**.

One mode in which such pins can be configured is **GPIO**. Then comes the Linux GPIO subsystem, which enables drivers to read signals on GPIO configured pins as high or low and to drive the signal high/low on GPIO configured pins. On the other hand, the **pin control** (abbreviated **pinctrl**) subsystem enables multiplexing of some pin/pin groups for different functions, and the capability to configure the electrical properties of pins, such as slew rate, pull up/down resistor, hysteresis, and so on.

To summarize, the pin controller mainly does two things: pin multiplexing, that is, reusing the same pin for different purposes, and pin configuration, that is, configuring electronic properties of pins. Then, the GPIO subsystem allows driving pins, provided these pins are configured to work in GPIO mode by the pin controller.

In this chapter, pin controller and GPIO subsystems will be introduced via the following topics:

- Introduction to some hardware terms
- Introduction to the pin control subsystem
- Dealing with the GPIO controller interface
- Getting the most out of the GPIO consumer interface
- Learning how not to write GPIO client drivers

Introduction to some hardware terms

The Linux kernel GPIO subsystem is not just about GPIO toggling. It is tightly coupled to the pin controller subsystem; they share some terms and concepts that we need to introduce:

- **Pin** and **pad**: A pin is a physical input or output wire/line that transports an electrical signal from or to a component. In schematics, the term "pin" is widely used. Contact pads, on the other hand, are the contact surface areas of a printed circuit board or an integrated circuit. As a result, a pin comes from a pad, and a pin is a pad by default.

- **GPIO**: Most MCUs and CPUs can share one pad among several functional blocks. This is accomplished by multiplexing the input and output signals of the pad. The different modes the pin/pad can operate in are known as **ALT modes** (or alternate modes), and it is common for CPUs to support up to eight settings (or modes) per pad. GPIO is one of these modes. It allows changing the pin direction and reading its value when it is configured as input or setting its value when it is configured as output. Other modes are ADC, UART Tx, UART Rx, SPI MOSI, SPI MISO, PWM, and so on.

- **Pin controller**: This is the underlying device or controller (or rather, a group of registers) allowing you to perform **pin multiplexing** (also referred to as **pinmux** or **pinmuxing**) to reuse the same pin for different purposes. Apart from pinmuxing, it allows pin configuration, that is, configuring electronic properties of pins. The following are some of these properties:

- **Biasing**, that is, setting the initial operating conditions, for example, grounding the pins or connecting them to Vdd. This is not to be confused with pull up and pull down, which is another property.

- **Pin debounce**, which is the time after which a state should be considered valid. This, for example, prevents multiple key pushes on keypads attached to GPIO lines.

- **Slew rate**, which determines how fast the pin toggles between the two logic states. It allows us to control the rise and fall time for the output signals. A trade-off has to be found because rapidly changing states consume more power and generate spikes, thus low slew rates are preferred, except for quick control signals such as parallel interfaces: EIM, EB&, SPI, or SDRAM, which need fast toggling.

- Pull Up/Down resistors

- **GPIO controller**: This is the device allowing you to drive pins when they are put in GPIO mode. It allows changing the GPIO direction and value.

Following the previous definitions, certain general rules have been established for writing pin controllers or GPIO controller drivers, and they are as follows:

- If your GPIO/pin controller can only do simple GPIO, implement just `struct gpio_chip` in `drivers/gpio/gpio-foo.c` and leave it there. Do not use the generic or old-style number-based GPIO.

- Keep your GPIO/pin controller in `drivers/gpio` if it can generate interrupts in addition to GPIO capabilities; simply fill in `struct irq_chip` and register it with the IRQ subsystem.

- Implement composite pin controller drivers in `drivers/pinctrl/pinctrl-foo.c` if this controller supports pinmuxing, advanced pin driver strength, complicated biasing, and so on.

- Maintain the `struct gpio_chip`, `struct irq_chip`, and `struct pinctrl_desc` interfaces.

Now that we are familiar with the terms related to the underlying hardware devices, let's introduce the Linux implementation, starting with the pin control subsystem.

Introduction to the pin control subsystem

The pin controller allows gathering pins, the modes these pins should operate in, and their configurations. The driver is responsible for providing the appropriate set of callbacks according to the features that are to be implemented, provided the underlying hardware supports these features.

The pin controller descriptor data structure is defined as follows:

```
struct pinctrl_desc {
    const char *name;
    const struct pinctrl_pin_desc *pins;
    unsigned int npins;
    const struct pinctrl_ops *pctlops;
    const struct pinmux_ops *pmxops;
    const struct pinconf_ops *confops;
    struct module *owner;
[...]
};
```

In that pin controller data structure, only relevant elements have been listed, and the following are their meanings:

- name is the name of the pin controller.

- pins: An array of pin descriptors that describe all the pins that this controller can handle. It has to be noted that the controller side represents each pin/pad as an instance of struct pinctrl_pin_desc, defined as follows:

```
struct pinctrl_pin_desc {
    unsigned number;
    const char *name;
[...]
};
```

In the preceding data structure, number represents the unique pin number from the global pin number space of the pin controller, and name is the name of this pin.

- npins: The number of descriptors in the pins array, usually obtained using ARRAY_SIZE() in the pins field.

- pctlops stores the pin control operation table, to support global concepts such as the grouping of pins. This is optional.

- pmxops represents the **pinmux** operations table if you support pinmuxing in your driver.

- confops: The pin configuration operations table if you support pin configuration in your driver.

Once the appropriate callbacks are defined and this data structure has been initialized, it can be passed to `devm_pinctrl_register()`, defined as the following:

```
struct pinctrl_dev *devm_pinctrl_register(
                    struct device *dev,
                    struct pinctrl_desc *pctldesc,
                    void *driver_data);
```

The preceding function will register the pin controller with the system, returning in the meantime a pointer to an instance of `struct pinctrl_dev`, representing the pin controller device, passed as a parameter to most (if not all) of the callback operations exposed by the pin controller driver. On error, the function returns an error pointer, which can be handled with `PTR_ERR`.

The controller's control, multiplexing, and configuration operation tables are to be set up according to the features supported by the underlying hardware. Their respective data structures are defined in the header files that must also be included in the driver, as follows:

```
#include <linux/pinctrl/pinconf.h>
#include <linux/pinctrl/pinconf-generic.h>
#include <linux/pinctrl/pinctrl.h>
#include <linux/pinctrl/pinmux.h>
```

When it comes to the pin control consumer interface, the following header must be used instead:

```
#include <linux/pinctrl/consumer.h>
```

Before being accessed by consumer drivers, pins must be assigned to the devices that need to control them. The recommended way to assign pins to devices is from the **device tree (DT)**. How pins groups are assigned in the DT closely depends on the platform, thus the pin controller driver and its binding.

Every pin control state is assigned a contiguous integer ID that starts at 0. A name property list can be used to map strings on top of these IDs to ensure that the same name always points to the same ID. It goes without saying that the set of states that must be defined in each device's DT node is determined by the binding of this device. This binding also determines whether to define the set of state IDs that must be provided, or whether to define the set of state names that must be provided. In any case, two properties can be used to assign a pin configuration node to a device:

- `pinctrl-<ID>`: This allows you to provide the list of pin configurations needed for particular states of the device. It is a list of `phandles` identified by `<ID>`, each of which points to a pin configuration node. These referenced pin configuration nodes must be child nodes of (or nested in) the pin controller node they belong to. This property can accept multiple entries so that multiple groups of pins may be configured and used for a particular device state, allowing in the meantime to specify pins from different pin controllers.

- `pinctrl-names`: This allows giving names to `pinctrl-<ID>` properties according to the state of the device owning the group(s) of pins. List entry 0 defines the name for the state whose ID is 0, list entry 1 for state ID 1, and so on. State ID 0 is commonly given the name `default`. A list of standardized states can be found in `include/linux/pinctrl/pinctrl-state.h`. However, clients or consumer drivers are free to implement any state they need, provided this state is documented in the device binding description.

Here is an excerpt of the DT, showing some device nodes, along with their pin control nodes. Let's name this excerpt `pinctrl-excerpt`:

```
&usdhc4 {
[...]
     pinctrl-0 = <&pinctrl_usdhc4_1>;
     pinctrl-names = "default";
};


gpio-keys {
    compatible = "gpio-keys";
    pinctrl-names = "default";
    pinctrl-0 = <&pinctrl_io_foo &pinctrl_io_bar>;
};


iomuxc@020e0000 { /* Pin controller node */
```

```
    compatible = "fsl,imx6q-iomuxc";
    reg = <0x020e0000 0x4000>;


    /* shared pinctrl settings */
    usdhc4 { /* first node describing the function */
        pinctrl_usdhc4_1: usdhc4grp-1 { /* second node */
            fsl,pins = <
                MX6QDL_PAD_SD4_CMD__SD4_CMD     0x17059
                MX6QDL_PAD_SD4_CLK__SD4_CLK     0x10059
                MX6QDL_PAD_SD4_DAT0__SD4_DATA0 0x17059
                MX6QDL_PAD_SD4_DAT1__SD4_DATA1 0x17059
                MX6QDL_PAD_SD4_DAT2__SD4_DATA2 0x17059
                MX6QDL_PAD_SD4_DAT3__SD4_DATA3 0x17059
                [...]
            >;
        };
    };
    [...]
    uart3 {
        pinctrl_uart3_1: uart3grp-1 {
            fsl,pins = <
                MX6QDL_PAD_EIM_D24__UART3_TX_DATA 0x1b0b1
                MX6QDL_PAD_EIM_D25__UART3_RX_DATA 0x1b0b1
            >;
        };
    };


    // GPIOs (Inputs)
    gpios {
        pinctrl_io_foo: pinctrl_io_foo {
            fsl,pins = <
                MX6QDL_PAD_DISP0_DAT15__GPIO5_IO09  0x1f059
                MX6QDL_PAD_DISP0_DAT13__GPIO5_IO07  0x1f059
            >;
        };
        pinctrl_io_bar: pinctrl_io_bar {
```

```
        fsl,pins = <
            MX6QDL_PAD_DISP0_DAT11__GPIO5_IO05    0x1f059
            MX6QDL_PAD_DISP0_DAT9__GPIO4_IO30     0x1f059
            MX6QDL_PAD_DISP0_DAT7__GPIO4_IO28     0x1f059
        >;
    };
  };
};
```

In the preceding example, a pin configuration is given in the form <PIN_FUNCTION> <PIN_SETTING>, where <PIN_FUNCTION> can be seen as the pin function or pin mode, and <PIN_SETTING> represents the pin's electrical properties:

```
MX6QDL_PAD_DISP0_DAT15__GPIO5_IO09 0x80000000
```

In the excerpt, MX6QDL_PAD_DISP0_DAT15__GPIO5_IO09 represents the pin function/mode, which is GPIO in this case, and 0x80000000 represents the pin settings or electrical properties.

Let's consider another example as follows:

```
MX6QDL_PAD_EIM_D25__UART3_RX_DATA 0x1b0b1
```

In that excerpt, MX6QDL_PAD_EIM_D25__UART3_RX_DATA represents the pin function, which is the RX line of UART3, and 0x1b0b1 represents its electrical settings.

The pin function is a macro whose value is meaningful for the pin controller driver only. These are generally defined in header files located in arch/<arch>/boot/dts/. If you use an UDOO quad, for example, which has an i.MX6 quad core (32-bit ARM), the pin function header would be arch/arm/boot/dts/imx6q-pinfunc.h. The following is the macro corresponding to the fifth line of the GPIO5 controller:

```
#define MX6QDL_PAD_DISP0_DAT11__GPIO5_IO05 0x19c 0x4b0 0x000
0x5 0x0
```

<PIN_SETTING> can be used to set up things such as pull-ups, pull-downs, keepers, drive strength, and so on. How it should be specified depends on the pin controller binding, and the meaning of its value depends on the SoC datasheet, generally in the IOMUX section. On i.MX6 IOMUXC, only the lower 17 bits are used for this purpose.

Back to `pinctrl-excerpt`, prior to selecting a pin group and applying its configuration, the driver must first obtain a handle to this group of pins using the `devm_inctrl_get()` function and then select the appropriate state using `pinctrl_lookup_state()` before finally applying the corresponding configuration state to hardware thanks to `pinctrl_select_state()`.

The following is an example that shows how to grab a pin control group and apply its default configuration:

```
#include <linux/pinctrl/consumer.h>

int ret;
struct pinctrl_state *s;
struct pinctrl *p;
foo_probe()
{
    p = devm_pinctrl_get(dev);
    if (IS_ERR(p))
        return PTR_ERR(p);

    s = pinctrl_lookup_state(p, name);
    if (IS_ERR(s))
        return PTR_ERR(s);

    ret = pinctrl_select_state(p, s);
    if (ret < 0) // on error
        return ret;
[...]
}
```

Like other resources (such as memory regions, clocks, and so on), it is a good practice to grab pins and apply their configuration from within the `probe()` function. However, this operation is so common that it has been integrated into the Linux device core as a step while probing devices. Thus, when a device is being probed, the device core will do the following:

- Grab the pins assigned to the device that is just about to probe using `devm_pinctrl_get()`.

- Look for the default pin state (`PINCTRL_STATE_DEFAULT`) using `pinctrl_lookup_state()`.

- Look in the meantime for an init (which means during the device initialization) pin state (`PINCTRL_STATE_INIT`) using the same API.

- Apply the init pin state if any, otherwise apply the default pin state.

- If power management is enabled, look for the optional sleep (`PINCTRL_STATE_SLEEP`) and idle (`PINCTRL_STATE_IDLE`) pin states for later use, during power management related operations.

See the `pinctrl_bind_pins()` function (defined in `drivers/base/pinctrl.c`), and the `really_probe()` function (defined in `drivers/base/dd.c`), which calls the former. These functions will help you understand how pins are bound to the device on its probing path.

> **Note**
>
> `pinctrl_select_state()` internally calls `pinmux_enable_setting()`, which in turn calls `pin_request()` on each pin in the pin control (group of pins) node.

The `pinctrl_put()` function can be used to release a pin control that has been requested using the non-managed API, that is, `pinctrl_get()`. That said, you can use `devm_pinctrl_get_select()`, given the name of the state to select, in order to configure pinmux in a single shot. This function is defined in `include/linux/pinctrl/consumer.h` as follows:

```
static struct pinctrl *devm_pinctrl_get_select(
                    struct device *dev, const char *name)
```

In the previous prototype, name is the name of the state as written in the `pinctrl-name` property. If the name of the state is `default`, the helper `devm_pinctr_get_select_default()` can be used, which is a wrapper around `devm_pinctl_get_select()` as follows:

```
static struct pinctrl * pinctrl_get_select_default(
                                    struct device *dev)
{
    return pinctrl_get_select(dev, PINCTRL_STATE_DEFAULT);
}
```

Now that we are familiar with the pin control subsystem (with both controller and consumer interfaces), we can learn how to deal with GPIO controllers, knowing that GPIO is an operating mode that a pin can work in.

Dealing with the GPIO controller interface

The GPIO controller interface is designed around a single data structure, `struct gpio_chip`. This data structure provides a set of functions, among which are methods to establish GPIO direction (input and output), methods used to access GPIO values (get and set), methods to map a given GPIO to IRQ and return the associated Linux interrupt number, and the `debugfs` dump method (showing extra state like pull-up config). Apart from these functions, that data structure provides a flag to determine the nature of the controller, that is, to allow checking whether this controller's accessors may sleep or not. Still from within this data structure, the driver can set the GPIO base number, from which GPIO numbering should start.

Back to the code, a GPIO controller is represented as an instance of `struct gpio_chip`, defined in `<linux/gpio/driver.h>` as follows:

```
struct gpio_chip {
    const char      *label;
    struct gpio_device    *gpiodev;
    struct device         *parent;
    struct module         *owner;

    int       (*request)(struct gpio_chip *gc,
                         unsigned int offset);
    void      (*free)(struct gpio_chip *gc,
                         unsigned int offset);
    int       (*get_direction)(struct gpio_chip *gc,
                         unsigned int offset);
    int       (*direction_input)(struct gpio_chip *gc,
                         unsigned int offset);
    int       (*direction_output)(struct gpio_chip *gc,
                       unsigned int offset, int value);
    int       (*get)(struct gpio_chip *gc,
                         unsigned int offset);
    int       (*get_multiple)(struct gpio_chip *gc,
                         unsigned long *mask,
                         unsigned long *bits);
    void      (*set)(struct gpio_chip *gc,
                       unsigned int offset, int value);
    void      (*set_multiple)(struct gpio_chip *gc,
```

```
                        unsigned long *mask,
                        unsigned long *bits);
    int        (*set_config)(struct gpio_chip *gc,
                        unsigned int offset,
                        unsigned long config);
    int        (*to_irq)(struct gpio_chip *gc,
                        unsigned int offset);

    int        (*init_valid_mask)(struct gpio_chip *gc,
                            unsigned long *valid_mask,
                            unsigned int ngpios);

    int        (*add_pin_ranges)(struct gpio_chip *gc);

    int        base;
    u16        ngpio;
    const char *const *names;
    bool       can_sleep;

#if IS_ENABLED(CONFIG_GPIO_GENERIC)
    unsigned long (*read_reg)(void __iomem *reg);
    void (*write_reg)(void __iomem *reg, unsigned long data);
    bool be_bits;
    void __iomem *reg_dat;
    void __iomem *reg_set;
    void __iomem *reg_clr;
    void __iomem *reg_dir_out;
    void __iomem *reg_dir_in;
    bool bgpio_dir_unreadable;
    int bgpio_bits;
    spinlock_t bgpio_lock;
    unsigned long bgpio_data;
    unsigned long bgpio_dir;
#endif /* CONFIG_GPIO_GENERIC */


#ifdef CONFIG_GPIOLIB_IRQCHIP
```

```
        struct gpio_irq_chip irq;
#endif /* CONFIG_GPIOLIB_IRQCHIP */

        unsigned long *valid_mask;

#if defined(CONFIG_OF_GPIO)
        struct device_node *of_node;
        unsigned int of_gpio_n_cells;
        int (*of_xlate)(struct gpio_chip *gc,
                    const struct of_phandle_args *gpiospec,
                    u32 *flags);
#endif /* CONFIG_OF_GPIO */
};
```

The following are the meanings of each element in the structure:

- label: This is the GPIO controller's functional name. It could be a part number or the name of the SoC IP block implementing it.

- gpiodev: This is the internal state container of the GPIO controller. It is also the structure through which the character device associated with this GPIO controller will be exposed.

- request is an optional hook for chip-specific activation. If provided, it is executed prior to allocating GPIO whenever you call gpio_request() or gpiod_get().

- free is an optional hook for chip-specific deactivation. If provided, it is executed before the GPIO is deallocated whenever you call gpiod_put() or gpio_free().

- get_direction is executed whenever you need to know the direction of the GPIO offset. The return value should be 0 to mean out, and 1 to mean in (the same as GPIOF_DIR_XXX) or a negative error.

- direction_input configures the signal offset as input, or returns an error.

- get returns the value of the GPIO offset; for output signals, this returns either the value actually sensed or zero.

- set assigns an output value to the GPIO offset.

- `set_multiple` is called when you need to assign output values for multiple signals defined by `mask`. If not provided, the kernel will install a generic hook that will walk through mask bits and execute `chip->set(i)` on each bit set. See here how you can implement this function:

```
static void gpio_chip_set_multiple(
                    struct gpio_chip *chip,
                    unsigned long *mask,
                    unsigned long *bits)
{
    if (chip->set_multiple) {
        chip->set_multiple(chip, mask, bits);
    } else {
        unsigned int i;
        /*
         * set outputs if the corresponding
         * mask bit is set
         */
        for_each_set_bit(i, mask, chip->ngpio)
            chip->set(chip, i, test_bit(i, bits));
    }
}
```

- `set_debounce` if supported by the controller, this hook is an optional callback provided to set the debounce time for the specified GPIO.

- `to_irq` is an optional hook to provide GPIO to IRQ mapping. This is called whenever you want to execute the `gpio_to_irq()` or `gpiod_to_irq()` function. This implementation may not sleep.

- `base` indicates the first GPIO number handled by this chip; or, if negative during registration, the kernel will automatically (dynamically) assign one.

- `ngpio` is the number of GPIOs this controller provides, starting from base to (`base + ngpio - 1`).

- `names`, if not NULL, must be a list (an array) of strings to use as alternative names for the GPIOs in this chip. The array must be `ngpio` sized, and any GPIO that does not need an alias may have its entry set to NULL in the array.

- `can_sleep` is a Boolean flag to be set if `get()`/`set()` methods may sleep. It is the case for GPIO controllers (also known as GPIO expanders) sitting on buses such as I2C or SPI, whose accesses may lead to sleep. This means that if the chip supports IRQs, these IRQs must be threaded because the chip access may sleep when, for example, reading out the IRQ status registers. For a GPIO controller mapped to memory (part of SoC), this can be set to `false`.

- Elements that are conditioned by the enabling of `CONFIG_GPIO_GENERIC` are related to a generic memory-mapped GPIO controller, with a standard register set.

- `irq`: The IRQ chip of this GPIO controller if the controller can map GPIOs to IRQs. In such cases, this field must be set up before the GPIO chip is registered.

- `valid_mask`: If not `NULL`, this element contains a bitmask of GPIOs that are valid to be used from the chip.

- `of_node` is a pointer to the device tree node representing this GPIO controller.

- `of_gpio_n_cells` is the number of cells used to form the GPIO specifier.

- `of_xlate` is a callback to translate a device tree GPIO specifier into a chip relative GPIO number and flags. This hook is invoked when a GPIO line from this controller is specified in the device tree and must be parsed. If not provided, the GPIO core will set it to `of_gpio_simple_xlate()` by default, which is a generic GPIO core helper that supports two cell specifiers. The first cell identifies the GPIO number, and the second one, the GPIO flags. Additionally, when setting the default callback, `of_gpio_n_cells` will be set to 2. If the GPIO controller needs one or more than two cell specifiers, you'll have to implement the corresponding translation callback.

Each GPIO controller exposes a number of signals that are identified in function calls by offset values in the range of 0 to (`ngpio` - 1). When those GPIO lines are referenced through calls such as `gpio_get_value(gpio)`, the offset is determined by subtracting `base` from the GPIO number and passed to the underlying driver function (`gpio_chip->get()` for example). The controller driver should then have the logic to map this offset to the control/status register associated with the GPIO.

Writing a GPIO controller driver

A GPIO controller needs nothing but the set of callbacks that correspond to the features it supports. After every callback of interest has been defined and other fields set, the driver should call devm_gpiochip_add_data() on the configured struct gpio_chip structure in order to register the controller with the kernel. You have probably guessed that you'd better use the managed API (the devm_ prefix) since it takes care of chip removal when necessary and releasing resources. If, however, you used the classical method, you will have to use gpiochip_remove() to remove the chip if necessary:

```
int gpiochip_add_data(struct gpio_chip *gc, void *data)
int devm_gpiochip_add_data(struct device *dev,
                        struct gpio_chip *gc, void *data)
```

In the preceding prototypes, gc is the chip to register and data is the driver's private data associated with this chip. They both return a negative error code if the chip can't be registered, such as because gc->base is invalid or already associated with a different chip. Otherwise, they return zero as a success code.

There are, however, pin controllers that are tightly coupled to the GPIO chip, and both are implemented in the same driver, much of which being in drivers/pinctrl/pinctrl-*.c. In such drivers, when gpiochip_add_data() is invoked, for device-tree-supported systems, the GPIO core will check the pin control's device node for the gpio-ranges property. If it is present, it will take care of adding the pin ranges for the driver.

However, the driver must call gpiochip_add_pin_range() in order to be compatible with older, existing device tree files that don't set the gpio-ranges attribute or systems that use ACPI. The following is an example:

```
if (!of_find_property(np, "gpio-ranges", NULL)) {
    ret = gpiochip_add_pin_range(chip,
                    dev_name(hw->dev), 0, 0, chip->ngpio);
    if (ret < 0) {
            gpiochip_remove(chip);
            return ret;
    }
}
```

Once more, it must be noted that the preceding is used just for backward compatibility for these old `pinctrl` nodes without the `gpio-ranges` property. Otherwise, calling `gpiochip_add_pin_range()` directly from a device tree-supported pin controller driver is deprecated. Please see *Section 2.1* of `Documentation/devicetree/bindings/gpio/gpio.txt` on how to bind pin controller and GPIO drivers via the `gpio-ranges` property.

We can see how easy it is to write a GPIO controller driver. In the book sources repository, you will find a working GPIO controller driver, for the MCP23016 I2C I/O expander from microchip, whose datasheet is available at `http://ww1.microchip.com/downloads/en/DeviceDoc/20090C.pdf`.

To write such drivers, the following header should be included:

```
#include <linux/gpio.h>
```

Following is an excerpt from the driver we have written for our controller:

```
#define GPIO_NUM 16
struct mcp23016 {
    struct i2c_client *client;
    struct gpio_chip gpiochip;
    struct mutex lock;
};


static int mcp23016_probe(struct i2c_client *client)
{
  struct mcp23016 *mcp;

  if (!i2c_check_functionality(client->adapter,
     I2C_FUNC_SMBUS_BYTE_DATA))
    return -EIO;

  mcp = devm_kzalloc(&client->dev, sizeof(*mcp),
                     GFP_KERNEL);
  if (!mcp)
    return -ENOMEM;

  mcp->gpiochip.label = client->name;
  mcp->gpiochip.base = -1;
```

```
    mcp->gpiochip.dev = &client->dev;
    mcp->gpiochip.owner = THIS_MODULE;
    mcp->gpiochip.ngpio = GPIO_NUM; /* 16 */
    /* may not be accessed from atomic context */
    mcp->gpiochip.can_sleep = 1;
    mcp->gpiochip.get = mcp23016_get_value;
    mcp->gpiochip.set = mcp23016_set_value;
    mcp->gpiochip.direction_output =
                            mcp23016_direction_output;
    mcp->gpiochip.direction_input =
                            mcp23016_direction_input;
    mcp->client = client;
    i2c_set_clientdata(client, mcp);

    return devm_gpiochip_add_data(&client->dev,
                                &mcp->gpiochip, mcp);
}
```

In the preceding excerpt, the GPIO chip data structure has been set up before being passed to devm_gpiochip_get_data(), which is called to register the GPIO controller with the system. As a result, a GPIO character device node will appear under /dev.

IRQ chip enabled GPIO controllers

IRQ chip support can be enabled in a GPIO controller by setting up the struct gpio_irq_chip structure embedded into this GPIO controller data structure. This struct gpio_irq_chip structure is used to group all fields related to interrupt handling in a GPIO chip and is defined as follows:

```
struct gpio_irq_chip {
    struct irq_chip *chip;
    struct irq_domain *domain;
    const struct irq_domain_ops *domain_ops;

    irq_flow_handler_t handler;
    unsigned int default_type;

    irq_flow_handler_t parent_handler;
```

```
    union {
        void *parent_handler_data;
        void **parent_handler_data_array;
    };

    unsigned int num_parents;
    unsigned int *parents;
    unsigned int *map;

    bool threaded;
    bool per_parent_data;

    int (*init_hw)(struct gpio_chip *gc);

    void (*init_valid_mask)(struct gpio_chip *gc,
                    unsigned long *valid_mask,
                    unsigned int ngpios);

    unsigned long *valid_mask;
    unsigned int first;

    void        (*irq_enable)(struct irq_data *data);
    void        (*irq_disable)(struct irq_data *data);
    void        (*irq_unmask)(struct irq_data *data);
    void        (*irq_mask)(struct irq_data *data);
};
```

There are architectures that may be multiple interrupt controllers involved in delivering an interrupt from the device to the target CPU. This feature is enabled in the kernel by setting the CONFIG_IRQ_DOMAIN_HIERARCHY config option.

In the previous data structure, some elements have been omitted. These are elements conditioned by CONFIG_IRQ_DOMAIN_HIERARCHY, that is, IRQ domain hierarchy related fields, which we won't discuss in this chapter. For the remaining elements, the following are their definitions:

- chip is the IRQ chip implementation.

- domain is the IRQ interrupt translation domain associated with chip; it is responsible for mapping between the GPIO hardware IRQ number and Linux IRQ number.

- domain_ops represents the set of interrupt domain operations associated with the IRQ domain.

- handler is the high-level interrupt flow handler (typically a predefined IRQ core function) for GPIO IRQs. There is a note on this field later in the section.

- default_type is the default IRQ triggering type applied during GPIO driver initialization.

- parent_handler is the interrupt handler for the GPIO chip's parent interrupts. It may be NULL if the parent interrupts are nested rather than chained. Moreover, setting this element to NULL will allow handling the parent IRQ in the driver. gpio_chip.can_sleep cannot be set to true if this handler is supplied because you cannot have chained interrupts on a chip that may sleep.

- parent_handler_data and parent_handler_data_array are data associated with, and passed to, the handler for the parent interrupt. This can either be a single pointer if per_parent_data is false, or an array of num_parents pointers otherwise. If per_parent_data is true, parent_handler_data_array cannot be NULL.

- num_parents is the number of interrupt parents for the GPIO chip.

- parents is a list of interrupt parents of the GPIO chip. Because the driver owns this list, the core will only refer to it, not edit it.

- map is a list of interrupt parents for each line of the GPIO chip.

- threaded indicates whether the interrupt handling is threaded (uses nested threads).

- per_parent_data tells whether parent_handler_data_array describes a num_parents sized array to be used as parent data.

- init_hw is an optional routine for initializing hardware before an IRQ chip will be added. This is extremely beneficial when a driver has to clear IRQ-related registers in order to avoid unwanted events.

- `init_valid_mask` is an optional callback that can be used to initialize `valid_mask`, which is used if not all GPIO lines are valid interrupts. There might be lines that just cannot fire interrupts, and this callback, when defined, is passed a bitmap in `valid_mask`, which will have `ngpios` bits from `0..(ngpios-1)` set to `1` as valid. The callback can then directly set some bits to `0` if they cannot be used for interrupts.

- `valid_mask`, if not `NULL`, contains a bitmask of GPIOs that are valid for inclusion in the IRQ domain of the chip.

- `first` is necessary in the case of static IRQ allocation. If set, `irq_domain_add_simple()` will allocate (starting from this value) and map all IRQs during initialization.

- `irq_enable`, `irq_disable`, `irq_unmask`, and `irq_mask` respectively store old `irq_chip.irq_enable`, `irq_chip.irq_disable`, `irq_chip.irq_unmask`, and `irq_chip.irq_mask` callbacks. See *Chapter 13, Demystifying the Kernel IRQ Framework*, for a detailed explanation.

Under the premise that your interrupts are 1-to-1 mapped to the GPIO line index, `gpiolib` will handle a significant portion of overhead code. In this 1-to-1 mapping, GPIO line offset 0 maps to hardware IRQ 0, GPIO line offset 1 maps to hardware IRQ 1, and so on until GPIO line offset `ngpio-1`, which maps to hardware IRQ `ngpio-1`. The bitmask `valid_mask` and the flag `need_valid_mask` in `gpio_irq_chip` can be used to mask off some lines as invalid for associating with IRQs, provided some GPIO lines do not have corresponding IRQs.

We can divide GPIO IRQ chips into two broad categories:

- **Cascaded interrupt chips**: This indicates that the GPIO chip has a single common interrupt output line that is triggered by any enabled GPIO line on that chip. The interrupt output line is subsequently routed to a parent interrupt controller one level up, which in the simplest case is the system's primary interrupt controller. An IRQ chip implements this by inspecting bits inside the GPIO controller to determine which line fired it. To figure this out, the IRQ chip part of the driver will need to inspect registers, and it will almost certainly need to acknowledge that it is handling the interrupt by clearing some bits (sometimes implicitly, by simply reading a status register), as well as setting up configurations such as edge sensitivity (rising or falling edge or a high/low level interrupt, for example).

- **Hierarchical interrupt chips**: This means that each GPIO line is connected to a parent interrupt controller one level up by a dedicated IRQ line. It is not necessary to query the GPIO hardware to determine which line has fired, but you might need to acknowledge the interrupt and configure edge sensitivity.

Cascaded GPIO IRQ chips usually fall in one of three categories:

- **Chained cascaded GPIO IRQ chips**: These are the types that are usually seen on SoCs. This means that the GPIOs have a fast IRQ flow handler that is called in a chain from the parent IRQ handler, which is usually the system interrupt controller. This means that the parent IRQ chip will immediately call the GPIO IRQ chip handler while holding the IRQs disabled. In its interrupt handler, the GPIO IRQ chip will then call something similar to this:

```
static irqreturn_t foo_gpio_irq(int irq, void *data)
    chained_irq_enter(...);
    generic_handle_irq(...);
    chained_irq_exit(...);
```

Because everything happens directly in the callback, chained GPIO IRQ chips cannot set the .can_sleep flag on struct gpio_chip to true. In this case, no slow bus traffic like I2C can be used.

- **Generic chained GPIO IRQ chips**: These are the same as CHAINED GPIO IRQCHIPS, but they don't use chained IRQ handlers. GPIO IRQs are instead dispatched via a generic IRQ handler, which is specified using request_irq(). In this IRQ handler, the GPIO IRQ chip will then end up calling something similar to the following sequence:

```
static irqreturn_t gpio_rcar_irq_handler(int irq,
                                        void *dev_id)
    /* go through the entire GPIOs and handle
     * all interrupts
     */
    for each detected GPIO IRQ
        generic_handle_irq(...);
```

- **Nested thread GPIO IRQ chips**: Off-chip GPIO expanders and any other GPIO IRQ chip sitting on a sleeping bus, such as I2C or SPI, fall under this category.

Of course, such drivers who require sluggish bus traffic to read out IRQ status
and other information, traffic which may result in further IRQs, cannot be
accommodated in a rapid IRQ handler with IRQs disabled. Instead, they must
create a thread and then mask the parent IRQ line until the interrupt is handled
by the driver. This driver's distinguishing feature is that it calls something like the
following in its interrupt handler:

```
static irqreturn_t pcf857x_irq(int irq,
                                void *data)
{
    struct pcf857x *gpio = data;
    unsigned long change, i, status;

    status = gpio->read(gpio->client);
    mutex_lock(&gpio->lock);
    change = (gpio->status ^ status) &
            gpio->irq_enabled;
    gpio->status = status;
    mutex_unlock(&gpio->lock);

    for_each_set_bit(i, &change, gpio->chip.ngpio)
     child_irq = irq_find_mapping(
                    gpio->chip.irq.domain, i);
        handle_nested_irq(child_irq);

    return IRQ_HANDLED;
}
```

Threaded GPIO IRQ chips are distinguished by the fact that they set the .can_
sleep flag on struct gpio_chip to true, indicating that the chip can sleep
when accessing the GPIOs.

> **Note**
>
> It is worth recalling that `gpio_irq_chip.handler` is the interrupt flow handler. It is the high-level IRQ-events handler, the one that calls the underlying handlers registered by client drivers using `request_irq()` or `request_threaded_irq()`. Its value depends on the IRQ being edge- or level-triggered. It is most often a predefined IRQ core function, one between `handle_simple_irq`, `handle_edge_irq`, and `handle_level_irq`. These are all kernel helper functions that do some operations before and after calling the real IRQ handler.
>
> When the parent IRQ handler calls `generic_handle_irq()` or `handle_nested_irq()`, the IRQ core will look for the IRQ descriptor structure (Linux's view of an interrupt) corresponding to the Linux IRQ number passed as an argument (`struct irq_desc *desc = irq_to_desc(irq)`) and calling `generic_handle_irq_desc()` on this descriptor, which will result in `desc->handle_irq(desc)`. You should note that `desc->handle_irq` corresponds to the high-level IRQ handler supplied earlier, which has been assigned to the IRQ descriptor using `irq_set_chip_and_handler()` during the mapping of this IRQ. Guess what, the mapping of these GPIO IRQs is done in `gpiochip_irq_map`, which is the `.map` callback of the default IRQ domain operation table (`gpiochip_domain_ops`) assigned by the GPIO core to the GPIO IRQ chip if not provided by the driver.
>
> To summarize, `desc->handle_irq = gpio_irq_chip.handler`, which may be `handle_level_irq`, `handle_simple_irq`, `handle_edge_irq`, or (rarely) a driver-provided function.

Example of adding IRQ chip support in a GPIO chip

In this section, we will demonstrate how to add the support of an IRQ chip into a GPIO controller driver. To do that, we will update our initial driver, more precisely, the probe method, as well as implementing an interrupt handler, which will hold the IRQ handling logic.

Let's consider the following figure:

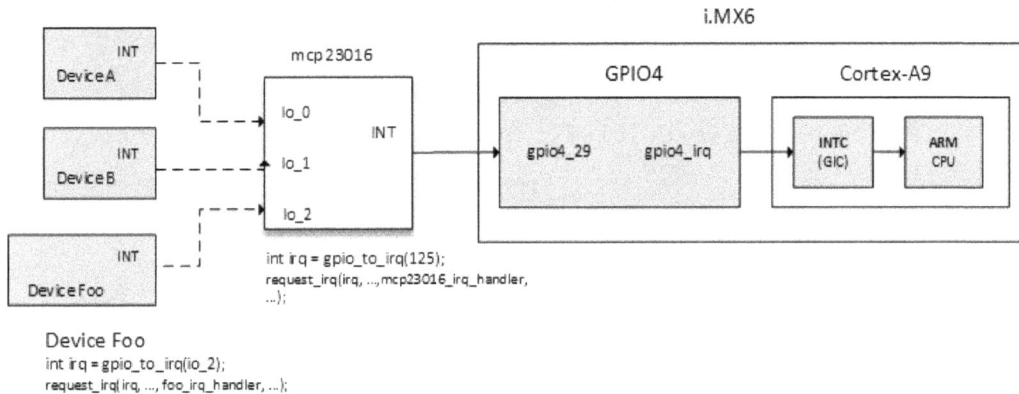

Figure 16.1 – Multiplexing IRQs

In the previous figure, let's consider that we have configured io_0 and io_1 as interrupt lines (this is what DeviceA and deviceB see).

Whether an interrupt happens on io_0 or io_1, the same parent interrupt line will be raised on the GPIO chip. At this step, the GPIO chip driver must figure out, by reading the GPIO status register of the GPIO controller, which GPIO line (io_0 or io_1) has really fired the interrupt. This is how, in the case of the MCP23016 chip, a single interrupt line (the parent actually) can be a multiplex for 16 GPIO interrupts.

Now let's update the initial GPIO controller driver. It must be noted that because the device sits on a slow bus, we have no choice but to implement nested (threaded) interrupt flow handling.

We start by defining our IRQ chip data structure, with a set of callbacks that can be used by the IRQ core. The following is an excerpt:

```
static struct irq_chip mcp23016_irq_chip = {
        .name = "gpio-mcp23016",
        .irq_mask = mcp23016_irq_mask,
        .irq_unmask = mcp23016_irq_unmask,
        .irq_set_type = mcp23016_irq_set_type,
};
```

In the preceding, the callbacks that have been defined depend on the need. In our case, we have only implemented interrupt (un)masking related callbacks, as well as the callback allowing you to set the IRQ type. To see the full description of a struct irq_chip structure, you can refer to *Chapter 13, Demystifying the Kernel IRQ Framework*.

Now that the IRQ chip data structure has been set up, we can modify the probe method as follows:

```
static int mcp23016_probe(struct i2c_client *client)
{
    struct gpio_irq_chip *girq;
    struct irq_chip *irqc;
[...]


    girq = &mcp->gpiochip.irq;
    girq->chip = &mcp23016_irq_chip;
    /* This will let us handling the parent IRQ in the driver
*/
    girq->parent_handler = NULL;
    girq->num_parents = 0;
    girq->parents = NULL;
    girq->default_type = IRQ_TYPE_NONE;
    girq->handler = handle_level_irq;
    girq->threaded = true;
[...]
    /*
     * Directly request the irq here instead of passing
     * a flow-handler.
     */
    err = devm_request_threaded_irq(
                    &client->dev,
                    client->irq,
                    NULL, mcp23016_irq,
                    IRQF_TRIGGER_RISING | IRQF_ONESHOT,
                    dev_name(&i2c->dev), mcp);
[...]


    return devm_gpiochip_add_data(&client->dev,
                            &mcp->gpiochip, mcp);
}
```

In the previous probe method update, we first initialized the `struct gpio_irq_chip` data structure embedded into `struct gpio_chip`, and then we registered an IRQ handler, which will act as the parent IRQ handler, responsible for enquiring the underlying GPIO chip for any IRQ-enabled GPIOs that have changed, and then running their IRQ handlers, if any.

Finally, the following is our IRQ handler, which must have been implemented before the `probe` function:

```
static irqreturn_t mcp23016_irq(int irq, void *data)
{
    struct mcp23016 *mcp = data;
    unsigned long status, changes, child_irq, i;
    status = read_gpio_status(mcp);

    mutex_lock(&mcp->lock);
    change = mcp->status ^ status;
    mcp->status = status;
    mutex_unlock(&mcp->lock);

    /* Do some stuff, may be adapting "change" according to
level */
    [...]
    for_each_set_bit(i, &change, mcp->gpiochip.ngpio) {
        child_irq =
            irq_find_mapping(mcp->gpiochip.irq.domain, i);
        handle_nested_irq(child_irq);
    }

    return IRQ_HANDLED;
}
```

In our interrupt handler, we simply read the current GPIO, and we compare it with the old status to determine GPIOs that have changes. All the tricks are handled by `handle_nested_irq()`, which is explained in *Chapter 13, Demystifying the Kernel IRQ Framework*, as well.

Now that we are done and are familiar with the implementation of IRQ chips in GPIO controller drivers, we can learn about the binding of these GPIO controllers, which will allow declaring the GPIO chip hardware in the device tree in a way the driver understands.

GPIO controller bindings

The device tree is the de facto standard to declare and describe devices on embedded platforms, especially on ARM architectures. It this then recommended for new drivers to provide the associated device bindings.

Back to GPIO controllers, there are mandatory properties that need to be provided:

- `compatible`: This is the list of strings to match the driver(s) that will handle this device.

- `gpio-controller`: This is a property that indicates to the device tree core that this node represents a GPIO controller.

- `gpio-cells`: Tells how many cells are used to describe a GPIO specifier. It should correspond to `gpio_chip.of_gpio_n_cells`, or to a value that `gpio_chip.of_xlate` can deal with. It is typically `<2>` for a less complex controller, with the first cell identifying the GPIO number, and the second defining the flags.

There are additional mandatory properties to define if the GPIO controller has IRQ chip support, that is, if this controller allows mapping its GPIO lines to IRQs. With such GPIO controllers, the following mandatory properties must be provided:

- `interrupt-controller`: Some controllers provide IRQs mapped to the GPIOs. In that case, the property `#interrupt-cells` should be set too and usually you use 2, but it depends on your needs. The first cell is the pin number, and the second represents the interrupt flags.

- `#interrupt-cells`: This must be defined as a value supported by the `xlate` hook of the IRQ domain, that is, `irq_domain_ops.xlate`. It is common for this hook to be set with `irq_domain_xlate_twocell`, which is a generic kernel IRQ core helper able to handle a two-cell specifier.

From the properties listed, we can declare our GPIO controller under its bus node, as follows:

```
&i2c1
    expander: mcp23016@20 {
        compatible = "microchip,mcp23016";
        reg = <0x20>;
        gpio-controller;
        #gpio-cells = <2>;
        interrupt-controller;
        #interrupt-cells = <2>;
```

```
                interrupt-parent = <&gpio4>;
                interrupts = <29 IRQ_TYPE_EDGE_FALLING>;
        };
};
```

This is all for the controller side. In order to demonstrate how clients can consume resources provided by the MCP23016, let's consider the following scenario: we have two devices, device A, named `foo`, and device B, named `bar`. The `bar` device consumes two GPIO lines from our controller (they will be used in output mode), and the `foo` device would like to map a GPIO to IRQ. This configuration could be declared in the device tree as follows:

```
parent_node {
    compatible = "simple-bus";

    foo_device: foo_device@1c {
        [...]
        reg = <0x1c>;
        interrupt-parent = <&expander>;
        interrupts = <2 IRQ_TYPE_EDGE_RISING>;
    };

    bar_device {
        [...]
        reset-gpios = <&expander 8 GPIO_ACTIVE_HIGH>;
        power-gpios = <&expander 12 GPIO_ACTIVE_HIGH>;
        [...]
    };
};
```

In the preceding excerpt, `parent_node` is given `simple-bus` as a `compatible` string in order to instruct the device tree core and the platform core to instantiate two platform devices, which correspond to our nodes. In that excerpt, we have also demonstrated how GPIOs, as well as IRQs from our controller, are specified. The number of cells used for each property matches the declaration in the controller binding.

GPIO- and pin-controller interaction

These two subsystems are closely related. A pin controller can route some or all of the GPIOs provided by a GPIO controller to pins on the package. This allows those pins to be muxed (also known as pinmuxing) between GPIO and other functions.

It may then be useful to know which GPIOs correspond to which pins on which pin controllers. The gpio-ranges property, which will be described below, represents this correspondence with a discrete set of ranges that map pins from the pin controller local number space to pins in the GPIO controller local number space.

The gpio-ranges format is the following:

```
<[pin controller phandle], [GPIO controller offset], [pin
controller offset], [number of pins]>;
```

The GPIO controller offset refers to the GPIO controller node containing the range definition. The bindings defined in Documentation/pinctrl/pinctrl-bindings.txt must be followed by the pin controller node referenced by phandle.

Each offset is a number between 0 and N. It is possible to stack any number of ranges with just one pin-to-GPIO line mapping if the ranges are concocted, but in practice, these ranges are generally gathered together as discrete sets.

The following is an example:

```
gpio-ranges = <&foo 0 20 10>, <&bar 10 50 20>;
```

This means the following:

- Ten pins (20..29) on pin controller foo are mapped to GPIO lines 0..9.
- Twenty pins (50..69) on pin controller bar are mapped to GPIO lines 10..29.

It must be noted that GPIOs have a global number space and the pin controller has a local number space, so we need to define a way to cross-reference them.

We want to map PIN GPIO_5_29 with PIN number 89 in the pin controller number space. The following is the device tree property to define the mapping between the GPIO and pin control subsystem:

```
&gpio5 {
    gpio-ranges = <&pinctrl 29 89 1> ;
}
```

In the previous excerpt, 1 GPIO line from the 29th GPIO line of GPIO bank5 will be mapped to pin ranges from 89 on pin controller `pinctrl`.

To illustrate this on a real platform, let's consider the i.MX6 SoC, which has 32 GPIOs per bank. The following is an excerpt from the pin controller node of i.MX6 SoCs, declared in `arch/arm/boot/dts/imx6qdl.dtsi`, and whose driver is `drivers/pinctrl/freescale/pinctrl-imx6dl.c`:

```
iomuxc: pinctrl@20e0000 {
    compatible = "fsl,imx6dl-iomuxc", "fsl,imx6q-iomuxc";
    reg = <0x20e0000 0x4000>;
};
```

Now that the pin controller has been declared, a GPIO controller (bank3) is declared in the same base device tree, `arch/arm/boot/dts/imx6qdl.dtsi`, as follows:

```
gpio3: gpio@20a4000 {
    compatible = "fsl,imx6q-gpio", "fsl,imx35-gpio";
    reg = <0x020a4000 0x4000>;
    interrupts = <0 70 IRQ_TYPE_LEVEL_HIGH>,
                 <0 71 IRQ_TYPE_LEVEL_HIGH>;
    gpio-controller;
    #gpio-cells = <2>;
    interrupt-controller;
    #interrupt-cells = <2>;
};
```

For information, the driver of this GPIO controller is `drivers/gpio/gpio-mxc.c`. After the GPIO controller node has been declared in the base device tree, this same GPIO controller node is overridden in a SoC-specific base device tree, `arch/arm/boot/dts/imx6q.dtsi`, as follows:

```
&gpio3 {
    gpio-ranges = <&iomuxc 0 69 16>, <&iomuxc 16 36 8>,
                  <&iomuxc 24 45 8>;
};
```

The preceding override of the GPIO controller node means the following:

- `<&iomuxc 0 69 16>` means that 16 pins, from pin 69 (to 84) on pin controller `iomuxc`, are mapped to GPIO lines starting from index 0 (to 15).

- `<&iomuxc 16 36 8>` means that 8 pins, from pin 36 (to 43) on pin controller `iomuxc`, are mapped to GPIO lines starting from index 16 (to 23).

- `<&iomuxc 24 45 8>` means that 8 pins, from pin 45 (to 52) on pin controller `iomuxc`, are mapped to GPIO lines starting from index 24 (to 31).

As expected, we have a 32-line GPIO bank, `16 + 8 + 8`.

As of now, we are able to both understand existing and instantiate new GPIO controllers from the device tree that interact with one or more pin controllers. As the last step in this GPIO controller binding learning curve, let's learn how to hog GPIOs in order to avoid writing a particular driver (or prevent existing ones) to control them.

Pin hogging

The GPIO chip can contain GPIO hog definitions. GPIO hogging is a mechanism providing automatic GPIO requests and configuration as part of the GPIO controller's driver probe function. This means that as soon as the pin control device is registered, the GPIO core will attempt to call `devm_pinctrl_get()`, `lookup_state()`, and `select_state()` on it.

The following are properties required for each GPIO hog definition, which is represented as a child node of the GPIO controller:

- `gpio-hog`: A property that indicates whether or not this child node represents a GPIO hog.

- `gpios`: Contains the GPIO information (ID, flags, ...) for each GPIO to affect. Will contain an integer multiple of the number of cells specified in its parent node (GPIO controller node).

- Only one of the following properties can be specified, scanned in the order they are listed. This means that when multiple properties are present, they will be searched in the order they are presented here, with the first match being considered as the intended configuration:

  - `input`: A property that specifies that the GPIO direction should be set to input.

  - `output-low`: A property that specifies that the GPIO should be configured as output, with an initial low state value.

  - `output-high`: A property that specifies that the GPIO direction should be set to output with an initial high value.

An optional property is line-name: the GPIO label name. If it's not present, the node name is used.

The following is an excerpt, where we first declare (as GPIO) the pins we are interested in in the pin controller node:

```
&iomuxc {
[...]
    pinctrl_gpio3_hog: gpio3hoggrp {
        fsl,pins = <
            MX6QDL_PAD_EIM_D19__GPIO3_IO19      0x1b0b0
            MX6QDL_PAD_EIM_D20__GPIO3_IO20      0x1b0b0
            MX6QDL_PAD_EIM_D22__GPIO3_IO22      0x1b0b0
            MX6QDL_PAD_EIM_D23__GPIO3_IO23      0x1b0b0
        >;
    };


[...]
}
```

After the pins of interest are declared, we can hog each GPIO under the node of the GPIO controller this GPIO belongs to, as follows:

```
&gpio3 {
    pinctrl-names = "default";
    pinctrl-0 = <&pinctrl_gpio3_hog>;

    usb-emulation-hog {
        gpio-hog;
        gpios = <19 GPIO_ACTIVE_HIGH>;
        output-low;
        line-name = "usb-emulation";
    };

    usb-model-hog {
        gpio-hog;
        gpios = <20 GPIO_ACTIVE_HIGH>;
        output-high;
```

```
            line-name = "usb-mode1";
    };

    usb-pwr-hog {
        gpio-hog;
        gpios = <22 GPIO_ACTIVE_LOW>;
        output-high;
      line-name = "usb-pwr-ctrl-en-n";
    };

    usb-mode2-hog {
        gpio-hog;
        gpios = <23 GPIO_ACTIVE_HIGH>;
        output-high;
        line-name = "usb-mode2";
    };
};
```

It must be noted that hogging pins should be used for those pins that are not controlled by any particular driver.

GPIO hogging was the last part on the GPIO controller side. Now that controllers have no more secrets for us, let's switch to the consumer interface.

Getting the most out of the GPIO consumer interface

The GPIO is a feature, or a mode in which a pin can operate, in terms of hardware. It is nothing more than a digital line that may be used as an input or output and has just two values (or states): 1 for high or 0 for low. The kernel's GPIO subsystem includes all of the functions you'll need to set up and manage GPIO lines from within your driver.

Before using a GPIO from within the driver, it must first be claimed by the kernel. It's a means to take control of a GPIO and prohibit other drivers from using it, as well as preventing the controller driver from being unloaded.

After claiming control of the GPIO, you can do the following:

- Set the direction and, if needed, set the GPIO configuration.

- If it's being used as an output, start toggling its output state (driving the line high or low).

- If used as input, set the debounce-interval if needed and read the state. For a GPIO line mapped to an IRQ, configure at what edge/level the interrupt should be triggered, and register a handler that will be run when the interrupt occurs.

In the Linux kernel, there are two different ways to deal with GPIOs:

- The legacy and deprecated integer-based interface, which uses integers to represent GPIOs.

- The new descriptor-based interface, where a GPIO is represented and described by an opaque structure, with a dedicated API. This is the recommended way to go.

While we will discuss the two approaches in this chapter, let's start with the legacy interface.

Integer-based GPIO interface – now deprecated

The integer-based interface is the most known usage of GPIOs in Linux systems, either in the kernel or in the user space. In this mode, the GPIO is identified by an integer, which is used for every operation that needs to be performed on this GPIO. The following is the header that contains the legacy GPIO access function:

```
#include <linux/gpio.h>
```

The integer-based interface relies on a set of functions, defined as follows:

```
bool gpio_is_valid(int number);
int  gpio_request(unsigned gpio, const char *label);
int  gpio_get_value_cansleep(unsigned gpio);
int  gpio_direction_input(unsigned gpio);
int  gpio_direction_output(unsigned gpio, int value);
void gpio_set_value(unsigned int gpio, int value);
int  gpio_get_value(unsigned gpio);
void gpio_set_value_cansleep(unsigned gpio, int value);
int gpio_get_value_cansleep(unsigned gpio);
void gpio_free(unsigned int gpio);
```

All the preceding functions are mapped to a set of callbacks provided by the GPIO controller through its `struct gpio_chip` structure, thanks to which it exposes a generic set of callback functions.

In all these functions, gpio represents the GPIO number we are interested in. Before using a GPIO, client drivers must call gpio_request() in order to take ownership of the line and, very importantly, to prevent this GPIO controller driver from being unloaded. In the same function, label is the label used by the kernel for labeling/describing the GPIO in sysfs as we can see in /sys/kernel/debug/gpio. gpio_request() returns 0 on success, and a negative error code on error. If in doubt, before requesting the GPIO, you can use the gpio_is_valid() function to check whether the specified GPIO number is valid on the system prior to it being requested.

Once a driver owes the GPIO, it can change its direction, depending on the need, whether it should be an input or output, using the gpio_direction_input() or gpio_direction_output() functions. In these functions, gpio is the GPIO number the driver needs to set the direction, which should have already been requested. There is a second parameter when it comes to configuring the GPIO as output, value, which is the initial state the GPIO should be in once the output direction is effective. Here again, in both functions, the return value is 0 on success or a negative error code on failure. Internally, these functions are mapped to lower-level callback functions exported by the driver of the GPIO chip that provides the GPIO line gpio.

Some GPIO controllers allow you to adjust the GPIO debounce-interval (this is only useful when the GPIO line is configured as input). This parameter can be set using gpio_set_debounce(). In this function, the debounce argument is the debounce time in milliseconds.

As it is a good practice to grab and configure resources in the driver's probing method, GPIO lines must respect this rule.

It has to be noticed that GPIO management (either configuration or getting/setting values) is not context agnostic; that is, there are memory-mapped GPIO controllers that can be accessed from any context (process and atomic contexts). On the other hand, there are GPIOs provided by discrete chips sitting on slow buses (such as I2C or SPI) that can sleep (because sending/receiving commands on such buses requires waiting to get to the head of a queue to transmit a command and get its response). Such GPIOs must be manipulated from a process context exclusively. A well-designed controller driver must be able to inform clients whether calls to its GPIO driving methods may sleep or not. This can be checked with the gpio_cansleep() function. This function returns true for GPIO lines whose controller sits on a slow bus, and false for GPIOs that belong to a memory-mapped controller.

Now that the GPIOs are requested and configured, we can set/get their values using the appropriate APIs. Here again, the APIs to use are context-dependent. For memory-mapped GPIO controllers, their GPIO lines can be accessed using `gpio_get_value()` or `gpio_set_value()`. The first function returns a value that represents the GPIO state, and the second one will set the value of the GPIO, which should have been configured as an output using `gpio_direction_output()`. `value` can be considered as Boolean for both functions, with zero indicating a low level and a non-zero value indicating a high level.

In case of doubt about the kind of GPIO controller from where the GPIO originates, the driver should use the context agnostic APIs, `gpio_get_value_cansleep()` and `gpio_set_value_cansleep()`. These APIs are safe to use in threaded contexts but also work in an atomic context.

> **Note**
> The legacy (that is, integer-based) interface supports specifying GPIOs from the device tree, in which case the APIs to be used to grab those GPIOs will be `of_get_gpio()`, `of_get_named_gpio()`, or similar APIs. These are mentioned here for studying purposes and won't be discussed in this chapter.

GPIO mapped to IRQ

There are GPIO controllers that allow their GPIO lines to be mapped to IRQs. These IRQs can be either edge- or level-triggered. The configuration depends on the needs. The GPIO controller is responsible for providing the mapping between the GPIO and its IRQ.

If the IRQ has been specified in the device tree and the underlying device is an I2C or SPI device, the consumer driver must just request the IRQ normally, since upon the device tree parsing, the GPIO mapped to IRQ specified in the device tree will be translated by the device tree core and assigned to your device structure, that is, `i2c_client.irq` or `spi_device.irq`. This is the case in `foo_device` from the example we have seen in the *GPIO controller bindings* section. For another device type, you'll have to call `irq_of_parse_and_map()` or a similar API.

If, however, the driver is given a GPIO (from module parameters, for example) or specified from the device tree without being mapped to IRQ there, the driver must use `gpio_to_irq()` to map the given GPIO number to its IRQ number:

```
int gpio_to_irq(unsigned gpio)
```

This function returns the corresponding Linux IRQ number, which can be passed to `request_irq()` (or the threaded counterpart, `request_threaded_irq()`) to register a handler for this IRQ:

```
int request_threaded_irq (unsigned int irq,
                irq_handler_t handler,
                irq_handler_t thread_fn,
                unsigned long irqflags,
                const char *devname,
                void *dev_id);
int request_any_context_irq (unsigned int irq,
                    irq_handler_t handler,
                    unsigned long flags,
                    const char * name,
                    void * dev_id);
```

`request_any_context_irq()` is smart enough to identify the underlying context supported by the IRQ chip integrated into the GPIO controller. If this controller's accessors can sleep, `request_any_context_irq()` will request a threaded IRQ, otherwise, it will request an atomic-context IRQ.

The following is a short example demonstrating what we have discussed so far:

```
static irqreturn_t my_interrupt_handler(int irq,
                                        void *dev_id)
{
    [...]
    return IRQ_HANDLED;
}


static int foo_probe(struct i2c_client *client)
{
    [...]
    struct device_node *np = client->dev.of_node;
    int gpio_int = of_get_gpio(np, 0);
    int irq_num = gpio_to_irq(gpio_int);
    int error =
        devm_request_threaded_irq(&client->dev, irq_num,
            NULL, my_interrupt_handler,
```

```
                    IRQF_TRIGGER_RISING | IRQF_ONESHOT,
                    input_dev->name, my_data_struct);
    if (error) {
        dev_err(&client->dev, "irq %d requested failed,
                %d\n", client->irq, error);
        return error;
    }
    [...]
    return 0;
}
```

In the previous excerpt, we have demonstrated how to use a legacy integer-based interface to grab a GPIO specified in the device tree, as well as the old API to translate this GPIO into a valid Linux IRQ number. These were the main points to highlight.

Though deprecated, we briefly introduced the legacy GPIO APIs. As is recommended, in the next section, we will deal with the new descriptor-based GPIO interface.

Descriptor-based GPIO interface: the new and recommended way

With the new descriptor-based GPIO interface, the subsystem has been oriented to the producer/consumer. The header required for the descriptor-based GPIO interface is the following:

```
#include <linux/gpio/consumer.h>
```

With the descriptor-based interface, a GPIO is described and characterized by a coherent data structure, struct gpio_desc, which is defined as follows:

```
struct gpio_desc {
    struct gpio_chip *chip;
    unsigned long    flags;
    const char       *label;
};
```

In the preceding data structure, chip is the controller providing this GPIO line; flags are the flags characterizing the GPIO and label is the name describing the GPIO.

Prior to requesting and acquiring ownership of GPIOs with the descriptor-based interface, these GPIOs must have been specified or mapped somewhere. It means they should be allocated to a driver, whereas with the legacy integer-based interface, a driver could just obtain a number from anywhere and request it as GPIO. Since descriptor-based GPIOs are represented by an opaque structure, such a method is not possible anymore.

With the new interface, GPIOs must exclusively be mapped to names or indexes, specifying at the same time the GPIO chips providing the GPIOs of interest. This gives us three ways to specify and assign GPIOs to drivers:

- **Platform data mapping**: In such cases, for example, the mapping is done in the board file.

- **Device tree**: The mapping is done in the device tree. This is the mapping we will discuss in this book.

- **Advanced Configuration and Power Interface mapping (ACPI)**: This is ACPI-style mapping. On x86-based systems, this is the most common configuration.

Now that we are done with the GPIO descriptor interface introduction, let's learn how it is mapped and assigned to devices.

GPIO descriptor mapping in the device tree and its APIs

GPIO descriptor mappings are defined in the device tree node of the consumer device. The GPIO descriptor mapping property must be named <name>-gpios or <name>-gpio, where <name> is meaningful enough to describe the function for which the GPIO(s) will be used. This is mandatory.

The reason is that descriptor-based GPIO lookup relies on the gpio_suffixes[] variable, a gpiolib variable defined in drivers/gpio/gpiolib.h as follows:

```
static const char * const gpio_suffixes[] =
                         { "gpios", "gpio" };
```

This variable is used in both device tree lookup and ACPI-based lookup. To see how it works, let's see how it is used in of_find_gpio(), the device tree's low-level GPIO lookup function defined as follows:

```
static struct gpio_desc *of_find_gpio(
                    struct device *dev,
                    const char *con_id,
                    unsigned int idx,
                    enum gpio_lookup_flags *flags)
```

```
{
    /* 32 is max size of property name */
    char prop_name[32];
    enum of_gpio_flags of_flags;
    struct gpio_desc *desc;
    unsigned int i;

    /* Try GPIO property "foo-gpios" and "foo-gpio" */
    for (i = 0; i < ARRAY_SIZE(gpio_suffixes); i++) {
        if (con_id)
            snprintf(prop_name, sizeof(prop_name),
                    "%s-%s", con_id,
                    gpio_suffixes[i]);
        else
            snprintf(prop_name, sizeof(prop_name), "%s",
                    gpio_suffixes[i]);

        desc = of_get_named_gpiod_flags(dev->of_node,
                                        prop_name, idx,
                                        &of_flags);

        if (!IS_ERR(desc) || PTR_ERR(desc) != -ENOENT)
            break;
[...]
}
```

Now let's consider the following node, which is an excerpt of `Documentation/gpio/board.txt`:

```
foo_device {
    compatible = "acme,foo";
    [...]
    led-gpios = <&gpio 15 GPIO_ACTIVE_HIGH>, /* red */
                <&gpio 16 GPIO_ACTIVE_HIGH>, /* green */
                <&gpio 17 GPIO_ACTIVE_HIGH>; /* blue */

    power-gpio = <&gpio 1 GPIO_ACTIVE_LOW>;
```

```
        reset-gpio = <&gpio 1 GPIO_ACTIVE_LOW>;
};
```

This is what a mapping should look like, with meaningful names, corresponding to the functions assigned to the GPIOs. This excerpt will be used as the basis for the rest of this section to demonstrate the use of the descriptor-based GPIO interface.

Now that the GPIOs have been specified in the device tree, the first thing to be done is to allocate GPIO descriptors and take the ownership of these GPIOs. This can be done using gpiod_get(), gpiod_getindex(), or gpiod_get_optional(), defined as follows:

```
struct gpio_desc *gpiod_get_index(struct device *dev,
                            const char *con_id,
                            unsigned int idx,
                            enum gpiod_flags flags)
struct gpio_desc *gpiod_get(struct device *dev,
                        const char *con_id,
                        enum gpiod_flags flags)
struct gpio_desc *gpiod_get_optional(struct device *dev,
                            const char *con_id,
                            enum gpiod_flags flags);
```

It must be noted that we can also use the device-managed variant of these APIs, defined as follows:

```
struct gpio_desc *devm_gpiod_get_index(
                            struct device *dev,
                            const char *con_id,
                            unsigned int idx,
                            enum gpiod_flags flags);
struct gpio_desc *devm_gpiod_get(struct device *dev,
                            const char *con_id,
                            enum gpiod_flags flags);
struct gpio_desc *devm_gpiod_get_optional(
                            struct device *dev,
                            const char *con_id,
                            enum gpiod_flags flags);
```

Both non _optional functions will return -ENOENT if no GPIO with the given function is assigned or a negative error if another error occurred. On success, the GPIO descriptor corresponding to the GPIO is returned. The first method returns the GPIO descriptor structure for the GPIO at a particular index (useful when the specifier is a list of GPIOs), whereas the second function always returns the GPIO at index 0 (single GPIO mapping). The _optional variant is useful for drivers that need to deal with optional GPIOs; it's the same as gpiod_get (), except that it returns NULL when no GPIO has been assigned to the device (that is, specified in the device tree).

In parameters, dev is the device to which the GPIO descriptor will belong. It is the underlying device structure the driver is responsible for; for example, i2c_client. dev, spi_device.dev, or platform_device.dev. con_id is the function of the GPIO within the consumer interface. It corresponds to the <name> prefix of the GPIO specifier property name in the device tree. idx is the index (starting from 0) of the GPIO in case the specifier contains a list of GPIOs. flags is an optional parameter that determines the GPIO initialization flags, to configure the direction and/or the initial output value. It is an instance of enum gpiod_flags, defined in include/linux/ gpio/consumer.h as follows:

```
enum gpiod_flags {
    GPIOD_ASIS   = 0,
    GPIOD_IN = GPIOD_FLAGS_BIT_DIR_SET,
    GPIOD_OUT_LOW = GPIOD_FLAGS_BIT_DIR_SET |
                    GPIOD_FLAGS_BIT_DIR_OUT,
    GPIOD_OUT_HIGH = GPIOD_FLAGS_BIT_DIR_SET |
                    GPIOD_FLAGS_BIT_DIR_OUT |
                    GPIOD_FLAGS_BIT_DIR_VAL,
};
```

Let's demonstrate in the following how these APIs can be used in drivers:

```
struct gpio_desc *red, *green, *blue, *power;

red = gpiod_get_index(dev, "led", 0, GPIOD_OUT_HIGH);
green = gpiod_get_index(dev, "led", 1, GPIOD_OUT_HIGH);
blue = gpiod_get_index(dev, "led", 2, GPIOD_OUT_HIGH);

power = gpiod_get(dev, "power", GPIOD_OUT_HIGH);
```

For the sake of readability, the preceding code does not perform error checking. The LED GPIOs will be active-high, but the power GPIO will be active-low (that is, `gpiod_is_active_low(power)` returns `true` in this case).

Since the `flags` argument is optional, there might be situations where either the initial flags are not specified or when the initial function of the GPIO needs to be changed. To address this, drivers can use `gpiod_direction_input()` or `gpiod_direction_output()` to change the GPIO direction. These APIs are defined as follows:

```
int gpiod_direction_input(struct gpio_desc *desc);
int gpiod_direction_output(struct gpio_desc *desc,
                          int value);
```

In the preceding APIs, `desc` is the GPIO descriptor of the GPIO of interest, and `value` is the initial value to apply to this GPIO when it is configured as output.

It must be noted that the same attention must be paid as with the integer-based interface. In other words, the driver must take care of whether the underlying GPIO chip is memory-mapped (and thus can be accessed in any context) or sits on a slow bus (which would require accessing the chip in process or threaded context exclusively). This can be achieved using the `gpiod_cansleep()` function, defined as follows:

```
int gpiod_cansleep(const struct gpio_desc *desc);
```

This function returns `true` if the underlying hardware can put the caller to sleep while it is accessed. In such cases, drivers should use dedicated APIs.

The following are APIs to get or set the GPIO value on a controller that sits on a slow bus, that is, a GPIO descriptor for which `gpiod_cansleep()` returned `true`:

```
int gpiod_get_value_cansleep(const struct gpio_desc *desc);
void gpiod_set_value_cansleep(struct gpio_desc *desc,
                             int value);
```

If the underlying chip is memory mapped, the following APIs can be used instead:

```
int gpiod_get_value(const struct gpio_desc *desc);
void gpiod_set_value(struct gpio_desc *desc, int value);
```

The context must be considered only if the driver is intended to access the GPIO(s) from within an interrupt handler or from within any other atomic context. Otherwise, you can just use the normal APIs, that is, the ones without the `_cansleep` suffix.

`gpiod_to_irq()` can be used to get the IRQ number that corresponds to a GPIO descriptor mapped to IRQ:

```
int gpiod_to_irq(const struct gpio_desc *desc);
```

The resulting IRQ number can be used with the `request_irq()` function (or the threaded variant `request_threaded_irq()`). If the driver does not need to bother with the context supported by the underlying hardware chip, `request_any_context_irq()` can be used instead. That said, the driver can use the device managed variant of these functions, that is, `devm_request_irq()`, `devm_request_threaded_irq()`, or `devm_request_any_context_irq()`.

If for any reason the module needs to switch back and forth between the descriptor-based interface and the legacy integer-based interface, the APIs `desc_to_gpio()` and `gpio_to_desc()` can be used for translations. They are defined as follows:

```
/* Convert between the old gpio_ and new gpiod_ interfaces */
struct gpio_desc *gpio_to_desc(unsigned gpio);
int desc_to_gpio(const struct gpio_desc *desc);
```

In the preceding, `gpio_to_desc()` takes a legacy GPIO number in the parameter and returns the associated GPIO descriptor, while `desc_to_gpio()` does the opposite.

The advantage of using the device-managed APIs is that drivers need not care about releasing the GPIO at all, since it will be handled by the GPIO core. If, however, non-managed APIs were used to request a GPIO descriptor, this descriptor must explicitly be released with `gpiod_put()`, defined as follows:

```
void gpiod_put(struct gpio_desc *desc);
```

Now that we are done with the consumer side's descriptor-based APIs, let's summarize what we have learned in a concrete example, from the mapping from the device tree to the consumer code based on consumer APIs.

Putting it all together

The following driver summarizes the concepts introduced in the descriptor-based interface. In this example, we need four GPIOs split as follows: two for LEDs (red and green, which are then configured as output) and two for buttons (thus configured as input). The logic to implement is that pushing button 1 toggles both LEDs only when button 2 is pushed as well.

To achieve that, let's consider the following mapping in the device tree:

```
foo_device {
    compatible = "packt,gpio-descriptor-sample";
    led-gpios = <&gpio2 15 GPIO_ACTIVE_HIGH>, // red
                <&gpio2 16 GPIO_ACTIVE_HIGH>, // green

    btn1-gpios = <&gpio2 1 GPIO_ACTIVE_LOW>;
    btn2-gpios = <&gpio2 31 GPIO_ACTIVE_LOW>;
};
```

Now that the GPIOs have been mapped in the device tree, let's write the platform driver that will leverage these GPIOs:

```
#include <linux/init.h>
#include <linux/module.h>
#include <linux/kernel.h>
#include <linux/platform_device.h> /* platform devices */
#include <linux/gpio/consumer.h>   /* GPIO Descriptor */
#include <linux/interrupt.h>       /* IRQ */
#include <linux/of.h>              /* Device Tree */

static struct gpio_desc *red, *green, *btn1, *btn2;
static unsigned int irq, led_state = 0;

static irq_handler_t btn1_irq_handler(unsigned int irq,
                                      void *dev_id)
{
    unsigned int btn2_state;

    btn2_state = gpiod_get_value(btn2);
    if (btn2_state) {
        led_state = 1 - led_state;
        gpiod_set_value(red, led_state);
        gpiod_set_value(green, led_state);
    }

    pr_info("btn1 interrupt: Interrupt! btn2 state is %d)\n",
```

```
                        led_state);
    return IRQ_HANDLED;
}
```

In the preceding, we have started with the IRQ handler. The toggling logic is implemented by `led_state = 1 - led_state`. Next, we implement the driver's `probe` method, as follows:

```
static int my_pdrv_probe (struct platform_device *pdev)
{
    int retval;
    struct device *dev = &pdev->dev;

    red = devm_gpiod_get_index(dev, "led", 0,
                            GPIOD_OUT_LOW);
    green = devm_gpiod_get_index(dev, "led", 1,
                            GPIOD_OUT_LOW);

    /* Configure GPIO Buttons as input */
    btn1 = devm_gpiod_get(dev, "led", 0, GPIOD_IN);
    btn2 = devm_gpiod_get(dev, "led", 1, GPIOD_IN);

    irq = gpiod_to_irq(btn1);
    retval = devm_request_threaded_irq(dev, irq, NULL,
                        btn1_pushed_irq_handler,
                        IRQF_TRIGGER_LOW | IRQF_ONESHOT,
                        "gpio-descriptor-sample", NULL);
    pr_info("Hello! device probed!\n");
    return 0;
}
```

The preceding probe method is quite simple. We first start requesting the GPIOs, then we translate the button 1 GPIO line into a valid IRQ number, and then we register a handler for this IRQ. You should pay attention to the fact that we have exclusively used device-managed APIs in that method.

Finally, we set up a device ID table before filling and registering our platform device driver, as follows:

```
static const struct of_device_id gpiod_dt_ids[] = {
    { .compatible = "packt,gpio-descriptor-sample", },
    { /* sentinel */ }
};

static struct platform_driver mypdrv = {
    .probe      = my_pdrv_probe,
    .driver     = {
        .name       = "gpio_descriptor_sample",
        .of_match_table = of_match_ptr(gpiod_dt_ids),
        .owner      = THIS_MODULE,
    },
};
module_platform_driver(mypdrv);
MODULE_AUTHOR("John Madieu <john.madieu@labcsmart.com>");
MODULE_LICENSE("GPL");
```

We may wonder why neither GPIO descriptors nor interrupts were released. This is because we exclusively used device-managed APIs in the probe function. Thanks to these, we do not need to release anything explicitly, thus we can get rid of the remove method of the platform driver.

If we use non-managed APIs, the remove method could look like the following:

```
static void my_pdrv_remove(struct platform_device *pdev)
{
    free_irq(irq, NULL);
    gpiod_put(red);
    gpiod_put(green);
    gpiod_put(btn1);
    gpiod_put(btn2);
    pr_info("good bye reader!\n");
}
static struct platform_driver mypdrv = {
    [...]
    .remove     = my_pdrv_remove,
```

```
    [...]
};
```

In the preceding, we can notice the use of regular `gpiod_put()` and `free_irq()` APIs to release GPIO descriptors and the IRQ line.

In this section, we have done with the kernel side of GPIO management, both on the controller and client sides. As we have learned all through this book, there are situations where we might want to avoid writing specific kernel code. Regarding GPIOs, the next section will teach us how not to write GPIO client drivers to control these GPIOs.

Learning how not to write GPIO client drivers

There are situations where writing user space code would achieve the same goals as writing kernel drivers. Moreover, the GPIO framework is one of the most used frameworks in the user space. It then goes without saying that there are several possibilities to deal with it in the user space, some of which we will introduce in this chapter.

Goodbye to the legacy GPIO sysfs interface

Sysfs has ruled GPIO management from the user space for quite a long time now. Though it is scheduled for removal, the sysfs GPIO interface still has a few days ahead of it. `CONFIG_GPIO_SYSFS` can still enable it, but its use is discouraged, and it will be removed from mainline Linux. This interface allows managing and controlling GPIOs through a set of files. It is located at `/sys/class/gpio/`, and the following are the common directory paths and attributes that are involved:

- `/sys/class/gpio/`: This is where it all starts. There are two special files in this directory, `export` and `unexport`, and as many directories as there are GPIO controllers registered with the system:

  - `export`: By writing the number of a GPIO to this file, we ask the kernel to export control of that GPIO to the user space. For example, typing `echo 21 > export` will create a `gpio21` node (resulting in a subdirectory of the same name) for GPIO #21, if this GPIO is not already requested by the kernel code.

  - `unexport`: The effect of exporting to the user space is reversed by writing the same GPIO number to this file. For example, the `gpio21` node exported using the `export` file will be removed by typing `echo 21 > unexport`.

On successful `gpio_chip` registration, a directory entry with a path such as `/sys/class/gpio/gpiochipX/` will be created, where `X` is the GPIO controller base (the controller providing GPIOs starting at #X), having the following attributes:

- `base`, whose value is the same as `X`, and which corresponds to `gpio_chip.base` (if assigned statically), and being the first GPIO managed by this chip.

- `label`, which is provided for diagnostics (not always unique).

- `ngpio`, which tells us how many GPIOs this controller provides (`N` to `N + ngpio - 1`). This is the same as defined in `gpio_chip.ngpios`.

- `/sys/class/gpio/gpioN/`: This directory corresponds to the GPIO line `N`, exported either using the `export` file or directly from the kernel. `/sys/class/gpio/gpio42/` (for GPIO #42) is an example. The following read/write attributes are contained in such directories:

  - `direction`: Use this file to get/set GPIO direction. Acceptable values are either `in` or `out` strings. This attribute will normally be written and writing the `out` value will initialize the GPIO value as `low` by default. To ensure glitch-free operation, values low and high may be written to configure the GPIO as an output with that initial value. If, however, the GPIO has been exported from the kernel (see the `gpiod_export()` or `gpio_export()` functions), then this attribute will be missing, disabling at the same time direction change.

  - `value`: This attribute can be used to get or set the state of the GPIO line based on its direction, input, or output. If the GPIO is configured as an output, any non-zero value written will set the output high, while writing 0 will set this output low. If the pin can be set up as an interrupt-generating line and is set to do so, then the `poll()` system function can be used on that file and will return when an interrupt occurs. Setting the events `POLLPRI` and `POLLERR` is required when using `poll()`. If, however, `select()` is used instead, the file descriptor should be set in `exceptfds`. After `poll()` returns, the user code should either `lseek()` to the beginning of the sysfs file and read the new value or close the file and re-open it to read the value. It is the same principle as we discussed for the pollable sysfs attribute.

  - `edge` determines the signal edge that will let the `poll()` or `select()` functions return. `none`, `rising`, `failing`, or `both` are acceptable values. This readable and writable file exists only if the GPIO can be configured as an interrupt generating input pin.

- `active_low`: When it is read, this attribute either returns 0 (for `false`) or 1 (meaning `true`). Writing any nonzero value will invert the `value` attribute for both reading and writing. Existing and subsequent `poll()`/`select()` support configuration through the `edge` attribute for rising and falling edges will follow this setting.

The following is a short sequence of commands demonstrating the use of the sysfs interface to drive GPIOs from the user space:

```
# echo 24 > /sys/class/gpio/export
# echo out > /sys/class/gpio/gpio24/direction
# echo 1 > /sys/class/gpio/gpio24/value
# echo 0 > /sys/class/gpio/gpio24/value
# echo high > /sys/class/gpio/gpio24/direction # shorthand for
out/1
# echo low > /sys/class/gpio/gpio24/direction # shorthand for
out/0
```

Let's not spend more time on this legacy interface. Without delay, let's switch to what kernel developers have provided as a new GPIO management interface from the user space, the `Libgpiod` library.

Welcome to the Libgpiod GPIO library

The Kernel Linux GPIO user space sysfs is deprecated and has been discontinued. That said, it was suffering from many ailments, some of which are as follows:

- State not tied to processes.

- A lack of concurrent access management to sysfs attributes.

- A lack of support for bulk GPIO operations, that is, performing operations on a set of GPIOs with a single command (in a single shot).

- A lot of operations were needed just to set a GPIO value (opening and writing into the export file, opening and writing into the direction file, opening and writing into the value file).

- Unreliable polling – user code had to poll on `/sys/class/gpio/gpioX/value`, and on each event, it was necessary to `close/re-open` or `lseek` in the file before re-reading the new value. This could lead to events being lost.

- It was not possible to set GPIO electrical properties.

- If the process crashed, the GPIOs remained exported; there was no context concept.

To address the limits of the sysfs interface, a new GPIO interface has been developed, the descriptor-based GPIO interface. It comes with GPIO character devices – a new user API, merged in Linux v4.8. This new interface has introduced the following improvements:

- One device file per GPIO chip: `/dev/gpiochip0`, `/dev/gpiochip1`, `/dev/gpiochipX` ....

- It's similar to other kernel interfaces: `open()` + `ioctl()` + `poll()` + `read()` + `close()`.

- It's possible to request multiple lines at once (for reading/setting values) using bulk-related APIs.

- It's possible to find GPIO lines and chips by name, which is much more reliable.

- Open source and open-drain flags, user/consumer strings, and uevents.

- Reliable polling, preventing the loss of events.

`Libgpiod` is shipped with a C API allowing you to get the most out of any GPIO chip registered on the system. That said, the C++ and Python languages are supported as well. The API is well documented, and too extensive to fully cover here. The basic use cases usually follow these steps:

1. Open the desired GPIO chip character device by calling one of the `gpiod_chip_open*` functions, such as `gpiod_chip_open_by_name()` or `gpiod_chip_open_lookup()`. This returns a pointer to `struct gpiod_chip`, which is used by subsequent API calls.

2. Retrieve the handle to the desired GPIO line by calling `gpiod_chip_get_line()`, which will return a pointer to an instance of `struct gpiod_line`. While the previous API returns the handle to a single GPIO line, the function `gpiod_chip_get_lines()` can be used if several GPIO lines are needed in a single shot. `gpiod_chip_get_lines()` will return a pointer to an instance of `struct gpiod_line_bulk`, which can be used later for bulk operations. The other API that can return a set of GPIO handles is `gpiod_chip_get_all_lines()`, which returns all the lines of a given GPIO chip in `struct gpiod_line_bulk`. When you have such a set of GPIO objects, you can request a GPIO line at a specific index local to this bulk object by using the `gpiod_line_bulk_get_line()` API.

3. Request the use of the line as an input or output by calling `gpiod_line_request_input()` or `gpiod_line_request_output()`. For bulk operations on a set of GPIO lines, `gpiod_line_request_bulk_input()` or `gpiod_line_request_bulk_output()` can be used instead.

4. Read the value of input GPIO lines by calling `gpiod_line_get_value()` for a single GPIO or `gpiod_line_get_value_bulk()` in the case of a set of GPIOs. For output GPIO lines, the level can be set by calling `gpiod_line_set_value()` for a single GPIO line or `gpiod_line_set_value_bulk()` on a set of output GPIOs.

5. When done, release the lines by calling `gpiod_line_release()` or `gpiod_line_release_bulk()`.

6. Once all the GPIO lines have been released, the associated chips can be released using `gpiod_chip_close()` on each.

`gpiod_line_release()` is to be called once done with a GPIO line. The GPIO line to release is passed as a parameter. If it is, however, a set of GPIOs that needs to be released, `gpiod_line_release_bulk()` should be used instead. It has to be noted that if the lines were not previously requested together (were not requested with `gpiod_line_request_bulk()`), the behavior of `gpiod_line_release_bulk()` is undefined.

There are sanity APIs it might worth mentioning, which are defined as follows:

```
bool gpiod_line_is_free(struct gpiod_line *line);
bool gpiod_line_is_requested(struct gpiod_line *line);
```

In the preceding APIs, `gpiod_line_is_requested()` can be used to check if the calling user owns this GPIO line. This function returns `true` if `line` was already requested, or `false` otherwise. It is different from `gpiod_line_is_free()`, which is used to check if the calling user has neither requested ownership `line` nor set up any event notifications on it. It returns `true` if `line` is free, and `false` otherwise.

Other APIs are available for more advanced functions such as configuring pin modes for pullup or pulldown resistors or registering a callback function to be called when an event occurs, such as the level of an input pin changing, as we will see in the next section.

Event- (interrupt-) driven GPIO

Interrupt-driven GPIO handling consists of grabbing one (`struct gpiod_line`) or more (`struct gpiod_line_bulk`) GPIO handles and listening for events on these GPIO lines, either infinitely or in a timed manner.

A GPIO line event is abstracted by a `struct gpiod_line_event` object, defined as follows:

```
struct gpiod_line_event {
    struct timespec ts;
```

```
    int event_type;
};
```

In the preceding data structure, `ts` is the time specifier data structure to represent the wait event timeout, and `event_type` is the type of event, which can be either `GPIOD_LINE_EVENT_RISING_EDGE` or `GPIOD_LINE_EVENT_FALLING_EDGE`, respectively for a rising edge event or a falling edge event.

After the GPIO handle(s) has been obtained using `gpiod_chip_get_line()` or `gpiod_chip_get_lines()` or `gpiod_chip_get_all_lines()`, the user code should request events of interest on these GPIO handles using one of the following APIs:

```
int gpiod_line_request_rising_edge_events(
                        struct gpiod_line *line,
                        const char *consumer);
int gpiod_line_request_bulk_rising_edge_events(
                        struct gpiod_line_bulk *bulk,
                        const char *consumer);


int gpiod_line_request_falling_edge_events(
                        struct gpiod_line *line,
                        const char *consumer);
int gpiod_line_request_bulk_falling_edge_events(
                        struct gpiod_line_bulk *bulk,
                        const char *consumer);


int gpiod_line_request_both_edges_events(
                        struct gpiod_line *line,
                        const char *consumer);
int gpiod_line_request_bulk_both_edges_events(
                        struct gpiod_line_bulk *bulk,
                        const char *consumer);
```

The preceding APIs request either rising edge, falling edge, or both edge events, respectively on a single GPIO line or on a set of GPIO (the bulk-related API).

After the events have been requested, the user code can wait on the GPIO lines of interest, waiting for the requested events to occur using one of the following APIs:

```
int gpiod_line_event_wait(struct gpiod_line *line,
                    const struct timespec *timeout);
int gpiod_line_event_wait_bulk(
                    struct gpiod_line_bulk *bulk,
                    const struct timespec *timeout,
                    struct gpiod_line_bulk *event_bulk);
```

In the preceding, gpiod_line_event_wait() waits for event(s) on a single GPIO line, while gpiod_line_event_wait_bulk() will wait on a set of GPIOs. In parameters, line is the GPIO line on which to wait events in the case of single GPIO monitoring, while bulk is the set of GPIO lines in the case of bulk monitoring. Finally, event_bulk is an output parameter, holding the set of GPIO lines on which the GPIO events of interest have occurred. These are all blocking APIs, which will continue execution flow only after the events of interest have occurred or after a timeout.

Once the blocking function returns, gpiod_line_event_read() must be used to read the events that occurred on the GPIO line(s) returned by the previously mentioned monitoring functions. This API has the following prototype:

```
int gpiod_line_event_read(struct gpiod_line *line,
                    struct gpiod_line_event *event);
```

On error, this API returns -1, otherwise, it returns 0. In parameters, line is the GPIO line to read the events on, and event is an output parameter, the event buffer to which the event data will be copied.

The following is an example of requesting an event and reading and processing that event:

```
char *chipname = "gpiochip0";

int ret;
struct gpiod_chip *chip;
struct gpiod_line *input_line;
struct gpiod_line_event event;
unsigned int line_num = 25;   /* GPIO Pin #25 */

chip = gpiod_chip_open_by_name(chipname);
if (!chip) {
```

```
        perror("Open chip failed\n");
        return -1;
}

input_line = gpiod_chip_get_line(chip, line_num);
if (!input_line) {
        perror("Get line failed\n");
        ret = -1;
        goto close_chip;
}

ret = gpiod_line_request_rising_edge_events(input_line,
                                        "gpio-test");
if (ret < 0) {
        perror("Request event notification failed\n");
        ret = -1;
        goto release_line;
}
while (1) {
  gpiod_line_event_wait(input_line, NULL); /* blocking */
  if (gpiod_line_event_read(input_line, &event) != 0)
        continue;

    /* should always be a rising event in our example */
    if (event.event_type != GPIOD_LINE_EVENT_RISING_EDGE)
        continue;

    [...]
}

release_line:
    gpiod_line_release(input_line);
close_chip:
    gpiod_chip_close(chip);
    return ret;
```

In the preceding snippet, we first look up the GPIO chip by its name and use the returned GPIO chip handle to grab a handle on GPIO line #25. Next, we request a rising events notification (interrupt-driven) on the GPIO line. After that, we loop on waiting for events to happen, read which event it was, and validate that it's a rising event.

Apart from the previous code example, let's now imagine a much complex example, where we monitor five GPIO lines, and let's start by feeding the required headers:

```
// file event-bulk.c
#include <gpiod.h>
#include <error.h>
#include <stdlib.h>
#include <stdio.h>
#include <string.h>
#include <unistd.h>
#include <sys/time.h>
```

Then, let's provide the static variables we will use in the program:

```
static struct gpiod_chip *chip;
static struct gpiod_line_bulk gpio_lines;
static struct gpiod_line_bulk gpio_events;


/* use GPIOs #4, #7, #9, #15, and #31 as input */
static unsigned int gpio_offsets[] = {4, 7, 9, 15, 31};
```

In the previous snippet, chip will hold the handle to the GPIO chip that we are interested in. gpio_lines will hold the handles of the event-driven GPIO lines, that is, the GPIO lines to be monitored. Finally, gpio_events will be given to the library so that upon monitoring, it is filled with the handles of GPIO lines on which events have occurred.

Finally, let's start implementing our main method:

```
int main(int argc, char *argv[])
{
    int err;
    int values[5] = {-1};
    struct timespec timeout;

    chip = gpiod_chip_open("/dev/gpiochip0");
```

```
    if (!chip) {
        perror("gpiod_chip_open");
        goto cleanup;
    }
```

In the previous snippet, we have simply opened the GPIO chip device and kept a pointer to it. Next, we will have to grab handles of the GPIO lines of interest and store them in `gpio_lines`:

```
    err = gpiod_chip_get_lines(chip, gpio_offsets, 5,
                                    &gpio_lines);
    if (err) {
        perror("gpiod_chip_get_lines");
        goto cleanup;
    }
```

Then, we use these GPIO line handles to request event monitoring on their underlying GPIO lines. Because we are interested in more than one GPIO, we use the `bulk` API variant, as follows:

```
    err = gpiod_line_request_bulk_rising_edge_events(
                    &gpio_lines, "rising edge example");
    if(err) {
        perror(
            "gpiod_line_request_bulk_rising_edge_events");
        goto cleanup;
    }
```

In the previous snippet, `gpiod_line_request_bulk_rising_edge_events()` will request rising edge event notifications. Now that we have requested event-driven monitoring for our GPIO, we can call the blocking monitoring API on these GPIO lines, as follows:

```
    /* Timeout of 60 seconds, pass in NULL to wait forever */
    timeout.tv_sec = 60;
    timeout.tv_nsec = 0;
    printf("waiting for rising edge event \n");
 marker1:
    err = gpiod_line_event_wait_bulk(&gpio_lines,
                                    &timeout, &gpio_events);
```

```
    if (err == -1) {
        perror("gpiod_line_event_wait_bulk");
        goto cleanup;
    } else if (err == 0) {
        fprintf(stderr, "wait timed out\n");
        goto cleanup;
    }
```

In the previous excerpt, since we need time-bounded event polling, we set up a `struct timespec` data structure with the desired timeout and we pass it to `gpiod_line_event_wait_bulk()`.

That said, reaching this step (passing the polling function) would mean that either the blocking monitoring API has timed out or that an event occurred on at least one of the GPIO lines that are monitored. The GPIO handles on which events occurred are stored in `gpio_events`, which is an output argument, and the list of monitored GPIO lines is passed in `gpio_lines`. It must be noted that both `gpio_lines` and `gpio_events` are bulk GPIO data structures.

If ever we are interested in reading the values of the GPIO lines on which events have occurred, we could do the following:

```
    err = gpiod_line_get_value_bulk(&gpio_events, values);
    if(err) {
        perror("gpiod_line_get_value_bulk");
        goto cleanup;
    }
```

If instead of reading the values of GPIO lines on which events occurred we needed to read the value of all the monitored GPIO lines, we would have replaced `gpio_events` with `gpio_lines` in the previous code.

Next, if we are interested in the type of event that occurred on each GPIO line in `gpio_events`, we can do the following:

```
for (int i = 0;
     i < gpiod_line_bulk_num_lines(&gpio_events);
     i++) {
    struct gpiod_line* line;
    struct gpiod_line_event event;
    line = gpiod_line_bulk_get_line(&gpio_events, i);
    if(!line) {
        fprintf(stderr, "unable to get line %d\n", i);
        continue;
    }
    if (gpiod_line_event_read(line, &event) != 0)
        continue;

    printf("line %s, %d\n", gpiod_line_name(line),
        gpiod_line_offset(line));
}
marker2:
```

In the preceding code, we iterate over each GPIO line in `gpio_events`, which represents the list of GPIO lines on which events have occurred. `gpiod_line_bulk_num_lines()` retrieves the number of GPIO lines held by the line bulk object, and `gpiod_line_bulk_get_line()` retrieves the line handle from a line bulk object at the given offset, local to this line bulk object. You should, however, note that to achieve the same goal, we could have used the `gpiod_line_bulk_foreach_line()` macro.

Then, on each GPIO line in the line bulk object, we invoke `gpiod_line_event_read()`, `gpiod_line_name()`, and `gpiod_line_offset()`. The first function will retrieve the event data structure corresponding to the event that occurred on that line. We could have then checked that the event type that occurred (especially when monitoring for both event types) is what we expected using something such as `if (event.event_type != GPIOD_LINE_EVENT_RISING_EDGE)`, for example. The second function is a helper that will retrieve the GPIO line name, while the third one, `gpiod_line_offset()`, will retrieve the GPIO line offset, global to the running system.

If we were interested in monitoring these GPIO lines infinitely or for a certain number of rounds, we could have wrapped the code between the `marker1` and `marker2` labels into a `while()` or a `for()` loop.

At the end of the execution flow, we do some cleaning, like the following:

```
cleanup:
    gpiod_line_release_bulk(&gpio_lines);
    gpiod_chip_close(chip);


    return EXIT_SUCCESS;
}
```

The previous cleaning code snippet first releases all the GPIO lines that we have requested, and then closes the associated GPIO chip.

> **Note**
>
> It must be noted that bulk GPIO monitoring must be done on a per GPIO chip basis. That is, it is not recommended to embed GPIO lines from different GPIO chips in the same line bulk object.

Now that we are done with API usage and have demonstrated it in a practical example, we can switch to command-line tools shipped with the libgpiod library.

Command-line tools

If you simply need to perform simple GPIO operations, the Gpiod library includes a collection of command-line tools that are particularly handy for interactively exploring GPIO functions and can be used in shell scripts to avoid the need to write C or C++ code. There are the following commands available:

- gpiodetect: Displays the list of all GPIO chips on the system, together with their names, labels, and the number of GPIO lines.

- gpioinfo: Displays the names, consumers, direction, active status, and other flags for all lines of the selected GPIO chips. gpioinfo gpiochip6 is an example. If no GPIO chip is given, the command will iterate through all GPIO chips on the system and list their associated lines.

- gpioget: Gets the values of GPIO lines specified.

- gpioset: Sets the values of specified GPIO lines, and potentially keeps them exported until a timeout, user input, or signal occurs.

- `gpiofind`: Given a line name, this command finds the associated GPIO chip name and line offset.

- `gpiomon`: Monitors GPIOs by waiting for events on these lines. This command allows you to specify which events to watch and how many of them should be processed before exiting or whether the events should be reported to the console.

Now that we have listed the available command-line tools, we can go on to learn about another mechanism offered by the GPIO subsystem, and that can be leveraged from the user space, thanks to which we can use the aforementioned tools.

The GPIO aggregator

GPIO access control now uses permissions on `/dev/gpiochip*` with the new interface. The typical Unix filesystem permissions enable all-or-nothing access control to these character devices. Compared to the earlier `/sys/class/gpio` interface, this new interface provides a number of advantages, which we listed at the beginning of the *Welcome to the Libgpiod GPIO library* section. One disadvantage, however, is that it creates one device file per GPIO chip, implying that access privileges are defined on a per GPIO chip basis, rather than per GPIO line.

As a result, the **GPIO aggregator** feature has been introduced and merged into version 5.8 of the Linux kernel. It allows you to combine a number of GPIOs into a virtual GPIO chip, which appears as an extra `/dev/gpiochip*` device.

This feature is handy for designating a set of GPIOs to a certain user and implementing access control. Furthermore, exporting GPIOs to a virtual machine is simplified and hardened because the virtual machine can just grab the entire GPIO controller and no longer has to worry about which GPIOs to grab and which not to, decreasing the attack surface. `Documentation/admin-guide/gpio/gpio-aggregator.rst` is where you'll find its documentation.

To have GPIO aggregator support in your kernel, you must have `CONFIG_GPIO_AGGREGATOR=y` in your kernel configuration. This feature can be configured either via sysfs or the device tree, as we will see in the next sections.

Aggregating GPIOs using sysfs

Aggregated GPIO controllers are instantiated and destroyed by writing to write-only attribute files in sysfs, mainly from the `/sys/bus/platform/drivers/gpio-aggregator/` directory.

This directory contains the followings attributes:

- `new_device`: Used to ask the kernel to instantiate an aggregated GPIO controller by writing a string describing the GPIOs to aggregate. The `new_device` file understands the format `[<gpioA>] [<gpiochipB> <offsets>] ...`:

  - `<gpioA>` is a GPIO line name.

  - `<gpiochipB>` is a GPIO chip label.

  - `<offsets>` is a comma-separated list of GPIO offsets and/or GPIO offset ranges denoted by dashes.

- `delete_device`: Used to ask the kernel to destroy an aggregated GPIO controller after use.

The following is an example that instantiated a new GPIO aggregator by aggregating GPIO line 19 of `e6052000.gpio` and GPIO lines 20-21 of `e6050000.gpio` into a new `gpio_chip`:

```
# echo 'e6052000.gpio 19 e6050000.gpio 20-21' > /sys/bus/
platform/drivers/gpio-aggregator/new_device
# gpioinfo gpio-aggregator.0
    gpiochip12 - 3 lines:
    line 0: unnamed unused input active-high
    line 1: unnamed unused input active-high
    line 2: unnamed unused input active-high
# chown geert /dev/gpiochip12
```

After use, the previously created aggregated GPIO controller can be destroyed using the following command, assuming it is named `gpio-aggregator.0`:

```
$ echo gpio-aggregator.0 > delete_device
```

From the previous example, the GPIO chip that resulted from the aggregation was `gpiochip12`, having three GPIO lines. Instead of `gpioinfo gpio-aggregator.0`, we could have used `gpioinfo gpiochip12`.

Aggregating GPIOs from the device tree

The device tree can also be used to aggregate GPIOs. To do so, simply define a node with `gpio-aggregator` as a compatible string and set the `gpios` property to the list of GPIOs that you want to be part of the new GPIO chip. A unique feature of this technique is that, like any other GPIO controller, the GPIO lines can be named and subsequently queried by user-space applications using the `libgpiod` library.

In the following, we will demonstrate the use of the GPIO aggregator with several GPIO lines from the device tree. First, we enumerate the pins we need to use GPIOs in our new GPIO chip. We do this under the pin controller node as follows:

```
&iomuxc {
[...]
    aggregator {
        pinctrl_aggregator_pins: aggretatorgrp {
            fsl,pins = <
                MX6QDL_PAD_EIM_D30__GPIO3_IO30       0x80000000
                MX6QDL_PAD_EIM_D23__GPIO3_IO23       0x80000000
                MX6QDL_PAD_ENET_TXD1__GPIO1_IO29     0x80000000
                MX6QDL_PAD_ENET_RX_ER__GPIO1_IO24    0x80000000
                MX6QDL_PAD_EIM_D25__GPIO3_IO25       0x80000000
                MX6QDL_PAD_EIM_LBA__GPIO2_IO27       0x80000000
                MX6QDL_PAD_EIM_EB2__GPIO2_IO30       0x80000000
                MX6QDL_PAD_SD3_DAT4__GPIO7_IO01      0x80000000
            >;
        };
    };
}
```

Now that our pins have been configured, we can declare our GPIO aggregator as follows:

```
gpio-aggregator {
    pinctrl-names = "default";
    pinctrl-0 = <&pinctrl_aggregator_pins>;
    compatible = "gpio-aggregator";

    gpios = <&gpio3 30 GPIO_ACTIVE_HIGH>,
            <&gpio3 23 GPIO_ACTIVE_HIGH>,
            <&gpio1 29 GPIO_ACTIVE_HIGH>,
```

```
                <&gpio1 25 GPIO_ACTIVE_HIGH>,
                <&gpio3 25 GPIO_ACTIVE_HIGH>,
                <&gpio2 27 GPIO_ACTIVE_HIGH>,
                <&gpio2 30 GPIO_ACTIVE_HIGH>,
                <&gpio7 1 GPIO_ACTIVE_HIGH>;

    gpio-line-names = "line_a", "line_b", "line_c",
            "line_d", "line_e", "line_f", "line_g",
            "line_h";
};
```

In this example, `pinctrl_aggregator_pins` is the GPIO pin node, which must have been instantiated under the pin controller node. `gpios` contains the list of GPIO lines the new GPIO chip must be made of. At the end, the meaning of `gpio-line-names` is line 30 of GPIO controller `gpio3` is used and is named `line_a`, line 23 of GPIO controller `gpio3` is used and is named `line_b`, line 29 of GPIO controller `gpio1` is used and named `line_c`, and so on up to line 1 of GPIO controller `gpio7`, which is named `line_h`.

From the user space, we can see the GPIO chip and its aggregated lines:

```
# gpioinfo
[...]
gpiochip9 - 8 lines:
    line 0: "line_a" unused input active-high
    line 1: "line_b" unused input active-high
[...]
    line 7: "line_g" unused input active-high
[...]
```

We can search a GPIO chip and a line number by the line name:

```
# gpiofind 'line_b'
gpiochip9 1
```

We can access a GPIO line by its name:

```
# gpioget $(gpiofind 'line_b')
1
#
# gpioset $(gpiofind 'line_h')=1
# gpioset $(gpiofind 'line_h')=0
```

We can change the GPIO chip device file ownership to allow user or group to access the attached lines:

```
# chown $user:$group /dev/gpiochip9
# chmod 660 /dev/gpiochip9
```

The GPIO chip created by the aggregator can be retrieved from sysfs in `/sys/bus/platform/devices/gpio-aggregator/`.

Aggregating GPIOs using a generic GPIO driver

Without a particular in-kernel driver, the GPIO aggregator can be used as a generic driver for a simple GPIO-operated device described in the device tree. Modifying the `gpio-aggregator` driver or writing to the `driver_override` file in sysfs are both options for binding a device to the GPIO aggregator.

Before we go further, let's talk about the `driver_override` file; this file is more precisely located in `/sys/bus/platform/devices/.../driver_override`. This file specifies the driver for a device, which will override the standard device tree, ACPI, ID table, and name matching, as we have seen in *Chapter 6, Introduction to Devices, Drivers, and Platform Abstraction*. It has to be noted that only a driver whose name matches the value written to `driver_override` will be able to bind to the device. The override is set by writing a string to the `driver_override` file (`echo vfio-platform > driver_override`), and it can be cleared by writing an empty string to the file (`echo > driver_override`). This reverts the device to its default binding of matching rules. It must, however, be noted that writing to driver override does not unbind the device from its existing driver or attempt to load the supplied driver automatically. The device will not bind to any driver if no driver with a matching name is currently loaded in the kernel. Devices can also use a `driver_override` name such as `none` to opt out of driver binding. There is no support for parsing delimiters, and only a single driver can be given in the override.

For example, given a `door` device, which is a GPIO-operated device described in the device tree, use its own compatible value as follows:

```
door {
        compatible = "myvendor,mydoor";

        gpios = <&gpio2 19 GPIO_ACTIVE_HIGH>,
                <&gpio2 20 GPIO_ACTIVE_LOW>;
        gpio-line-names = "open", "lock";
};
```

It can be bound to the GPIO aggregator with either of the following methods:

- Adding its compatible value to `gpio_aggregator_dt_ids[]` in `drivers/gpio/gpio-aggregator.c`
- Binding manually using `driver_override`

The first method is quite straightforward:

```
$ echo gpio-aggregator > /sys/bus/platform/devices/door/driver_
override
$ echo door > /sys/bus/platform/drivers/gpio-aggregator/bind
```

In the previous commands, we have written the driver's name (`gpio-aggregator` in this case) in the `driver_override` file present in the device directory, `/sys/bus/platform/devices/<device-name>/`. After that, we have bound the device to the driver by writing the device name in the `bind` file present in the driver's directory, `/sys/bus/<bus-name>/drivers/<driver-name>/`. It has to be noted that `<bus-name>` corresponds to the bus framework the driver belongs to. It could be `i2c`, `spi`, `platform`, `pci`, `isa`, `usb`, and so on.

After the binding, a new GPIO chip, `door`, will be created. Its information can then be carried out as follows:

```
$ gpioinfo door
gpiochip12 - 2 lines:
        line   0:       "open"       unused   input   active-high
        line   1:       "lock"       unused   input   active-high
```

Next, the library APIs can be used on this GPIO chip like any other normal (non-virtual) GPIO chip.

We are now done with GPIO aggregation from the user space in particular, and with GPIO management from the user space in general. We have learned how to create virtual GPIO chips to isolate a set of GPIOs, and we have learned how to use the GPIO library to drive these GPIOs.

Summary

In this chapter, we introduced the pin control framework and described its interaction with the GPIO subsystem. We learned how to deal with GPIOs, either as a controller or consumer, from both the kernel and the user space. Though the legacy integer-based interface is deprecated, it was introduced because it is still widely used. Additionally, we introduced some advanced topics such as IRQ chip support in the GPIO chip and the mapping of GPIOs to IRQs. We ended this chapter by learning how to deal with GPIOs from the user space, by writing C code or by using dedicated command-line tools provided by the standard Linux GPIO library, `libgpiod`.

In the next chapter, we deal with input devices, which can be implemented using GPIOs.

17
Leveraging the Linux Kernel Input Subsystem

Input devices are devices that you can use to interact with the system. Such devices include buttons, keyboards, touchscreens, mice, and more. They work by sending events that are caught and broadcast over the system by the input core. This chapter will explain each structure that's used by the input core to handle input devices, as well as how to manage events from the user space.

In this chapter, we will cover the following topics:

- Introduction to the Linux kernel input subsystem – its data structures and APIs
- Allocating and registering an input device
- Using polled input devices

- Generating and reporting input events
- Handling input devices from the user space

Introduction to the Linux kernel input subsystem – its data structures and APIs

The main data structures and APIs of this subsystem can be found in the `include/linux/input.h` files. The following line is required in any input device driver:

```
#include <linux/input.h>
```

Whatever type of input device it is, whatever type of event it sends, an input device is represented in the kernel as an instance of the struct `input_dev`:

```
struct input_dev {
  const char *name;
  const char *phys;

  unsigned long evbit[BITS_TO_LONGS(EV_CNT)];
  unsigned long keybit[BITS_TO_LONGS(KEY_CNT)];
  unsigned long relbit[BITS_TO_LONGS(REL_CNT)];
  unsigned long absbit[BITS_TO_LONGS(ABS_CNT)];
  unsigned long mscbit[BITS_TO_LONGS(MSC_CNT)];

  unsigned int repeat_key;

  int rep[REP_CNT];
  struct input_absinfo *absinfo;
  unsigned long key[BITS_TO_LONGS(KEY_CNT)];

  int (*open)(struct input_dev *dev);
  void (*close)(struct input_dev *dev);

  unsigned int users;
  struct device dev;

  unsigned int num_vals;
```

```
    unsigned int max_vals;
    struct input_value *vals;

    bool devres_managed;
};
```

For the sake of readability, some elements in the structure have been omitted. Let's look at these fields in more detail:

- `name` represents the name of the device.
- `phys` is the physical path to the device in the system hierarchy.
- `evbit` is a bitmap of the types of events that are supported by the device. The following are some events:

  - `EV_KEY` is for devices that support sending key events (for example, keyboards, buttons, and so on)
  - `EV_REL` is for devices that support sending relative positions (for example, mice, digitizers, and so on)
  - `EV_ABS` is for devices that support sending absolute positions (for example, joysticks) The list of events is available in the kernel source in the `include/linux/input-event-codes.h` file. You can use the `set_bit()` macro to set the appropriate bit, depending on your input device's capabilities. Of course, a device can support more than one type of event. For example, a mouse driver will set both `EV_KEY` and `EV_REL`, as shown here:

    ```
    set_bit(EV_KEY, my_input_dev->evbit);
    set_bit(EV_REL, my_input_dev->evbit);
    ```

- `keybit` is for `EV_KEY` enabled devices and consists of a bitmap of keys/buttons that this device exposes; for example, `BTN_0`, `KEY_A`, `KEY_B`, and so on. The complete list of keys/buttons can be found in the `include/linux/input-event-codes.h` file.
- `relbit` is for `EV_REL` enabled devices and consists of a bitmap of relative axes for the device; for example, `REL_X`, `REL_Y`, `REL_Z`, and so on. Have a look at `include/linux/input-event-codes.h` for the complete list.
- `absbit` is for `EV_ABS` enabled devices and consists of a bitmap of absolute axes for the device; for example, `ABS_Y`, `ABS_X`, and so on. Have a look at the same previous file for the complete list.

- `mscbit` is for `EV_MSC` enabled devices and consists of a bitmap of miscellaneous events that are supported by the device.

- `repeat_key` stores the key code of the last key pressed; it is used when the autorepeat feature is implemented by the software.

- `rep` stores the current values for auto repeat parameters, typically the delay and rate.

- `absinfo` is an array of `&struct input_absinfo` elements that holds information about the absolute axes (the current value, `min`, `max`, `flat`, `fuzz`, and the resolution). You should use the `input_set_abs_params()` function to set those values:

```
void input_set_abs_params(struct input_dev *dev,
                          unsigned int axis, int min,
                          int max, int fuzz, int flat)
```

- `min` and `max` specify the lower and upper bound values, respectively. `fuzz` indicates the expected noise on the specified channel of the specified input device. In the following examples, we're setting each channel's bound:

```
#define ABSMAX_ACC_VAL        0x01FF
#define ABSMIN_ACC_VAL        -(ABSMAX_ACC_VAL)
[...]
set_bit(EV_ABS, idev->evbit);
input_set_abs_params(idev, ABS_X, ABSMIN_ACC_VAL,
                ABSMAX_ACC_VAL, 0, 0);
input_set_abs_params(idev, ABS_Y, ABSMIN_ACC_VAL,
                ABSMAX_ACC_VAL, 0, 0);
input_set_abs_params(idev, ABS_Z, ABSMIN_ACC_VAL,
                ABSMAX_ACC_VAL, 0, 0);
```

- `key` reflects the current state of the device's keys/buttons.

- `open` is a method that's called when the very first user calls `input_open_device()`. Use this method to prepare the device, such as to interrupt a request, poll a thread start, and so on.

- `close` is called when the very last user calls `input_close_device()`. Here, you can stop polling (which consumes a lot of resources).

- `users` stores the number of users (input handlers) that opened this device. It is used by `input_open_device()` and `input_close_device()` to ensure that `dev->open()` is only called when the first user opens the device and that `dev->close()` is only called when the very last user closes the device.

- `dev` is the struct device associated with this device (for device model).

- `num_vals` is the number of values that are queued in the current frame.

- `max_vals` is the maximum number of values that are queued in a frame.

- `Vals` is the array of values that are queued in the current frame.

- `devres_managed` indicates that the devices are managed with the **devres** framework and don't need to be explicitly unregistered or freed.

Now that you're familiar with the main input device's data structure, we can start registering such devices within the system.

Allocating and registering an input device

Before the events that are supported by an input device can be seen by the system, memory needs to be allocated for this device first using the `devm_input_allocate_device()` API. Then, the device needs to be registered with the system using `input_device_register()`. The former API will take care of freeing up the memory and unregistering the device when it leaves the system. However, non-managed allocation is still available but not recommended, `input_allocate_device()`. By using non-managed allocation, the driver becomes responsible for making sure that `input_unregister_device()` and `input_free_device()` are called to unregister the device and free its memory when they're on the unloading path of the driver, respectively. The following are the respective prototypes of these APIs:

```
struct input_dev *input_allocate_device(void)
struct input_dev *devm_input_allocate_device(
                                    struct device *dev)
void input_free_device(struct input_dev *dev)
int input_register_device(struct input_dev *dev)
void input_unregister_device(struct input_dev *dev)
```

Device allocation may sleep, so it must not be called in the atomic context or with a spinlock being held. The following is an excerpt of the `probe` function of an input device sitting on the I2C bus:

```c
struct input_dev *idev;
int error;
/*
 * such allocation will take care of memory freeing and
 * device unregistering
 */
idev = devm_input_allocate_device(&client->dev);
if (!idev)
    return -ENOMEM;

idev->name = BMA150_DRIVER;
idev->phys = BMA150_DRIVER "/input0";
idev->id.bustype = BUS_I2C;
idev->dev.parent = &client->dev;

set_bit(EV_ABS, idev->evbit);
input_set_abs_params(idev, ABS_X, ABSMIN_ACC_VAL,
                     ABSMAX_ACC_VAL, 0, 0);
input_set_abs_params(idev, ABS_Y, ABSMIN_ACC_VAL,
                     ABSMAX_ACC_VAL, 0, 0);
input_set_abs_params(idev, ABS_Z, ABSMIN_ACC_VAL,
                     ABSMAX_ACC_VAL, 0, 0);

error = input_register_device(idev);
if (error)
    return error;
error = devm_request_threaded_irq(&client->dev,
            client->irq,
            NULL, my_irq_thread,
            IRQF_TRIGGER_RISING | IRQF_ONESHOT,
            BMA150_DRIVER, NULL);
if (error) {
    dev_err(&client->dev, "irq request failed %d,
```

```
            error %d\n", client->irq, error);
    return error;
}
return 0;
```

As you may have noticed, in the preceding code, no memory freeing nor device unregistering is performed when an error occurs because we have used the managed allocation for both the input device and the IRQ. That said, the input device has an IRQ line so that we're notified of a state change on the underlying device. This is not always the case as the system may lack available IRQ lines, in which case the input core will have to poll the device frequently so that it doesn't miss events. We discuss this in the next section.

Using polled input devices

Polled input devices are special input devices that rely on polling to sense device state changes; the generic input device type relies on IRQ to sense changes and send events to the input core.

A polled input device is described in the kernel as an instance of struct input_polled_dev structure, which is a wrapper around the generic struct input_dev structure. The following is its declaration:

```
struct input_polled_dev {
    void *private;

    void (*open)(struct input_polled_dev *dev);
    void (*close)(struct input_polled_dev *dev);
    void (*poll)(struct input_polled_dev *dev);
    unsigned int poll_interval; /* msec */
    unsigned int poll_interval_max; /* msec */
    unsigned int poll_interval_min; /* msec */

    struct input_dev *input;
    bool devres_managed;
};
```

Let's take a look at the elements in this structure:

- `private` is the driver's private data.

- `open` is an optional method that prepares the device for polling (enables the device and sometimes flushes the device's state).

- `close` is an optional method that is called when the device is no longer being polled. It is used to put devices into low power mode.

- `poll` is a mandatory method that's called whenever the device needs to be polled. It is called at the frequency of `poll_inteval`.

- `poll_interval` is the frequency at which the `poll()` method should be called. It defaults to 500 milliseconds unless it's overridden when you're registering the device.

- `poll_interval_max` specifies the upper bound for the poll interval. It defaults to the initial value of `poll_interval`.

- `poll_interval_min` specifies the lower bound for the poll interval. It defaults to 0.

- `input` is the input device that the polled device is built around. It must be initialized by the driver (by its ID, name, and bits). The polled input device just provides an interface to use polling instead of IRQ, to sense device state change.

Memory can be allocated for a polled input device using `devm_input_allocate_polled_device()`. This is a managed allocation API that takes care of freeing memory and unregistering the device as appropriate. Similarly, the non-managed API can be used for allocation, `input_allocate_polled_device()`, in which case you must take care of calling `input_free_polled_device()` by yourself. The following code shows the prototypes of those APIs:

```
struct input_polled_dev
    *devm_input_allocate_polled_device(
                             struct device *dev)
struct input_polled_dev *input_allocate_polled_device(void)
void input_free_polled_device(struct
                          input_polled_dev *dev)
```

For resource-managed devices, the input_dev->devres_managed field will be set to true by the input core. Then, you should take care of initializing the mandatory fields of the underlying struct input_dev, as we saw in the previous section. The polling interval must be set too; otherwise, it will default to 500 ms.

Once the fields have been allocated and initialized, the polled input device can be registered using input_register_polled_device(), which returns 0 on success. For managed allocation, unregistering is handled by the system; you need to call input_unregister_polled_device() by yourself to perform the reverse operation. The following are their prototypes:

```
int input_register_polled_device(
                    struct input_polled_dev *dev)
void  input_unregister_polled_device(
                    struct input_polled_dev *dev)
```

A typical example of the probe() function for such a device may look as follows. First, we define the driver data structure, which will gather all the necessary resources:

```
struct my_struct {
    struct input_pulled_dev *polldev;
    struct gpio_desc *gpio_btn;
    [...]
}
```

Once the driver data structure has been defined, the probe() function can be implemented. The following is its body:

```
static int button_probe(struct platform_device *pdev)
{
    struct my_struct *ms;
    struct input_dev *input_dev;
    int error;
    struct device *dev = &pdev->dev;
    ms = devm_kzalloc(dev, sizeof(*ms), GFP_KERNEL);
    if (!ms)
        return -ENOMEM;
    ms->polldev = devm_input_allocate_polled_device(dev);
    if (!ms->polldev)
        return -ENOMEM;
```

```
    /* This gpio is not mapped to IRQ */
    ms->gpio_btn = devm_gpiod_get(dev, "my-btn", GPIOD_IN);
    ms->polldev->private = ms;
    ms->polldev->poll = my_btn_poll;
    ms->polldev->poll_interval = 200;/* Poll every 200ms */
    ms->polldev->open = my_btn_open;

     /* Initializing the underlying input_dev fields */
    input_dev = ms->poll_dev->input;
    input_dev->name = "System Reset Btn";
    /* The gpio belongs to an expander sitting on I2C */
    input_dev->id.bustype = BUS_I2C;
    input_dev->dev.parent = dev;
    /* Declare the events generated by this driver */
    set_bit(EV_KEY, input_dev->evbit);
    set_bit(BTN_0, input_dev->keybit); /* buttons */

    retval = input_register_polled_device(ms->poll_dev);
    if (retval) {
        dev_err(dev, "Failed to register input device\n");
        return retval;
    }
    return 0;
}
```

Once again, neither unregistering nor freeing are handled by ourselves when an error occurs because we have used managed allocations.

The following is what our open callback may look like:

```
static void my_btn_open(struct input_polled_dev *poll_dev)
{
    struct my_strut *ms = poll_dev->private;
    dev_dbg(&ms->poll_dev->input->dev, "reset open()\n");
}
```

In our example, it does nothing. However, the open method is used to prepare the resources that are needed by the device.

Deciding whether you should implement a polled input device is straightforward. The usual way is to use classic input devices if an IRQ line is available; alternatively, you can fall back to the polled device:

```
if(client->irq > 0){
    /* Use generic input device */
} else {
    /* Use polled device */
}
```

Other elements may need to be considered when you're choosing between implementing a polled input device or an IRQ-based one; the preceding code is just a suggestion.

Now that we are familiar with this subset of input devices, we can consider registering and unregistering input devices. That said, even though the input device has been registered, we can't interact with it yet. In the next section, we will learn how the input device can report events to the kernel.

Generating and reporting input events

Device allocation and registration are essential, but they are useless if the device is unable to report events to the input core, which is what input devices are designed to do. Depending on the type of event our device can support, the kernel provides the appropriate APIs to report them to the core.

Given an `EV_XXX` capable device, the corresponding report function would be `input_report_xxx()`. The following table shows the mappings between the most important event types and their report functions:

Even Type	Report Function	Code Example

EV_KEY	input_report_key()	input_report_key(
		    poll_dev->input,
		    BTN_0,
		    gpiod_get_value(
		    ms->reset_btn_desc) & 1);

EV_REL	input_report_rel()	input_report_rel(
		    nunchuk->input,
		    REL_X,
		    nunchuk->report.joy_x -
		        128)/10);

EV_ABS	input_report_abs()	• input_report_abs(
		        bma150->input,
		        ABS_X, x_value);
		• input_report_abs(
		        bma150->input,
		        ABS_Y, y_value);
		• input_report_abs(
		        bma150->input,
		        ABS_Z, z_value);

Table 18.1 – Mapping the input device's capabilities and the report APIs

The prototypes for these report APIs are as follows:

```
void input_report_abs(struct input_dev *dev,
                unsigned int code, int value)
void input_report_key(struct input_dev *dev,
                unsigned int code, int value)
void input_report_rel(struct input_dev *dev,
                unsigned int code, int value)
```

The list of available report functions can be found in `include/linux/input.h` in the kernel source file. They all have the same skeleton:

- `dev` is the input device that's responsible for the event.

- `code` represents the event code; for example, `REL_X` or `KEY_BACKSPACE`. The complete list can be found in `include/linux/input-event-codes.h`.

- `value` is the value the event carries. For an `EV_REL` event type, it carries the relative change. For an `EV_ABS` (joysticks and so on) event type, it contains an absolute new value. For an `EV_KEY` event type, it should be set to 0 for key release, 1 for a keypress, and 2 for auto-repeat.

Once all these changes have been reported, the driver should call `input_sync()` on the input device to indicate that this event is complete. The input subsystem will collect these events into a single packet and send it through `/dev/input/event<X>`, which is the character device that represents our `struct input_dev` on the system. Here, `<X>` is the interface number that's been assigned to the driver by the input core:

```
void input_sync(struct input_dev *dev)
```

Let's look at an example of this. The following is an excerpt from the **bma150** digital acceleration sensors drivers in `drivers/input/misc/bma150.c`:

```
static void threaded_report_xyz(struct bma150_data *bma150)
{
    u8 data[BMA150_XYZ_DATA_SIZE];
    s16 x, y, z;
    s32 ret;

    ret = i2c_smbus_read_i2c_block_data(bma150->client,
                        BMA150_ACC_X_LSB_REG,
                        BMA150_XYZ_DATA_SIZE, data);
    if (ret != BMA150_XYZ_DATA_SIZE)
        return;

    x = ((0xc0 & data[0]) >> 6) | (data[1] << 2);
    y = ((0xc0 & data[2]) >> 6) | (data[3] << 2);
    z = ((0xc0 & data[4]) >> 6) | (data[5] << 2);

    /* sign extension */
```

```
x = (s16) (x << 6) >> 6;
y = (s16) (y << 6) >> 6;
z = (s16) (z << 6) >> 6;

    input_report_abs(bma150->input, ABS_X, x);
    input_report_abs(bma150->input, ABS_Y, y);
    input_report_abs(bma150->input, ABS_Z, z);
    /* Indicate this event is complete */
    input_sync(bma150->input);
}
```

In the preceding excerpt, `input_sync()` tells the core to consider the three reports as the same event. This makes sense since the position has three axes (*X*, *Y*, and *Z*) and we do not want *X*, *Y*, or *Z* to be reported separately.

The best place to report the event is inside the poll function for a polled device or the IRQ routine (threaded part or not) for an IRQ-enabled device. If you perform some operations that may sleep, you should process your report inside the threaded part of the IRQ handler. The following code shows how our initial example could implement the `poll` method:

```
static void my_btn_poll(struct input_polled_dev *poll_dev)
{
    struct my_struct *ms = polldev->private;
    struct i2c_client *client = mcp->client;

    input_report_key(polldev->input, BTN_0,
                        gpiod_get_value(ms->rgpio_btn) & 1);
    input_sync(poll_dev->input);
}
```

In the preceding code, our input device reports the 0 key code. In the next section, we will discuss how the user space can handle those report events and codes.

Handling input devices from the user space

A node will be created in the /dev/input/ directory for each input device (polled or not) that has been successfully registered with the system. In my case, the node corresponds to event 0 because it is the first and only input device on my target board. You can use the udevadm tool to display information about the device:

```
# udevadm info /dev/input/event0
P: /devices/platform/input-button.0/input/input0/event0
N: input/event0
S: input/by-path/platform-input-button.0-event
E: DEVLINKS=/dev/input/by-path/platform-input-button.0-event
E: DEVNAME=/dev/input/event0
E: DEVPATH=/devices/platform/input-button.0/input/input0/event0
E: ID_INPUT=1
E: ID_PATH=platform-input-button.0
E: ID_PATH_TAG=platform-input-button_0
E: MAJOR=13
E: MINOR=64
E: SUBSYSTEM=input
E: USEC_INITIALIZED=74842430
```

Another tool that you can use, which allows you to print the keys that are supported by the device, is evtest. It can also catch and print events when they are reported by the device. The following code shows its usage on our input device:

```
# evtest /dev/input/event0
input device opened()
Input driver version is 1.0.1
Input device ID: bus 0x0 vendor 0x0 product 0x0 version 0x0
Input device name: "Packt Btn"
Supported events:
    Event type 0 (EV_SYN)
    Event type 1 (EV_KEY)
        Event code 256 (BTN_0)
```

Not only the input devices we have written drivers for can be managed with `evtest`. In the following example, I am using the USB-C headset that's connected to my computer. It has input device capabilities since it provides volume-related keys:

```
jma@labcsmart-sqy:~$ sudo evtest /dev/input/event4
Input driver version is 1.0.1
Input device ID: bus 0x3 vendor 0x12d1 product 0x3a07 version
0x111
Input device name: "Synaptics HUAWEI USB-C HEADSET"
Supported events:
  Event type 0 (EV_SYN)
  Event type 1 (EV_KEY)
    Event code 114 (KEY_VOLUMEDOWN)
    Event code 115 (KEY_VOLUMEUP)
    Event code 164 (KEY_PLAYPAUSE)
    Event code 582 (KEY_VOICECOMMAND)
  Event type 4 (EV_MSC)
    Event code 4 (MSC_SCAN)
Properties:
Testing ... (interrupt to exit)


Event: time 1640231369.347093, type 4 (EV_MSC), code 4 (MSC_
SCAN), value c00e9
Event: time 1640231369.347093, type 1 (EV_KEY), code 115 (KEY_
VOLUMEUP), value 1
Event: time 1640231369.347093, -------------- SYN_REPORT -----
-------
Event: time 1640231369.487017, type 4 (EV_MSC), code 4 (MSC_
SCAN), value c00e9
Event: time 1640231369.487017, type 1 (EV_KEY), code 115 (KEY_
VOLUMEUP), value 0
Event: time 1640231369.487017, -------------- SYN_REPORT -----
-------
```

In the preceding code, I pushed the up volume key to see how it is reported. `evtest` can even be used with your keyboard, with the only condition being that you identify the corresponding input device node in `/dev/input/`.

As we have seen, every registered input device is represented by a /dev/input/
event<X> character device, which we can use to read the event from the user space.
An application that's reading this file will receive event packets in the struct input_
event format, which has the following declaration:

```
struct input_event {
  struct timeval time;
  __u16 type;
  __u16 code;
  __s32 value;
}
```

Let's look at the meaning of each element in the structure:

- time is a timestamp that corresponds to the time when the event happened.

- type is the event type; for example, EV_KEY for a keypress or release, EV_REL
  for a relative moment, or EV_ABS for an absolute one. More types are defined in
  include/linux/input-event-codes.h.

- code is the event code; for example, REL_X or KEY_BACKSPACE. Again, a
  complete list can be found in include/linux/input-event-codes.h.

- value is the value that the event carries. For an EV_REL event type, it carries the
  relative change. For an EV_ABS (joysticks and so on) event type, it contains the
  absolute new value. For an EV_KEY event type, it is set to 0 for a key release, 1 for a
  keypress, and 2 for auto-repeat.

A user space application can use blocking and non-blocking reads, but also poll() or
select() system calls, to be notified of events after opening this device. The following is
an example of the select() system call. Let's start by enumerating the headers we need
to implement our example:

```
#include <unistd.h>
#include <fcntl.h>
#include <stdio.h>
#include <stdlib.h>
#include <linux/input.h>
#include <sys/select.h>
```

Then, we must define our input device path as a macro as it will be used often:

```
#define INPUT_DEVICE "/dev/input/event0"

int main(int argc, char **argv)
{
    int fd, ret;
    struct input_event event;
    ssize_t bytesRead;
    fd_set readfds;
```

Next, we must open the input device and keep its file descriptor for later use. Failing to open the input device is considered as an error, so we must exit the program:

```
    fd = open(INPUT_DEVICE, O_RDONLY);
    if(fd < 0){
        fprintf(stderr,
            "Error opening %s for reading", INPUT_DEVICE);
        exit(EXIT_FAILURE);
    }
```

Now, we have a file descriptor representing our opened input device. We can use the `select()` system call to sense any key press or release:

```
    while(1){
        FD_ZERO(&readfds);
        FD_SET(fd, &readfds);
        ret = select(fd + 1, &readfds, NULL, NULL, NULL);
        if (ret == -1) {
            fprintf(stderr,
                    "select call on %s: an error ocurred",
                    INPUT_DEVICE);
            break;
        }
        else if (!ret) { /* If we used timeout */
            fprintf(stderr,
                    "select on %s: TIMEOUT", INPUT_DEVICE);
            break;
        }
```

At this point, we have done the necessary sanity checks on the return path of select().
Note that select() returns zero if it timed out before any file descriptors became ready,
hence else if in the preceding code.

The change is effective now, let's read the data to see what it corresponds to:

```
/* File descriptor is now ready */
if (FD_ISSET(fd, &readfds)) {
    bytesRead = read(fd, &event,
                        sizeof(struct input_event));
    if(bytesRead == -1)
        /* Process read input error*/
        [...]
    if(bytesRead != sizeof(struct input_event))
        /* Read value is not an input event */
        [...] /* handle this error */
```

If the execution flow reaches this mean, it means that everything went well. Now, we can
walk through the events that are supported by our input device and compare them to the
event that is reported by the input core before a decision is made:

```
/* We could have done a switch/case if we had
 * many codes to look for */
if (event.code == BTN_0) {
    /* it concerns our button */
    if (event.value == 0) {
        /* Process keyRelease if need be */
        [...]
    }
    else if(event.value == 1){
        /* Process KeyPress */
        [...]
    }
    }
    }
    }
    close(fd);
    return EXIT_SUCCESS;
}
```

For further debugging purposes, if your input device is based on GPIOs, you can successively push/release the button and check whether the GPIO's state has changed:

```
# cat /sys/kernel/debug/gpio   | grep button
 gpio-195 (gpio-btn           ) in  hi
# cat /sys/kernel/debug/gpio   | grep button
 gpio-195 (gpio-btn           ) in  lo
```

Moreover, if the input device has an IRQ line, it may make sense to check the statistic for this IRQ line to make sure it is coherent. For example, here, we must check whether the request has succeeded and how many times it has been fired:

```
# cat /proc/interrupts | grep packt
160:        0       0       0       0  gpio-mxc   0  packt-input-
button
```

In this section, we learned how to deal with the input device from the user space and provided some debugging tips for when something goes wrong. We used the `select()` system call to sense input events, though we could have used `poll()` as well.

Summary

This chapter described the input framework and highlighted the difference between polled and interrupt-driven input devices. At this point, you should have the necessary knowledge to write a driver for any input driver, whatever its type, and whatever input event it supports. The user space interface was also discussed, and an example was provided.

Index

V

W

X

Packt>

Packt.com

Subscribe to our online digital library for full access to over 7,000 books and videos, as well as industry leading tools to help you plan your personal development and advance your career. For more information, please visit our website.

Why subscribe?

- Spend less time learning and more time coding with practical eBooks and Videos from over 4,000 industry professionals
- Improve your learning with Skill Plans built especially for you
- Get a free eBook or video every month
- Fully searchable for easy access to vital information
- Copy and paste, print, and bookmark content

Did you know that Packt offers eBook versions of every book published, with PDF and ePub files available? You can upgrade to the eBook version at packt.com and as a print book customer, you are entitled to a discount on the eBook copy. Get in touch with us at customercare@packtpub.com for more details.

At www.packt.com, you can also read a collection of free technical articles, sign up for a range of free newsletters, and receive exclusive discounts and offers on Packt books and eBooks.

Other Books You May Enjoy

If you enjoyed this book, you may be interested in these other books by Packt:

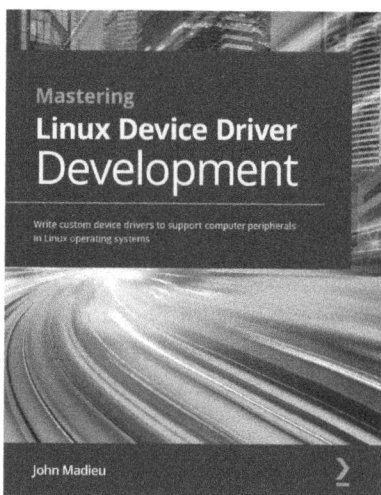

Mastering Linux Device Driver Development

John Madieu

ISBN: 9781789342048

- Explore and adopt Linux kernel helpers for locking, work deferral, and interrupt management
- Understand the Regmap subsystem to manage memory accesses and work with the IRQ subsystem
- Get to grips with the PCI subsystem and write reliable drivers for PCI devices
- Write full multimedia device drivers using ALSA SoC and the V4L2 framework
- Build power-aware device drivers using the kernel power management framework
- Find out how to get the most out of miscellaneous kernel subsystems such as NVMEM and Watchdog

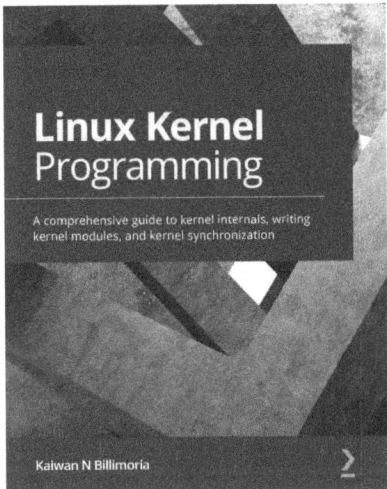

Linux Kernel Programming

Kaiwan N Billimoria

ISBN: 9781789953435

- Write high-quality modular kernel code (LKM framework) for 5.x kernels
- Configure and build a kernel from source
- Explore the Linux kernel architecture
- Get to grips with key internals regarding memory management within the kernel
- Understand and work with various dynamic kernel memory alloc/dealloc APIs
- Discover key internals aspects regarding CPU scheduling within the kernel
- Gain an understanding of kernel concurrency issues
- Find out how to work with key kernel synchronization primitives

Packt is searching for authors like you

If you're interested in becoming an author for Packt, please visit `authors.packtpub.com` and apply today. We have worked with thousands of developers and tech professionals, just like you, to help them share their insight with the global tech community. You can make a general application, apply for a specific hot topic that we are recruiting an author for, or submit your own idea.

Share Your Thoughts

Now you've finished *Linux Device Driver Development - Second Edition*, we'd love to hear your thoughts! Scan the QR code below to go straight to the Amazon review page for this book and share your feedback or leave a review on the site that you purchased it from.

`https://packt.link/r/1803240067`

Your review is important to us and the tech community and will help us make sure we're delivering excellent quality content.